STRUCTURAL VIBRATION

Exact Solutions for Strings, Membranes, Beams, and Plates

STRUCTURAL VIBRATION

Exact Solutions for Strings, Membranes, Beams, and Plates

C.Y. Wang and C.M. Wang

CRC Press
Taylor & Francis Group
Boca Raton London New York

CRC Press is an imprint of the
Taylor & Francis Group, an **informa** business

CRC Press
Taylor & Francis Group
6000 Broken Sound Parkway NW, Suite 300
Boca Raton, FL 33487-2742

Printed on acid-free paper
Version Date: 20130703

International Standard Book Number-13: 978-1-4665-7684-1 (Hardback)

Visit the Taylor & Francis Web site at
http://www.taylorandfrancis.com

and the CRC Press Web site at
http://www.crcpress.com

Contents

Preface

There is a staggering number of research studies on the vibration of structures. Based on a simple search using the Science Citation Index, the numbers of references associated with the following words are 1,000 for "vibration and string," 2,000 for "vibration and membrane," 7,000 for "vibration and plate," and 16,000 for "vibration and beam, bar or rod." This clearly illustrates the importance of the subject of free and forced vibrations for analysis and design of structures and machines.

The free vibration of a structural member eventually ceases due to energy dissipation, either from the material strains or from the resistance of the surrounding fluid. The frequency of such a system will be lowered by damping. But since damping also causes the amplitude to decay, the resonance with a forced excitation of a strongly damped system will not be as important as the weakly damped system. In this book, we shall consider the undamped system, which models the weakly damped system, and only focus on the exact solutions for free transverse vibration of strings, bars, membranes, and plates because these solutions elucidate the intrinsic, fundamental, and unexpected features of the solutions. They also serve as benchmarks to assess the validity, convergence, and accuracy of numerical methods and approximate analytical methods. We define exact solutions to mean solutions in terms of known functions as well as those solutions determined from exact characteristic equations. However, this book will not cover longitudinal in-plane/translational vibrations, shear waves, torsional oscillations, infinite domains (wave propagation), discrete systems (such as linked masses), and frames. The exact solutions for a wide range of differential equations are useful to academics teaching differential equations, as they may draw the practical problems associated with the differential equations.

There are seven chapters in this book. Chapter 1 gives the introduction to structural vibration and the importance of the natural frequencies in design. Chapter 2 presents the vibration solutions for strings. Chapter 3 presents the vibration solutions for membranes. Chapter 4 deals with vibration of bars and beams. Chapter 5 gives the vibration solutions for isotropic plates with uniform thickness. Chapter 6 deals with plates with complicating effects such as the presence of in-plane forces, internal spring support, internal hinge, elastic foundation, and nonuniform thickness distribution. Chapter 7 presents vibration solutions for nonisotropic plates, such as orthotropic, sandwich, laminated, and functionally graded plates.

Owing to the vastness of the literature, there may be relevant papers that escaped our search in the Science Citation Index. To these authors, we offer our sincere apology. Such omissions shall be rectified in a future edition.

Finally, we wish to express our thanks to Dr. Tay Zhi Yung and Mr. Ding Zhiwei of the National University of Singapore for checking the manuscript and plotting the vibration mode shapes and also to Dr. Liu Bo of The Solid Mechanics Research Centre, Beihang University, China, for contributing the sections on rectangular isotropic and orthotropic Mindlin plates.

C. Y. Wang and C. M. Wang

About the Authors

C. Y. Wang is a professor in the Department of Mathematics with a joint appointment in mechanical engineering at the Michigan State University, East Lansing, Michigan. He obtained his BS from Taiwan University and PhD from Massachusetts Institute of Technology. Prof. Wang has published about 170 papers in solid mechanics (elastica, torsion, buckling, and vibrations of structural members), 170 papers in fluid mechanics (exact Navier-Stokes solutions, Stokes flow, unsteady viscous flow), and 120 papers in other areas (biological, thermal, electromechanics). Prof. Wang wrote a monograph *Perturbation Methods* (Taiwan University Press) and is a coauthor of *Exact Solutions for Buckling of Structural Members* (CRC Press). He has served as a technical editor for *Applied Mechanics Reviews*.

C. M. Wang is a professor in the Department of Civil and Environmental Engineering and the director of the Engineering Science Programme, Faculty of Engineering, National University of Singapore. He is a chartered structural engineer, a fellow of the Singapore Academy of Engineering, a fellow of the Institution of Engineers Singapore, and a fellow of the Institution of Structural Engineers. His research interests are in the areas of structural stability, vibration, optimization, plated structures, and Mega-Floats. He has published more than 400 scientific publications, co-edited three books, *Analysis and Design of Plated Structures: Stability and Dynamics: Volumes 1 and 2* (Woodhead Publishing) and *Very Large Floating Structures* (Taylor & Francis) and co-authored three books: *Vibration of Mindlin Plates* (Elsevier), *Shear Deformable Beams and Plates: Relationships with Classical Solutions* (Elsevier), and *Exact Solutions for Buckling of Structural Members* (CRC Press). He is the editor-in-chief of the *International Journal of Structural Stability and Dynamics* and the *IES Journal Part A: Civil and Structural Engineering* and an editorial board member of *Engineering Structures, Advances in Applied Mathematics and Mechanics*, *Ocean Systems Engineering*, and *International Journal of Applied Mechanics*.

1 Introduction to Structural Vibration

1.1 WHAT IS VIBRATION?

Vibration may be regarded as any motion that repeats itself after an interval of time, or one may define vibrations as oscillations of a system about a position of equilibrium (Kelly 2007). Examples of vibratory motion include the swinging of a pendulum, the motion of a plucked guitar string, tidal motion, the chirping of a male cicada by rubbing its wings, the flapping of airplane wings in turbulence, the soothing motion of a massage chair, or the swaying of a slender tall building due to wind or an earthquake.

The key parameters in describing vibration are amplitude, period, and frequency. The amplitude of vibration is the maximum displacement of a vibrating particle or body from its position of equilibrium, and this is related to the applied energy. The period is the time taken for one complete cycle of the motion. The frequency is the number of cycles per unit time or the reciprocal of the period. The angular (or circular) frequency is the product of the frequency and 2π, and hence its unit is radians per unit time.

Vibrations may be classified as either *free vibration* or *forced vibration*. Free vibration takes place when a system oscillates under the action of forces inherent within the system itself—when externally imposed forces are absent. A system under free vibration will vibrate at one or more of its natural frequencies, which are dependent on the mass and stiffness distributions as well as the boundary conditions. In contrast, forced vibration occurs when an external periodic force is applied to the system.

When the effects of friction can be neglected, the vibrations are referred to as undamped. Realistically, all vibrations are damped to some degree. If a free vibration is only slightly damped, its amplitude gradually decreases until the motion comes to an end after a certain time. If the damping is sufficiently large, vibration is suppressed, and the system then quickly regains its original equilibrium position. A damped forced vibration is maintained so long as the periodic force that causes the vibration is applied. The amplitude of the vibration is affected by the magnitude of the damping forces.

From an energy viewpoint, vibration may be defined as a phenomenon that involves alternating interchange of potential energy and kinetic energy. If the system is damped, then some energy is dissipated in each cycle of the vibration, and the vibratory motion will ultimately come to an end. If a steady motion of vibration is to be maintained, then the energy dissipated due to damping has to be compensated by an external source.

1.2 BRIEF HISTORICAL REVIEW ON VIBRATION OF STRINGS, MEMBRANES, BEAMS, AND PLATES

According to Rao (1986, 2005), it is likely that the interest in vibration dates back to the time of the discovery of early musical instruments such as whistles, strings, or drums, which produce sound from vibration. Drawings of stringed instruments have been found on the walls of Egyptian tombs that were built around 3000 BC.

In the course of seeking why some notes sounded more pleasant than others, the Greek mathematician and philosopher Pythagoras (582–507 BC) conducted experiments on vibrating strings, and he observed that the pitch of the note (the frequency of the sound) was dependent on the tension and length of the string. Galileo (1638), the Italian physicist and astronomer, took measurements to establish a relationship between the length and frequency of vibration for a simple pendulum and for strings; he also observed the resonance of two connecting bodies. Marinus Mersenne (1636), a French mathematician and theologian, also studied the behavior of vibrating strings. English scientist Robert Hooke (1635–1703) and French mathematician and physicist Joseph Sauveur (1653–1716) performed further studies on the relationship between the pitch and frequency of a vibrating taut string. Sauveur is noted for introducing the terms *nodes* (stationary points), *loops*, *fundamental frequency*, and *harmonics*, and he is the first scientist to record the phenomenon of *beats*.

The breakthrough in formulating the governing equations for structural vibration problems may be attributed to Sir Isaac Newton (1687), who was the first to formulate the laws of classical mechanics, and to Gottfried Leibniz (1693) as well as Newton for creating calculus. Euler (1744) and Bernoulli (1751) discovered the differential equation governing the lateral vibration of prismatic bars and investigated its solution for the case of small deflections. Lagrange (1759) also made important contributions to the theory of vibrating strings. Euler (1766) derived the equations for the vibration of rectangular membranes under uniform tension as well as for the vibration of a ring. Poisson (1829) derived the governing equation for vibrating circular membranes and gave the solutions for the axisymmetric vibration mode. Pagani (1829) worked out the nonaxisymmetric vibration solution for circular membranes. Coulomb (1784) investigated the torsional oscillations of a metal cylinder suspended by a wire.

The German physicist Chladni observed nodal patterns on flat square plates at their resonant frequencies using sand spread evenly on the plate surface. The sand formed regular patterns as the sand accumulated along the nodal lines of zero vertical displacements upon induction of vibration. Figure 1.1 shows the patterns of square plates that were originally published in Chladni's book (Chladni 1802). In 1816, Sophie Germain successfully derived the differential equation for the vibration of plates by means of calculus of variations. However, she made a mistake in neglecting the strain energy due to the twisting of the plate mid-plane. The correct version of the governing differential equation, without its derivation, was found posthumously among Lagrange's notes in 1813. Thus, Lagrange has been credited as being the first to present the correct equation for thin plates. By using trigonometric series introduced by Fourier around that time, Navier (1823) was able to readily determine the exact vibration solutions for rectangular plates with simply supported

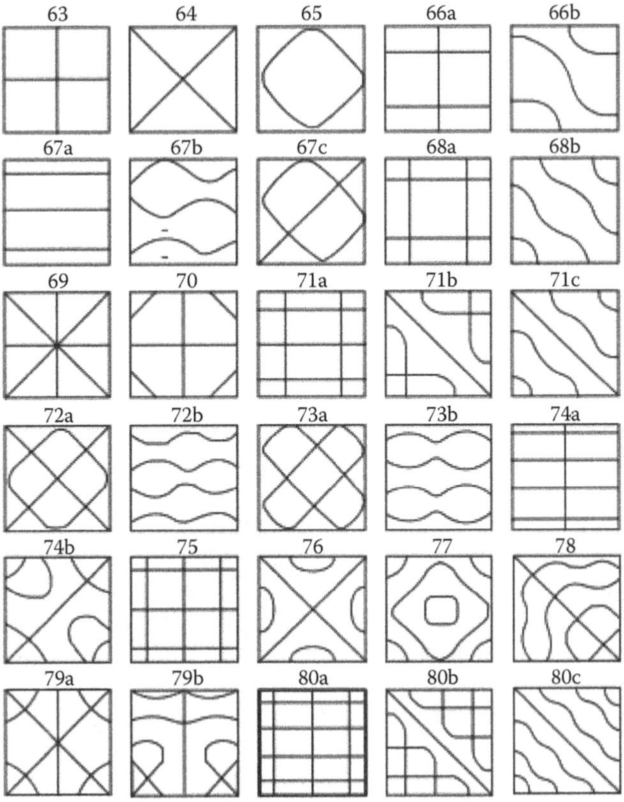

FIGURE 1.1 Chladni's original figures of vibrating square plates showing nodal lines. *Source*: http://en.wikipedia.org/wiki/File:Chladini.Diagrams.for.Quadratic.Plates.svg.

edges. Poisson (1829) extended Navier's work to circular plates. The extended plate theory that considered the combined bending and stretching actions of a plate has been attributed to Kirchhoff (1850). His other significant contribution is the application of a virtual displacement method for solving plate problems.

Lord Rayleigh (1877) presented a theory to explain the phenomenon of vibration that to this day is still used to determine the natural frequencies of vibrating structures. Based on the plate assumptions made by Kirchhoff (1850) and Rayleigh's theory, early researchers used analytical techniques to solve the vibration problems of plates. For example, Voigt (1893) and Carrington (1925) successfully derived the exact vibration frequency solutions for a simply supported rectangular plate and a fully clamped circular plate, respectively. Ritz (1909) was one of the early researchers to solve the problem of the freely vibrating plate, which does not have an exact solution. He demonstrated how to reduce the upper-bound frequencies by including more than a single trial (admissible) function and performing a minimization with respect to the unknown coefficients of these trial functions. The method became known as the Ritz method. Liew and Wang (1992, 1993) automated the Ritz method for analysis of arbitrarily shaped plates.

The theories of vibration of beams and plates were investigated further by Timoshenko (1921) and Mindlin (1951), and their theories allow for the effects of transverse shear deformation and rotary inertia. Other, more refined beam and plate theories that do away with the need for a shear correction factor were developed by Bickford (1982), Reddy (1984), and Reddy and Phan (1985), who employed higher-order polynomials in the expansion of the displacement components through the beam or plate thickness. Leissa (1969) produced an excellent monograph entitled "Vibration of Plates," which contains a wealth of vibration solutions for a wide range of plate shapes and boundary conditions. Originally published by NASA in 1969, Leissa's monograph was reprinted in 1993 by the Acoustical Society of America due to popular demand.

1.3 IMPORTANCE OF VIBRATION ANALYSIS IN STRUCTURAL DESIGN

When designing structures, the effect of vibration on them is a very important factor to consider. Obviously, structures used to support heavy centrifugal machines like motors and turbines are subjected to vibration. Vibration causes excessive wear of bearings, material cracking, fasteners to become loose, noise, and abrasion of insulation around electrical conductors, resulting in short circuiting (Wowk 1991). When cutting a metal, vibration can cause chatter, which affects the quality of the surface finish. Structural vibration may cause discomfort and even fear in the occupants working in the building, make it difficult to operate machinery, and cause malfunctioning of equipment.

The natural frequencies of a structure are very important to structural and mechanical engineers when designing for human comfort, structural serviceability and operational requirements, and against the occurrence of resonance. Resonance occurs when the natural frequency of the structure coincides with the excitation frequency. This resonance phenomenon has to be avoided so as to prevent excessive deformation, fatigue cracks, and even the collapse of the entire structure. For example, the spectacular collapse of the Tacoma Narrows suspension bridge (that spanned the Tacoma Narrows strait of Puget Sound between Tacoma and the Kitsap Peninsula in the U.S. state of Washington) in 1940 was a result of resonance caused by strong wind gusts. Therefore, structural engineers design their structures to have a fundamental natural frequency of vibration that satisfies a specific minimum frequency given in design codes. For instance, the American Association of State Highway and Transportation Officials (AASHTO) specifies the minimum frequency for a pedestrian bridge to be 3 Hz. For office buildings, it is recommended that the natural frequency of floor structures be kept to within 4 Hz, whereas for performance stages and dance floors, this minimum limit of natural frequency may be raised to 8.4 Hz (Technical Guidance Note 2012).

Given the undesirable and devastating effects that vibrations can have on machines and structures, vibration analysis and testing have become a standard procedure in the design of structures (Richardson and Ramsey 1981; McConnell and Varoto 2008). Vibration may be reduced by using the illustrative vibrating mechanical

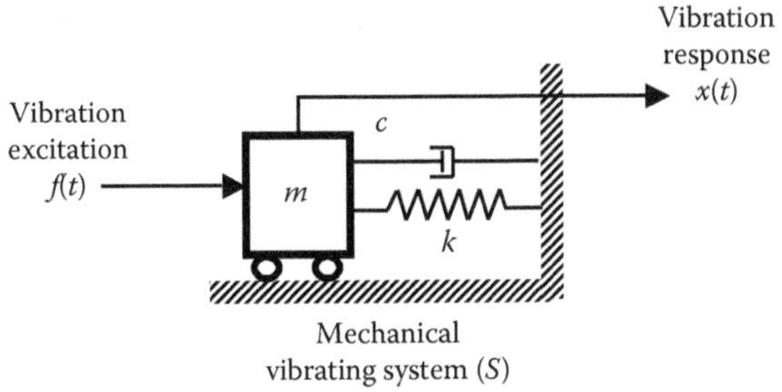

FIGURE 1.2 A vibrating mechanical system.

system shown in Figure 1.2, where the forcing excitations *f*(*t*) to the mechanical system *S* cause the vibration response *x*(*t*). The problem at hand is to suppress *x*(*t*) to an acceptable level. The three general ways to do this are:

1. *Isolation.* Suppress the excitations of the vibration. This method deals with the forcing excitation *f*(*t*)
2. *Design modification.* Modify or redesign the mechanical system so that for the same levels of excitation, the resulting vibrations are acceptable. This method deals with the mechanical system *S*, which has a mass *m*, stiffness *k*, and damping coefficient *c*.
3. *Control.* Absorb or dissipate the vibrations using external devices, through implicit or explicit sensing and control. This method deals with the vibration response *x*(*t*).

Within each category, there are several approaches for mitigating vibration. Actually, each of these approaches needs either redesign or modification. It is to be noted that the removal of faults (e.g., misalignments and malfunctions by repair or parts replacement) can also reduce vibrations. This approach may be included in any of the three categories listed here (De Silva 2007).

In order to understand isolation well, we need to know the concept of mechanical impedance (Wowk 1991). When vibrations travel through different materials and metal interfaces, they get reduced or attenuated. With the concept of impedance, we can insert materials into the force transmission path so as to reduce the amplitude of the vibration. Generally, any material with a lower stiffness than the adjacent material will function well to attenuate the force, and it works in both directions. Mechanical springs, air springs, cork, fiberglass, polymer, and rubber are the typical isolator materials. The performance of the isolator is a function of frequency.

On the other hand, vibration can also be useful in several industrial applications. For example, compactors, vibratory conveyors, hoppers, sieves, and washing machines take advantage of vibration to do the job. More interestingly, vibrations are found to be able to improve the efficiency of certain machining, casting, forging, and welding processes. Vibration is also used in nondestructive testing of materials and structures, in vibratory finishing processes, and in electronic circuits to filter out the unwanted frequencies (Rao 1986). It is also employed in shake tables to simulate earthquakes for testing structural designs against seismic action. Of course, most people enjoy the vibration of a massaging chair/device on their bodies.

1.4 SCOPE OF BOOK

In this book, we focus our attention on the free, harmonic, and flexural vibration of strings, membranes, beams, and plates. Damping is assumed to be small, and hence it is neglected. In each of the many structural vibration problems treated herein, we present the exact natural angular (or circular) frequencies and their accompanying mode shapes. Exact solutions are very important, as they clearly reveal the intrinsic features of the solutions and provide benchmarks to assess the validity, convergence, and accuracy of numerical solutions. Here, we define an exact solution as one that can be expressed in terms of a finite number of terms, and the proposed solution may contain elementary or common functions such as harmonic or Bessel functions. Special functions, such as hypergeometric functions, are excluded. Analytical solutions that are not exact, such as infinite series solutions and asymptotic solutions, are also excluded.

The governing differential equations of motion for the problems treated herein are obtained by using the method of elementary analysis, and the equations are solved for different boundary conditions. Analytical vibration solutions of structures with complicated geometries and boundary conditions are difficult or impossible to obtain. In such cases, numerical methods are required. However, for some cases of structural geometries and boundary conditions, it is possible to solve the differential equations exactly in a closed form. In this book, the authors present as many analytical vibration solutions as possible in one single volume for ready use by engineers, academicians, and researchers in structural dynamic analysis and design. This book addresses a variety of boundary conditions, restraints, and mass and stiffness distributions in the hope that the reader may better understand the effects of shape, restraints, and boundary conditions on vibration frequencies and mode shapes.

The numerous differential equations and their solutions presented in this book are also useful for academicians, especially when they wish to provide practical problems to the differential equations that they present to students of engineering science.

REFERENCES

American Association of State Highway and Transportation Officials. 2012. Technical guidance note: Floor vibration. *Structural Engineer* 90 (7): 32–34.

Bernoulli, D. 1751. De vibrationibus et sono laminarum elasticarum commentationes physicogeometricae. *Commentari Academiae Scientiarum Imperialis Petropolitanae* 13 (1741–43): 105–20.

Bickford, W. B. 1982. A consistent higher order beam theory. *Developments in Theoretical and Applied Mechanics* 11:137–50.

Carrington, H. 1925. The frequencies of vibration of flat circular plates fixed at the circumference. *Phil. Mag.* 50 (6): 1261–64.

Chladni, E. F. F. 1802. *Die akustik*. Leipzig: Breitkopf & Härtel.

Coulomb, C. A. 1784. *Recherches theoretiques et experimentales sur la force de torsion et sur l'elasticite des fils de metal*. Paris: Memoirs of the Paris Academy.

De Silva, C. W. 2007. *Vibration damping, control, and design*. Boca Raton, FL: Taylor & Francis.

Euler, L. 1744. De curvis elasticis. In *Methodus inveniendi lineas curvas maximi minimive proprietate gaudentes, sive solutio problematis isoperimetrici lattissimo sensu accepti.* Lausanne-Geneva, Switzerland: Bousquet.

———. 1766. De motu vibratorio tympanorum. *Novi Commentarii Academiae Scientiarum Imperialis Petropolitanae* 10:243–60.

Galileo Galilei. 1638. *Dialogues concerning two new sciences.*

Kelly, S. G. 2007. *Advanced vibration analysis.* Boca Raton, FL: Taylor & Francis.

Kirchhoff, G. 1850. Uber das gleichgwich und die bewegung einer elastsichen scheibe. *J. Angew. Math.* 40:51–88.

Lagrange, J. L. de. 1759. Recherches sur la méthod de maximis et minimis. *Misc. Taurinensia, Torino* 1, *Oeuvres* 1, 3–20.

Leibniz, G. W. 1693. Supplementum geometriae dimensoriae, seu generalissima omnium tetragonismorum effectio per motum: Similiterque multiplex constructio lineae ex data tangentium conditione. *Acta Eruditorum* 110:294–301 and 166:282–84.

Leissa, A. W. 1969. *Vibration of plates.* NASA SP-160. Washington, DC: U.S. Government Printing Office. Repr. Sewickley, PA: Acoustical Society of America, 1993.

Liew, K. M., and C. M. Wang. 1992. Vibration analysis of plates by pb-2 Rayleigh-Ritz method: Mixed boundary conditions, reentrant corners and curved internal supports. *Mechanics of Structures and Machines* 20 (3): 281–92.

———. 1993. pb-2 Rayleigh-Ritz method for general plate analysis. *Engineering Structures* 15 (1): 55–60.

Marinus Mersenne. 1636. *Harmonicorum liber.* Lvtetiae Parisiorvum: Sumptibus Gvillielmi Bavdry.

McConnell, K. G., and P. S. Varoto. 2008. *Vibration testing: Theory and practice.* 2nd ed. Hoboken, NJ: John Wiley & Sons.

Mindlin, R. D. 1951. Influence of rotary inertia and shear on flexural motions of isotropic, elastic plates. *Journal of Applied Mechanics* 18:31–38.

Navier, C. L. M. H. 1823. Extrait des recherches sur la flexion des plans elastiques. *Bull. Sci. Soc. Philomarhique de Paris* 5:95–102.

Newton, I. 1687. *Philosophiae naturalis principia mathematica* [Mathematical principles of natural philosophy]. Cambridge, England.

Pagani, M. 1829. *Note sur le mouvement vibratoire d'une membrane elastique de forme circulaire.* Brussels: Royal Academy of Science.

Poisson, S. D. 1829. Memoire sur l'équilibre et le mouvement des corps élastiques. *Mem. Acad. Roy. Des Sci. de L'Inst. France* Ser. 2 8:357.

Rao, S. S. 1986. *Mechanical vibrations.* 2nd ed. Reading, MA: Addison-Wesley.

———. 2005. *Vibration of continuous systems.* Hoboken, NJ: John Wiley & Sons.

Rayleigh, J. W. 1877. *Theory of sound.* Vol. 1. New York: Macmillan. Repr. New York: Dover, 1945.

Reddy, J. N. 1984. A simple higher-order theory for laminated composite plates. *Journal of Applied Mechanics* 51:745–52.

Reddy, J. N., and N. D. Phan. 1985. Stability and vibration of isotropic, orthotropic and laminated plates according to a higher-order shear deformation theory. *Journal of Sound and Vibration* 98 (2): 157–70.

Richardson, M. H., and K. A. Ramsey. 1981. Integration of dynamic testing into the product design cycle. *Sound and Vibration* 15 (11): 14–27.

Ritz, W. 1909. Uber eine neue methode zur lösung gewisser variations: Probleme der mathematischen physik. *Journal fur Reine und Angewandte Mathematik* 135:1–61.

Timoshenko, S. P. 1921. On the correction for shear of the differential equation for transverse vibration of prismatic bars. *Philosophical Magazine* 41:744–46.

Voigt, W. 1893. Bemerkungen zu dem problem der transversalen schwingungen rechteckiger platten. *Nachr. Ges. Wiss* (Göttingen) 6:225–30.

Wowk, V. 1991. *Mechanical vibrations: Measurement and analysis.* New York: McGraw-Hill.

2 Vibration of Strings

2.1 INTRODUCTION

Strings are basic structural elements that support only tension. They also approximate cables, with negligible bending, and chains, with numerous links. There seems to be no exact solution to the vibration of strings where the static shape deviates from a straight line. We assume that the tension, density, and deflection are continuous functions of position along the string, except perhaps at a single point in the interior span of the string. Thus, we limit our presentation to a string with at most two connected segments. For example, we include a string composed of two segments of different constant densities, but not multiple segments, since the solution of the latter can be extended similarly.

In the tables and figures of this chapter, we present the first five natural frequencies of vibration. The lowest one is the fundamental frequency, below which no natural vibration would occur.

2.2 ASSUMPTIONS AND GOVERNING EQUATIONS FOR STRINGS

A string is slender, i.e., its lateral dimensions are infinitesimal compared to the longitudinal length. It does not admit any bending moment, shear, or axial compression. Rotational inertia is negligible. We assume that there is a stable equilibrium straight state. The vibrations, mainly lateral, are small compared to the string length, which is finite.

One can derive the string equations by considering the dynamic balance on an elemental segment as shown in Figure 2.1, or if damping is absent, by the energy method. The governing equation of a string, derived in many texts (e.g., see Magrab 2004), is given by

$$\frac{\partial}{\partial x'}\left(T'(x')\frac{\partial w'}{\partial x'}\right)=\rho(x')\frac{\partial^2 w'}{\partial t'^2} \tag{2.1}$$

Here w' is the lateral deflection, x' is the distance from one end, $T'(x')>0$ is the tension, $\rho(x')>0$ is the density (mass per unit length), and t' is the time. The tension is governed by the static force balance, i.e.,

$$\frac{dT'}{dx'}+f'(x')=0 \tag{2.2}$$

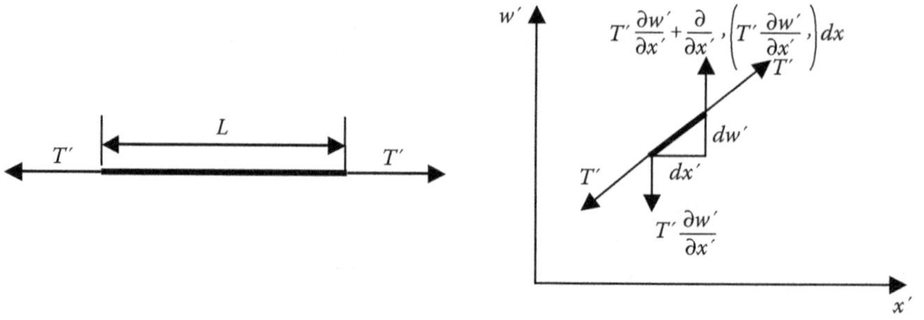

FIGURE 2.1 String under tension and an elemental segment of string.

where f' is the force per length acting along the string. Of particular interest is when the string is hanging (i.e., f' is constant).

By normalizing all lengths by the string length L, the tension by the maximum tension T_0, the density by the maximum density ρ_0, and the time by $L\sqrt{\rho_0/T_0}$ and dropping the primes, Equation (2.1) becomes

$$\frac{\partial}{\partial x}\left(T\frac{\partial w'}{\partial x}\right)=\rho\frac{\partial^2 w'}{\partial t^2} \tag{2.3}$$

For free vibrations, we can assume that

$$w'(x',t')=\bar{w}(x)e^{i\bar{\omega}t'} \tag{2.4}$$

where $i=\sqrt{-1}$ and $\bar{\omega}$ is the angular frequency of vibration. Let $w=\bar{w}/L, T=T'/T_0$, $t=t'/(L\sqrt{\rho_0/T_0})$, $\omega=\bar{\omega}L\sqrt{\rho_0/T_0}$, and recognizing that only the real part of w has significance, Equation (2.3) becomes

$$\frac{d}{dx}\left(T\frac{dw}{dx}\right)+\rho\omega^2 w=0 \tag{2.5}$$

2.3 BOUNDARY CONDITIONS

The boundary conditions at an end of a string include

- Fixed end, where

$$w=0 \tag{2.6}$$

- Sliding end, where there is no transverse resistance

$$\frac{dw}{dx}=0 \tag{2.7}$$

- Massed end, where, by transverse force balance,

$$T'\frac{\partial w'}{\partial x'} = \mp m\frac{\partial^2 w'}{\partial t'^2} \tag{2.8}$$

or in a nondimensional form

$$T\frac{dw}{dx} \mp \alpha\omega^2 w = 0 \tag{2.9}$$

where m is the point mass at the end, $\alpha = m/\rho_0 L$ is a mass ratio, the top sign is for an end with the normal in the x-direction, and the bottom sign otherwise.

- Elastically lateral supported end, where

$$T'\frac{\partial w'}{\partial x'} = \mp kw' \tag{2.10}$$

or in a nondimensional form

$$T\frac{dw}{dx} = \mp \beta w \tag{2.11}$$

where k is the spring constant and $\beta = kL/T_0$ is a normalized spring constant.

There are other boundary conditions, such as viscous dashpots, which are not as important. The aforementioned boundary conditions can be combined into a canonical form, i.e.,

$$T\frac{dw}{dx} \mp (\alpha\omega^2 + \beta)w = 0 \tag{2.12}$$

For a fixed end, α or β is infinite, and for a sliding end, $\alpha = \beta = 0$.

2.4 CONSTANT PROPERTY STRING

In this case, the tension and density are constants. By setting $T = \rho = 1$, Equation (2.5) becomes

$$\frac{d^2 w}{dx^2} + \omega^2 w = 0 \tag{2.13}$$

(a) String with fixed ends

(b) String with sliding ends

(c) String with one end fixed and the other end sliding

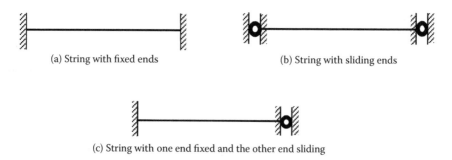

FIGURE 2.2 Strings with different end conditions.

Let α_0,β_0 be the values at $x = 0$, and α_1,β_1 be the values at $x = 1$. The solution of Equation (2.13) is

$$w = C_1 \sin(\omega x) + C_2 \cos(\omega x) \tag{2.14}$$

In view of the boundary conditions in Equation (2.12), one obtains

$$\omega C_1 - (\alpha_0\omega^2 - \beta_0)C_2 = 0 \tag{2.15}$$

$$\omega(\cos\omega\, C_1 - \sin\omega\, C_2) - (\alpha_1\omega^2 - \beta_1)(\sin\omega\, C_1 + \cos\omega\, C_2) = 0 \tag{2.16}$$

For nontrivial C_1,C_2, the exact characteristic equation for the frequency ω is

$$\omega[(\alpha_1\omega^2 - \beta_1)\cos\omega + \omega\sin\omega] - (\alpha_0\omega^2 - \beta_0)[\omega\cos\omega - (\alpha_1\omega^2 - \beta_1)\sin\omega] = 0 \tag{2.17}$$

If both ends are fixed, set α_0 or β_0, and α_1 or β_1 to infinity. Thus, we obtain $\sin\omega = 0$ or $\omega = n\pi$, where n is a positive integer. The fundamental frequency, or the frequency below which the string would not vibrate, is $\omega = \pi$. If both ends are sliding, set $\alpha_0 = \beta_0 = 0$ and $\alpha_1 = \beta_1 = 0$, and the frequencies are the same, i.e., $\omega = n\pi$. For one end fixed and one end sliding, we find $\cos\omega = 0$ or $\omega = (n - 1/2)\pi$. The frequencies for other combinations can be generated from Equation (2.17). Strings with different end conditions are shown in Figure 2.2. Mode shapes for strings with different end conditions are shown in Figures 2.3a, 2.3b, and 2.3c. Since the vibration amplitudes are arbitrary, they are made equal in the figures.

2.5 TWO-SEGMENT CONSTANT PROPERTY STRING

We consider a composite string composed of two connected constant-property segments. Let a subscript 1 denote the segment $0 \le x \le b$ and a subscript 2 denote the segment $b \le x \le 1$. At the joint, the string is continuous

$$w_1(b) = w_2(b) \tag{2.18}$$

Also there may be a point mass and a supporting spring at the joint. By carrying out a transverse force balance at $x = b$, one obtains

$$T_2 \frac{dw_2}{dx} - T_1 \frac{dw_1}{dx} + (\alpha\omega^2 - \beta)w = 0 \tag{2.19}$$

Three important cases will be illustrated. In each case we assume that the tension is the same, i.e., $T_1 = T_2 = 1$, and the ends are fixed.

2.5.1 Different Densities

Figure 2.4 shows a two-segment composite string. Segment 1 has the maximum density ($\rho_1 = 1$), whereas segment 2 has the smaller density ($\rho_2 < \rho_1$).

Let $\gamma = \rho_2/\rho_1 \leq 1$. The governing equations are

$$\frac{d^2 w_1}{dx^2} + \omega^2 w_1 = 0 \tag{2.20}$$

$$\frac{d^2 w_2}{dx^2} + \gamma\omega^2 w_2 = 0 \tag{2.21}$$

The boundary conditions are that the deflections are zero at the ends, i.e.,

$$w_1(0) = 0, \quad w_2(1) = 0 \tag{2.22}$$

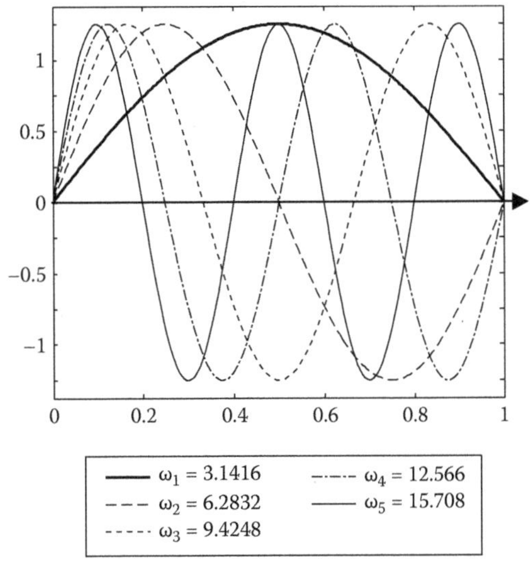

FIGURE 2.3 (a) Mode shapes for a string with fixed ends. (*continued*)

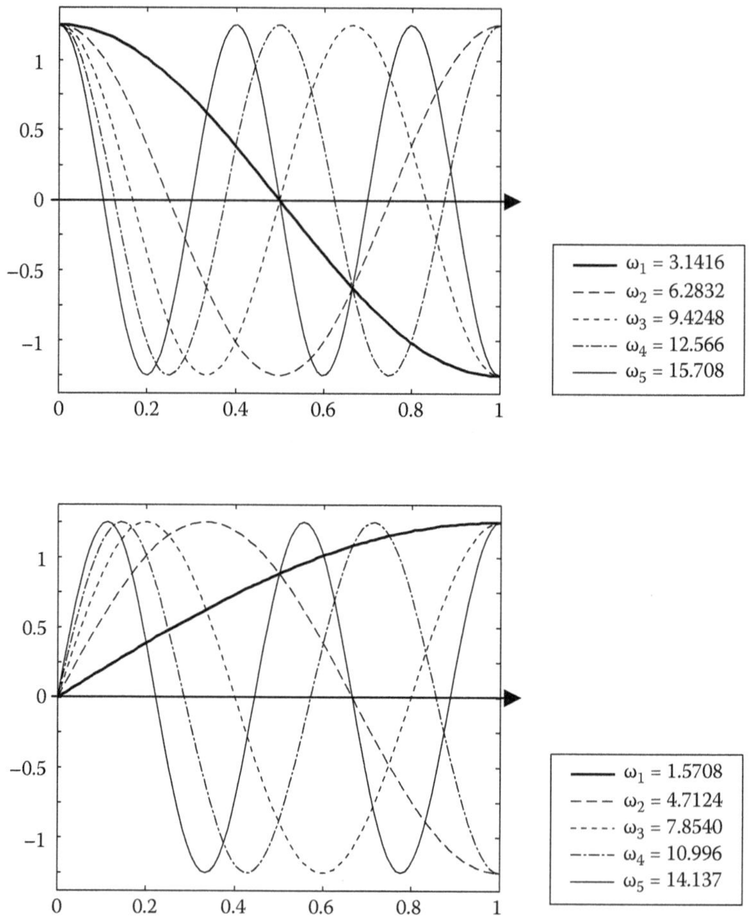

FIGURE 2.3 (b) Mode shapes for a string with sliding ends. (c) Mode shapes with one end fixed and one end sliding.

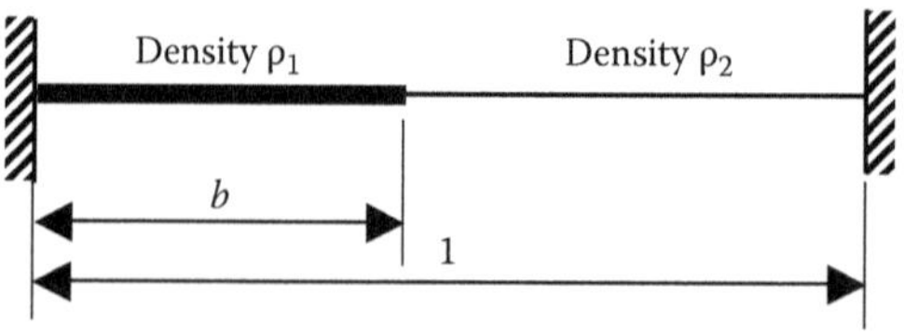

FIGURE 2.4 Composite string with two segments of different densities.

The solutions for the foregoing governing equations and boundary conditions are

$$w_1 = C_1 \sin(\omega x), \quad w_2 = C_2 \sin\left[\sqrt{\gamma}\omega(x-1)\right] \tag{2.23}$$

At the joint, we have from Equation (2.19)

$$\frac{dw_1}{dx}(b) = \frac{dw_2}{dx}(b) \tag{2.24}$$

Equations (2.18) and (2.24) yield the characteristic equation

$$\sqrt{\gamma}\sin(\omega b)\cos\left[\sqrt{\gamma}\omega(1-b)\right] + \cos(\omega b)\sin\left[\sqrt{\gamma}\omega(1-b)\right] = 0 \tag{2.25}$$

Equation (2.25) is equivalent to that found by Levinson (1976).

The first five frequencies for various γ and b are given in Table 2.1. Notice that the higher the average density, the lower is the frequency. Figure 2.5 shows sample mode shapes of a two-segment string with $\gamma = 0.5$, $b = 0.5$.

TABLE 2.1
Frequencies for Two-Segment String

b	ω	$\gamma = 0.1$	$\gamma = 0.3$	$\gamma = 0.5$	$\gamma = 0.7$	$\gamma = 0.9$
0.1	ω_1	9.5592	5.6896	4.4281	3.7497	3.3103
	ω_2	16.112	11.063	8.7665	7.4690	6.6140
	ω_3	23.223	15.872	12.941	11.136	9.9067
	ω_4	32.944	20.714	16.994	14.754	13.189
	ω_5	42.260	26.121	21.109	18.358	16.464
0.3	ω_1	5.5976	4.7628	4.1072	3.6338	3.2841
	ω_2	12.929	8.9029	7.6873	7.0030	6.4982
	ω_3	17.060	14.192	12.906	10.651	9.7650
	ω_4	25.060	17.955	15.794	14.270	13.062
	ω_5	29.373	23.387	19.622	17.641	16.284
0.5	ω_1	3.9648	3.7728	3.5799	3.3945	3.2221
	ω_2	9.4712	8.4791	7.5415	6.8937	6.4531
	ω_3	15.065	12.024	10.840	10.191	9.6663
	ω_4	19.354	16.060	14.888	13.771	12.906
	ω_5	23.038	20.681	18.314	17.010	16.111
0.7	ω_1	3.3401	3.2984	3.2552	3.2106	3.1648
	ω_2	7.2563	7.0657	6.8509	6.6226	6.3939
	ω_3	11.455	11.029	10.506	10.002	9.5921
	ω_4	15.743	14.480	13.864	13.181	12.741
	ω_5	20.042	18.299	17.052	16.434	15.945
0.9	ω_1	3.1505	3.1486	3.1466	3.1446	3.1426
	ω_2	6.3462	6.3331	6.3196	6.3054	6.2908
	ω_3	9.6043	9.5694	9.5318	9.4913	9.4477
	ω_4	12.918	12.854	12.782	12.702	12.614
	ω_5	16.275	16.175	16.061	15.930	15.785

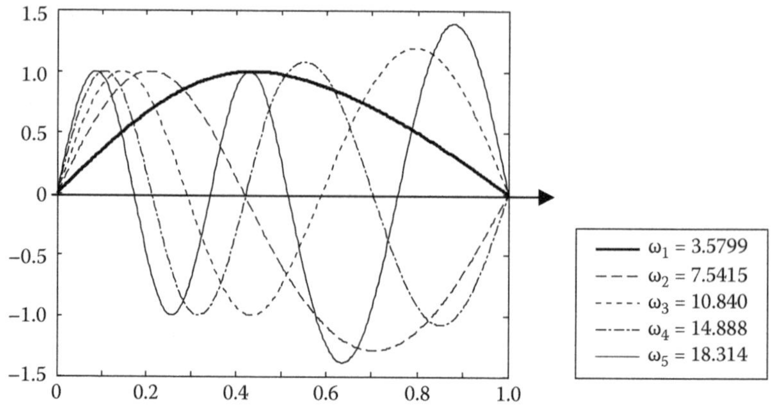

FIGURE 2.5 Mode shapes for two-segment string ($\gamma = 0.5, b = 0.5$).

2.5.2 A Mass Attached on the Span

Consider a string of constant property with a mass m attached on the span as shown in Figure 2.6. The solution is given by Equation (2.23) with $\gamma = 1$. The condition at the joint is

$$\frac{dw_2}{dx} - \frac{dw_1}{dx} + \alpha\omega^2 w = 0 \tag{2.26}$$

Equations (2.18) and (2.26) yield

$$\alpha\omega \sin(\omega b) \sin[\omega(1 - b)] - \sin \omega = 0 \tag{2.27}$$

Equation (2.27) is equivalent to that found by Chen (1963).

Table 2.2 shows the results. Since symmetry is present, the frequencies are evaluated only for $b \le 0.5$. Frequency decreases with increased mass and more centered location. Notice that the fifth frequency of $b = 0.2$ and $b = 0.4$ is 15.708 regardless of the attached mass. This is because the mass is at one of the nodes of the fifth mode. Similarly for the centered mass ($b = 0.5$), the second and the fourth frequencies are independent of mass. Sample mode shapes for a string with an attached mass at its mid-span are shown in Figure 2.7.

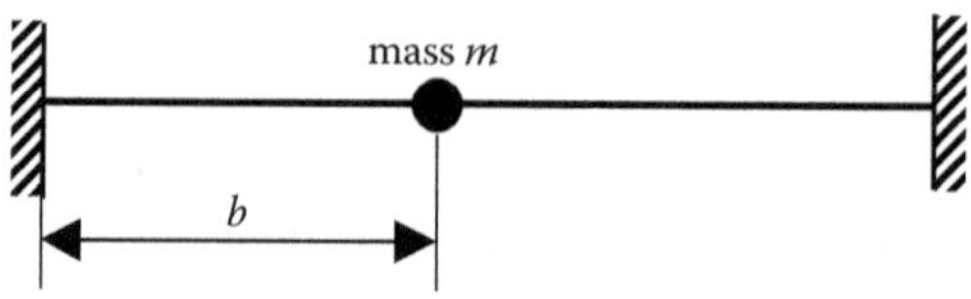

FIGURE 2.6 String with an attached mass m.

TABLE 2.2
Frequencies for String with a Mass

b	ω	$\alpha = 0.5$	$\alpha = 1$	$\alpha = 1.5$	$\alpha = 2$	$\alpha = 2.5$	$\alpha = 3$
0.1	ω_1	2.9370	2.6444	2.3602	2.1306	1.9503	1.8066
	ω_2	4.7680	4.0682	3.8227	3.7159	3.6593	3.6249
	ω_3	7.3897	7.1643	7.0981	7.0669	7.0488	7.0370
	ω_4	10.700	10.583	10.545	10.526	10.515	10.508
	ω_5	14.123	14.043	14.016	14.003	13.995	13.989
0.2	ω_1	2.5525	2.1099	1.8226	1.6241	1.4777	1.3644
	ω_2	4.6365	4.2830	4.1589	4.0980	4.0622	4.0387
	ω_3	8.1506	8.0087	7.9583	7.9326	7.9170	7.9066
	ω_4	11.957	11.878	11.847	11.832	11.822	11.815
	ω_5	15.708	15.708	15.708	15.708	15.708	15.708
0.3	ω_1	2.3116	1.8662	1.6013	1.4230	1.2930	1.1930
	ω_2	5.0394	4.7902	4.6943	4.6442	4.6136	4.5930
	ω_3	9.1778	9.01010	9.0661	9.0463	9.0337	9.0248
	ω_4	11.780	10.789	10.685	10.632	10.560	10.579
	ω_5	13.694	13.575	13.537	13.518	13.507	13.500
0.4	ω_1	2.1906	1.7539	1.5014	1.3330	1.2106	1.1167
	ω_2	5.6780	5.5013	5.4242	5.3816	5.3546	5.3360
	ω_3	8.3965	8.01525	8.0583	8.0090	7.9787	7.9584
	ω_4	10.802	10.636	10.580	10.553	10.536	10.526
	ω_5	15.708	15.708	15.708	15.708	15.708	15.708
0.5	ω_1	2.1538	1.7207	1.4720	1.3065	1.1865	1.0943
	ω_2	6.2832	6.2832	6.2832	6.2832	6.2832	6.2832
	ω_3	7.2872	6.8512	6.6774	6.5846	6.5271	6.4880
	ω_4	12.566	12.566	12.566	12.566	12.566	12.566
	ω_5	13.157	12.875	12.774	12.723	12.692	12.672

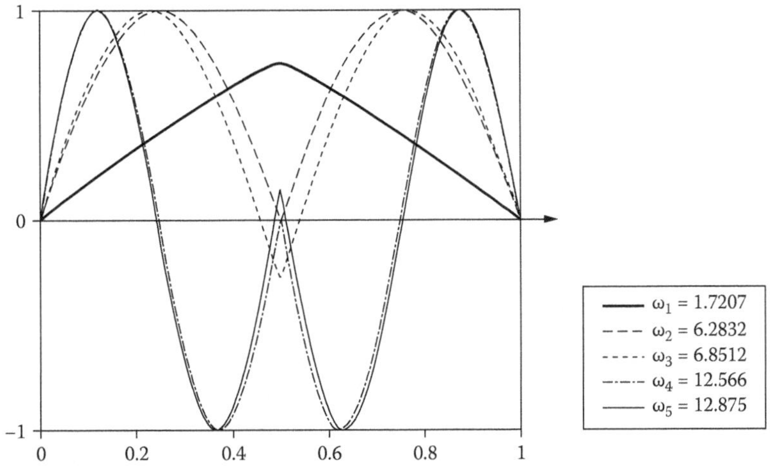

FIGURE 2.7 Mode shapes of string with an attached mass at mid-span (i.e., $\alpha = 1, b = 0.5$).

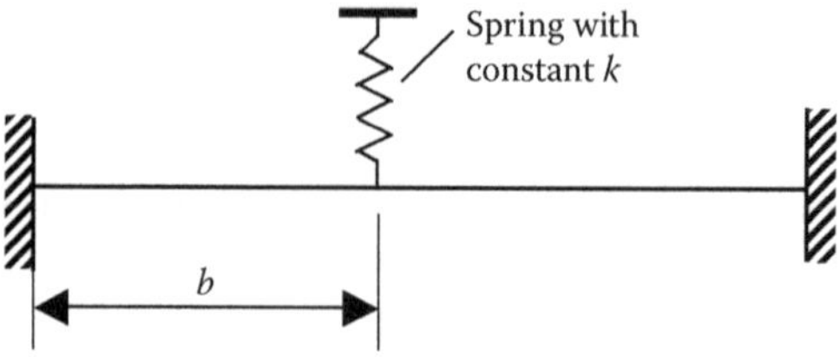

FIGURE 2.8 Fixed-ended string with lateral spring support.

2.5.3 A Supporting Spring on the Span

Figure 2.8 shows a string being supported by an elastic spring with a spring constant k. The condition at the joint is

$$\frac{dw_2}{dx} - \frac{dw_1}{dx} - \beta w = 0 \tag{2.28}$$

Equations (2.18), (2.23), and (2.28) with $\gamma = 1$ give

$$\beta \sin(\omega b) \sin[\omega(1 - b)] + \omega \sin \omega = 0 \tag{2.29}$$

The frequencies are shown in Table 2.3. In this case, the frequency is increased by the stiffness of the spring and by a more centered location. Again at the nodal points, the spring has no effect on the frequency. Sample mode shapes for string supported by a lateral spring with $b = 0.5$, $\beta = 20$ are shown in Figure 2.9.

2.6 TRANSFORMATION FOR NONUNIFORM TENSION AND DENSITY

If tension and/or density vary along the string, Equation (2.5) may not have an exact solution. The following transformation, somewhat more general than that suggested by Horgan and Chan (1999), reduces it to the Liouville normal form. Let

$$w(x) = W(z) \tag{2.30}$$

$$z(x) \equiv \frac{1}{z_1}\int_0^x \frac{1}{T(s)}ds, \qquad z_1 \equiv \int_0^1 \frac{1}{T(s)}ds \tag{2.31}$$

We see that z is also in [0,1]. Then Equation (2.5) becomes

$$\frac{d^2W}{dz^2} + (z_1^2 \rho T)\omega^2 W = 0 \tag{2.32}$$

TABLE 2.3
Frequencies for Spring-Supported String

b	ω	$\beta = 5$	$\beta = 10$	$\beta = 15$	$\beta = 20$	$\beta = 25$	$\beta = 30$
0.1	ω_1	3.2482	3.3053	3.3408	3.3650	3.3824	3.3957
	ω_2	6.4855	6.6006	6.6735	6.7234	6.7596	6.7870
	ω_3	9.7004	9.8734	9.9874	10.067	10.125	10.168
	ω_4	12.880	13.106	13.266	13.381	13.466	13.530
	ω_5	16.013	16.273	16.480	16.638	16.759	16.852
0.2	ω_1	3.4742	3.6134	3.6881	3.7343	3.7657	3.7883
	ω_2	6.8328	7.1362	7.3121	7.4225	7.4970	7.5503
	ω_3	9.9048	10.325	10.648	10.884	11.053	11.176
	ω_4	12.718	12.894	13.086	13.280	13.468	13.642
	ω_5	15.708	15.708	15.708	15.708	15.708	15.708
0.3	ω_1	3.7640	4.0104	4.1355	4.2097	4.2585	4.2930
	ω_2	6.9506	7.4562	7.8130	8.0633	8.2411	8.3700
	ω_3	9.4782	9.5360	9.5964	9.6569	9.7154	9.7703
	ω_4	12.692	12.795	12.878	12.946	13.001	13.047
	ω_5	16.017	16.295	16.535	16.736	16.901	17.036
0.4	ω_1	4.0621	4.4690	4.6813	4.8065	4.8874	4.9434
	ω_2	6.5559	6.7925	6.9765	7.1153	7.2205	7.3017
	ω_3	9.5936	9.7311	9.8394	9.9244	9.9915	10.045
	ω_4	12.919	13.244	13.528	13.770	13.974	14.146
	ω_5	15.708	15.708	15.708	15.708	15.708	15.708
0.5	ω_1	4.2128	4.7613	5.0906	5.3073	5.4596	5.5719
	ω_2	6.2832	6.2832	6.2832	6.2832	6.2832	6.2832
	ω_3	9.9186	10.327	10.652	10.909	11.113	11.277
	ω_4	12.566	12.566	12.566	12.566	12.566	12.566
	ω_5	16.018	16.303	16.559	16.783	16.977	17.146

With the boundary condition at any end, Equation (2.12) becomes

$$\frac{dW}{dz} \mp (z_1\alpha\omega^2 - z_1\beta)W = 0 \tag{2.33}$$

In the special case,

$$z_1^2\rho(x)T(x) = 1 \tag{2.34}$$

and if α and β is adjusted to $z_1\bar{\alpha}$ and $z_1\bar{\beta}$, then the nonuniform problem reduces exactly to the uniform string problem. The frequencies ω are exactly the same. Two different problems having exactly the same eigenvalues or frequencies are called isospectral problems. Therefore, for tension inversely proportional to density (there are infinite different combinations), the frequency of the uniform string can be applied. There are, however, other cases where the product ρT is not constant.

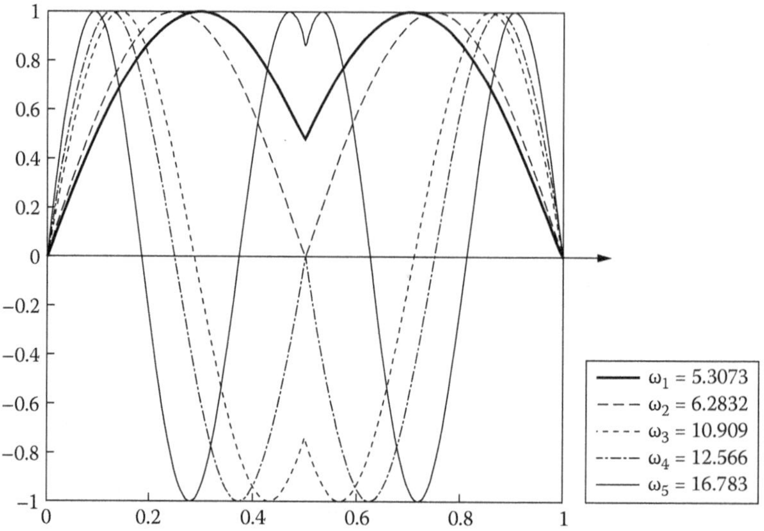

FIGURE 2.9 Mode shapes of a string supported by a spring at mid-length ($b = 0.5$, $\beta = 20$).

2.7 CONSTANT TENSION AND VARIABLE DENSITY

By setting $T = 1$, Equation (2.5) gives

$$\frac{d^2w}{dx^2} + \rho(x)\omega^2 w = 0 \tag{2.35}$$

Depending on the form of the density, there are many exact solutions to Equation (2.35). We shall discuss the more important cases, namely, the power law density distribution and the exponential density distribution. Some specific examples were given by Horgan and Chan (1999).

2.7.1 Power Law Density Distribution

In this case,

$$\rho = (1 - cx)^m \tag{2.36}$$

Let the maximum density be at $x = 1$. If m is positive, $1 > c > 0$, whereas if m is negative, $c < 0$. If $m \neq -2$, let

$$\eta = \left(\frac{\omega}{c}\right)^{\frac{2}{m+2}} (1 - cx) \tag{2.37}$$

Then Equation (2.35) becomes

$$\frac{d^2w}{d\eta^2} + \eta^m w = 0 \tag{2.38}$$

The solution is given by (e.g., Murphy 1960)

$$w = C_1\sqrt{\eta}J_{1/p}\left(\frac{2}{p}\eta^{p/2}\right) + C_2\sqrt{\eta}J_{-1/p}\left(\frac{2}{p}\eta^{p/2}\right) \tag{2.39}$$

where $p = m + 2$ and J's are Bessel functions. In the case of $1/p$ being an integer, $J_{-1/p}$ is replaced by the second kind $Y_{1/p}$. From the end conditions, we can determine the constants C_1, C_2. Suppose the ends are fixed (other boundary conditions may apply). At the ends, let

$$\eta_0 = \left(\frac{\omega}{c}\right)^{\frac{2}{m+2}}, \qquad \eta_1 = \left(\frac{\omega}{c}\right)^{\frac{2}{m+2}}(1-c) \tag{2.40}$$

Then for nontrivial values of C_1,C_2, Equation (2.39) gives the characteristic equation

$$J_{1/p}\left(\frac{2}{p}\eta_0^{p/2}\right)J_{-1/p}\left(\frac{2}{p}\eta_1^{p/2}\right) - J_{-1/p}\left(\frac{2}{p}\eta_0^{p/2}\right)J_{1/p}\left(\frac{2}{p}\eta_1^{p/2}\right) = 0 \tag{2.41}$$

If $m = 1$, the Bessel functions in Equation (2.41) can be expressed in Airy functions. This special case was given by Fulcher (1985). If $m = -4$, the Bessel functions reduce to harmonic functions

$$\sin\left(\frac{1}{\eta_0} - \frac{1}{\eta_1}\right) = 0 \tag{2.42}$$

or

$$\omega = n\pi(1-c) \tag{2.43}$$

This solution was found by Borg (1946).

If $m = -2$, let

$$\eta = 1 - cx \tag{2.44}$$

Equation (2.35) becomes

$$\frac{d^2w}{d\eta^2} + \left(\frac{\omega}{c}\right)^2 \frac{1}{\eta^2} w = 0 \tag{2.45}$$

The solutions are in terms of η powers. If $\omega < |c|/2$, the solution is

$$w = C_1 \eta^{\frac{1+\sqrt{1-4\omega^2/c^2}}{2}} + C_2 \eta^{\frac{1+\sqrt{1-4\omega^2/c^2}}{2}} \tag{2.46}$$

If $\omega > |c|/2$, the solution is

$$w = C_1 \sqrt{\eta} \sin\left(\frac{\sqrt{4\omega^2/c^2 - 1}}{2} + \ln \eta \right) + C_2 \sqrt{\eta} \cos\left(\frac{\sqrt{4\omega^2/c^2 - 1}}{2} + \ln \eta \right) \tag{2.47}$$

If $\omega = |c|/2$, the solution is

$$w = C_1 \sqrt{\eta} + C_2 \sqrt{\eta} \ln \eta \tag{2.48}$$

If both ends are fixed, only Equation (2.47) can have nontrivial solutions. The frequency is governed by

$$\frac{\sqrt{4\omega^2/c^2 - 1}}{2} \ln(1-c) = n\pi \tag{2.49}$$

or the frequency is

$$\omega = \frac{|c|}{2} \sqrt{1 + \left(\frac{2n\pi}{\ln(1-c)} \right)^2} \tag{2.50}$$

Tables 2.4 and 2.5 show the vibration frequencies for various m and c values. All entries are from Equation (2.41), except for the $m = -2$ case, which is from Equation (2.50).

In particular, if $m = 1$ the density describes a homogeneous string with a cross section of constant width and a height that tapers linearly along the axis. If $m = 2$, the cross section is similar along the axis, and both width and height taper linearly. Sample mode shapes for a string with a constant tension and variable density are shown in Figure 2.10.

2.7.2 Exponential Density Distribution

Consider the case when the density decreases exponentially, i.e.,

$$\rho = e^{-bx} \tag{2.51}$$

Let

$$\eta = e^{-bx/2} \tag{2.52}$$

TABLE 2.4
Frequencies for Power Law Density $m \geq 0$

m	ω	$c = 0.2$	$c = 0.4$	$c = 0.6$	$c = 0.8$	$c = 1$
0.5	ω_1	3.2256	3.3224	3.4364	3.5743	3.7486
	ω_2	6.4528	6.6537	6.9000	7.2179	7.6664
	ω_3	9.6797	9.9832	10.359	10.854	11.590
	ω_4	12.907	13.312	13.816	14.486	15.515
	ω_5	16.133	16.641	17.272	18.116	19.441
1	ω_1	3.3106	3.5077	3.7402	4.0181	4.3537
	ω_2	6.6252	7.0359	7.5434	8.1920	9.0486
	ω_3	9.9388	10.560	11.336	12.352	13.755
	ω_4	13.252	14.083	15.126	16.505	18.464
	ω_5	16.566	17.605	18.913	20.654	23.174
1.5	ω_1	3.3969	3.6971	4.0519	4.4703	4.9579
	ω_2	6.8002	7.4294	8.2109	9.1962	10.430
	ω_3	10.202	11.155	12.354	13.905	15.919
	ω_4	13.604	14.878	16.491	18.604	21.413
	ω_5	17.005	18.601	20.626	23.297	26.907
2	ω_1	3.4841	3.8905	4.3702	4.9286	5.5614
	ω_2	6.9780	7.8338	8.8998	10.223	11.811
	ω_3	10.470	11.767	13.408	15.499	18.083
	ω_4	13.961	15.698	17.909	20.764	24.361
	ω_5	17.452	19.627	22.404	26.022	30.640

and Equation (2.35) becomes the Bessel equation

$$\eta \frac{d^2 w}{d\eta^2} + \frac{dw}{d\eta} + \frac{4\omega^2}{b^2}\eta w = 0 \tag{2.53}$$

The solution is

$$w = C_1 J_0\left(\frac{2\omega}{b}e^{-bx/2}\right) + C_2 Y_0\left(\frac{2\omega}{b}e^{-bx/2}\right) \tag{2.54}$$

The characteristic equation is

$$J_0\left(\frac{2\omega}{b}\right)Y_0\left(\frac{2\omega}{b}e^{-b/2}\right) - Y_0\left(\frac{2\omega}{b}\right)J_0\left(\frac{2\omega}{b}e^{-b/2}\right) = 0 \tag{2.55}$$

The results are shown in Table 2.6. Sample mode shapes for a string with an exponential density distribution are shown in Figure 2.11.

TABLE 2.5
Frequencies for Power Law Density $m \leq 0$

m	ω	$c = -0.5$	$c = -1$	$c = -1.5$	$c = -2$	$c = -2.5$
−0.5	ω_1	3.3178	3.4649	3.5927	3.7064	3.8095
	ω_2	6.6312	6.9164	7.1617	7.3786	7.5743
	ω_3	9.9455	10.371	10.736	11.058	11.347
	ω_4	13.260	13.826	14.311	14.738	15.123
	ω_5	16.575	17.281	17.886	18.420	18.900
−1	ω_1	3.5001	3.8094	4.0862	4.3394	4.5745
	ω_2	6.9920	7.5932	8.1252	8.6079	9.0531
	ω_3	10.486	11.383	12.174	12.891	13.552
	ω_4	13.980	15.173	16.226	17.179	18.055
	ω_5	17.474	18.965	20.279	21.468	22.561
−1.5	ω_1	3.6802	4.1746	4.6216	5.0401	5.4367
	ω_2	7.3657	8.3139	9.1755	9.9751	10.727
	ω_3	11.046	12.461	13.744	14.933	16.050
	ω_4	14.726	16.610	18.317	14.897	21.380
	ω_5	18.407	20.760	22.891	24.863	26.714
−2	ω_1	3.8821	4.5599	5.1973	5.8060	6.3927
	ω_2	7.7522	9.0785	10.313	11.482	12.601
	ω_3	11.625	13.606	15.447	17.187	18.850
	ω_4	15.498	18.136	20.585	22.899	25.108
	ω_5	19.372	22.667	25.725	28.614	31.372

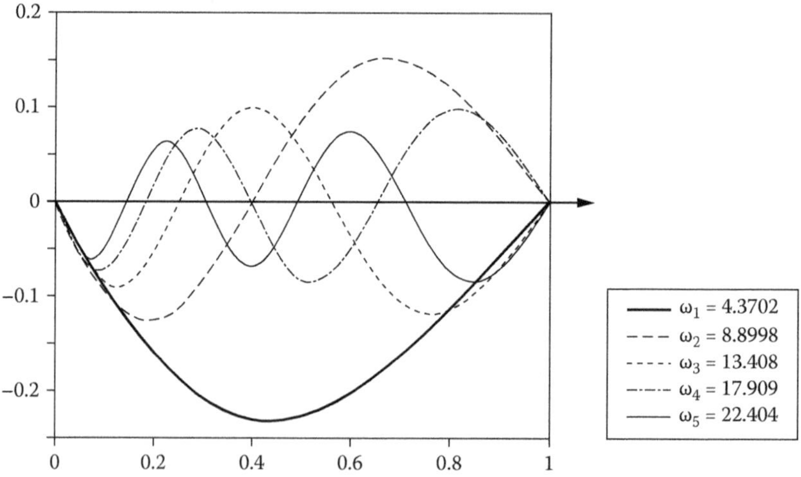

FIGURE 2.10 Constant tension and variable density with $m = 2$, $c = 0.6$.

TABLE 2.6
Frequencies for String with Exponential Density Distribution

b	0.5	1.0	1.5	2.0	2.5	3.0
ω_1	3.5478	3.9797	4.4349	4.9107	5.4046	5.9142
ω_2	7.0999	7.9780	8.9149	9.9072	10.951	12.040
ω_3	10.651	11.972	13.386	14.888	16.472	18.133
ω_4	14.202	15.966	17.854	19.863	21.985	24.214
ω_5	17.753	19.958	22.321	24.836	27.495	30.289

2.8 VARIABLE TENSION AND CONSTANT DENSITY

We shall only consider the most important variable tension case, i.e., the self weight acting on a vertically hanging string. The integration of Equation (2.2) gives the linear distribution

$$T' = T_1 + \rho_0 g(L - x') \tag{2.56}$$

where T_1 is the tension at the lower end and g is the gravitational acceleration. The maximum tension is at the upper end

$$T_0 = T_1 + \rho_0 gL \tag{2.57}$$

Thus

$$T = \frac{T'}{T_0} = 1 - ax \tag{2.58}$$

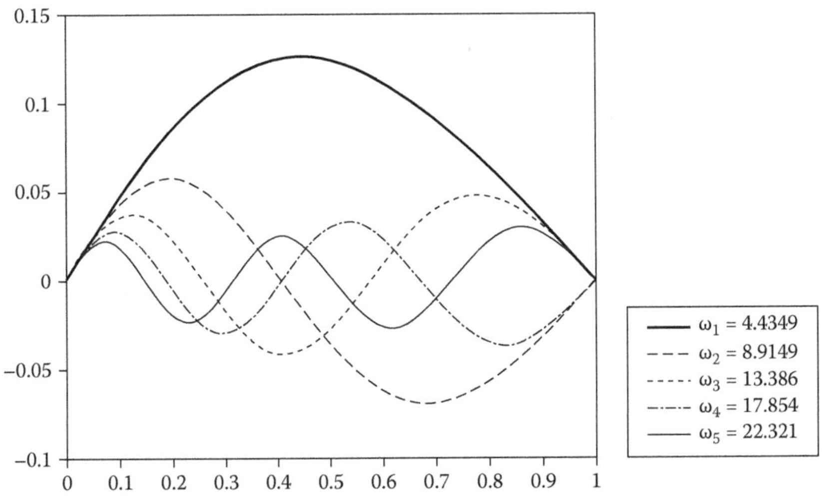

FIGURE 2.11 Mode shapes of string with exponential density distribution $b = 1.5$.

where

$$a = \frac{1}{\frac{T_1}{\rho_0 g L} + 1} < 1 \tag{2.59}$$

By integrating, Equation (2.31) gives

$$z = \frac{1}{a z_1} \ln\left(\frac{1}{1-ax}\right), \qquad z_1 = \frac{1}{a} \ln\left(\frac{1}{1-a}\right) \tag{2.60}$$

from which we find

$$T = e^{-a z_1 z} \tag{2.61}$$

By letting $y = z_1 z$, Equation (2.32) gives

$$\frac{d^2 W}{dy^2} + e^{-ay} \omega^2 W = 0 \tag{2.62}$$

This is exactly the governing equation for the exponential density distribution in Section 2.7.2. The solution is

$$W = C_1 J_0(\upsilon) + C_2 Y_0(\upsilon) \tag{2.63}$$

where

$$\upsilon = \frac{2\omega}{a} e^{-a z_1 z/2} \tag{2.64}$$

2.8.1 Vertical String Fixed at Both Ends

Consider the case when a string with fixed ends is oriented vertically as shown in Figure 2.12. Denote

$$\upsilon_1 = \upsilon\big|_{z=1} = \frac{2\omega}{a} e^{-a z_1/2} = \frac{2\omega}{a}\sqrt{1-a} \tag{2.65}$$

Since the deflection is zero at $z = 0$ and $z = 1$, Equation (2.63) yields the characteristic equation

$$J_0\left(\frac{2\omega}{a}\right) Y_0\left(\frac{2\omega}{a}\sqrt{1-a}\right) - Y_0\left(\frac{2\omega}{a}\right) J_0\left(\frac{2\omega}{a}\sqrt{1-a}\right) = 0 \tag{2.66}$$

The results are shown in Table 2.7. Note that frequencies decrease due to self weight. Sample vibration mode shapes for a vertical string fixed at both ends with self weight $a = 0.5$ are shown in Figure 2.13

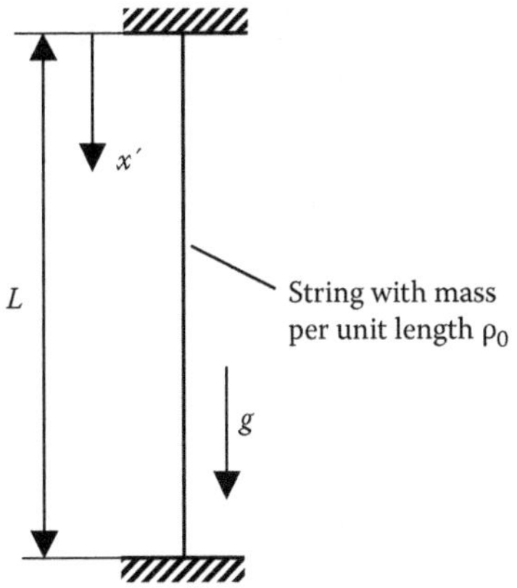

FIGURE 2.12 Vertical string with fixed ends.

TABLE 2.7
Frequencies for a Vertical String Fixed at Both Ends

a	0.1	0.3	0.5	0.7	0.9
ω_1	3.0609	2.8839	2.6775	2.4202	2.0355
ω_2	6.1219	5.7695	5.3610	4.8567	4.1169
ω_3	9.1829	8.6547	8.0432	7.2897	6.1901
ω_4	12.244	11.540	10.725	9.7218	8.2606
ω_5	15.305	14.425	13.407	12.154	10.330

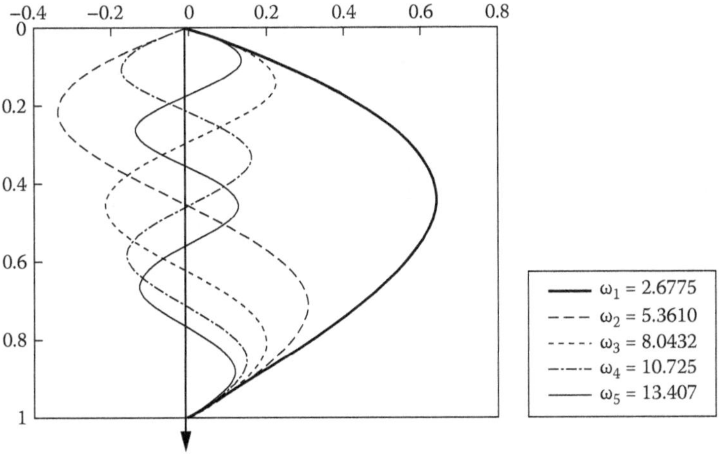

FIGURE 2.13 Mode shapes of string with self weight $a = 0.5$.

2.8.2 Vertical String with Sliding Spring on Top and a Free Mass at the Bottom

Figure 2.14 shows a hanging string with a spring-supported sliding top end and mass m hanging at the lower end (Wang and Wang 2010). This situation models a cantilever crane lift.

Now $T_1 = mg$ and

$$a = \frac{1}{\alpha + 1} \tag{2.67}$$

and α is defined as in Section 2.3. The solution is still Equation (2.63). The boundary condition at the top is

$$\frac{dW}{dz} + z_1 \beta W = 0 \tag{2.68}$$

where β is the normalized spring constant. The bottom boundary condition is

$$\frac{dW}{dz} - z_1 \alpha \omega^2 W = 0 \tag{2.69}$$

Let $v_0 = 2\omega/a$. The characteristic equation is

$$\begin{aligned}&[a\upsilon_0 J_1(\upsilon_0)/2 + \beta J_0(\upsilon_0)][a\upsilon_1 Y_1(\upsilon_1)/2 - \alpha\omega^2 Y_0(\upsilon_1)] - \\ &[a\upsilon_0 Y_1(\upsilon_0)/2 + \beta Y_0(\upsilon_0)][a\upsilon_1 J_1(\upsilon_1)/2 - \alpha\omega^2 J_0(\upsilon_1)] = 0\end{aligned} \tag{2.70}$$

Table 2.8 shows the results. In the special case of $\beta = \infty$, Equation (2.70) reduces to a characteristic equation derived by Sujith and Hodges (1995), who used a

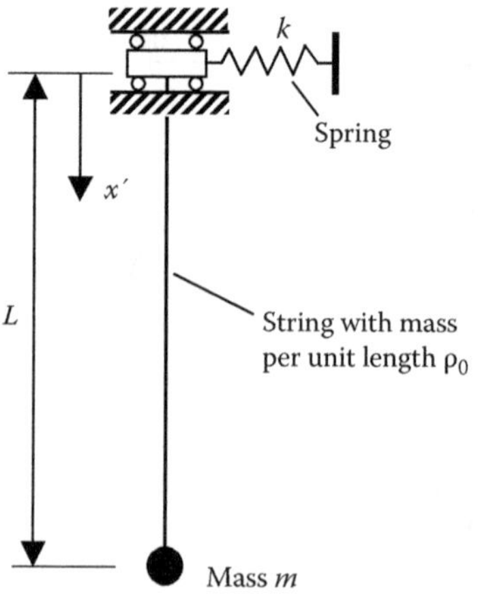

FIGURE 2.14 Hanging string with a tip mass.

TABLE 2.8
Frequencies of the Hanging String with End Mass

α	ω	$\beta = 0$	$\beta = 5$	$\beta = 10$	$\beta = 15$	$\beta = 20$	$\beta = \infty$
0	ω_1	1.9159	1.3260	1.2637	1.2431	1.2328	1.2024
	ω_2	3.5078	3.0210	2.8978	2.8525	2.8295	2.7600
	ω_3	5.0867	4.6193	4.5351	4.4694	4.4347	4.3269
	ω_4	6.6619	6.3350	6.1665	6.0856	6.0407	5.8958
	ω_5	8.2353	7.9592	7.7904	7.6992	7.6460	7.4655
0.5	ω_1	1.6757	0.9743	0.9293	0.9150	0.8873	0.8873
	ω_2	3.9089	3.1905	2.9976	2.9286	2.7923	2.7923
	ω_3	6.3137	5.7719	5.5078	5.3889	5.1279	5.1279
	ω_4	8.7581	8.3408	8.0722	7.9247	7.5504	7.5504
	ω_5	11.217	10.882	10.631	10.472	9.9994	9.9994
1	ω_1	1.6349	0.8232	0.7833	0.7708	0.7647	0.7470
	ω_2	4.1395	3.3526	3.1264	3.0453	3.0048	2.8862
	ω_3	6.7755	6.2103	5.9145	5.7770	5.7026	5.4712
	ω_4	9.4368	9.0096	8.7194	8.5531	8.4541	8.1175
	ω_5	12.107	11.767	11.502	11.327	11.213	10.781
1.5	ω_1	1.6171	0.7282	0.6917	0.6804	0.6749	0.6590
	ω_2	4.2663	3.4459	3.2009	3.1129	3.0690	2.9409
	ω_3	7.0205	6.4450	6.1330	5.9854	5.9051	5.6549
	ω_4	9.7934	9.3620	9.0613	8.8854	8.7795	8.4163
	ω_5	12.573	12.230	11.959	11.776	11.655	11.191
2	ω_1	1.6071	0.6605	0.6267	0.6163	0.6112	0.5967
	ω_2	4.3469	3.5064	3.2493	3.1568	3.1107	2.9764
	ω_3	7.1739	6.5924	6.2705	6.1165	6.0325	5.7702
	ω_4	10.016	9.5820	9.2750	9.0931	8.9829	8.6026
	ω_5	12.863	12.519	12.244	12.056	11.931	11.445
2.5	ω_1	1.6007	0.6090	0.5773	0.5676	0.5629	0.5494
	ω_2	4.4027	3.5487	3.2832	3.1875	3.1399	3.0012
	ω_3	7.2793	6.6939	6.3653	6.2069	6.1203	5.8495
	ω_4	10.168	9.7328	9.4217	9.2358	9.1225	8.7305
	ω_5	13.061	12.717	12.440	12.248	12.120	11.620
3	ω_1	1.5962	0.5680	0.5381	0.5290	0.5246	0.5119
	ω_2	4.4437	3.5801	3.3083	3.2103	3.1614	3.0196
	ω_3	7.3562	6.7682	6.4346	6.2731	6.1845	5.9075
	ω_4	10.279	9.8429	9.5289	9.3399	9.2245	8.8237
	ω_5	13.206	12.861	12.582	12.388	12.258	11.747

slightly different normalization. When $\alpha = 0$, it corresponds to the simple hanging string problem that was first solved by Bernoulli. The frequencies are the roots of $J_0(2\omega) = 0$. Notice that the fundamental frequency (the simple pendulum mode) decreases while the higher frequencies (string vibrations) increase with increased end mass.

2.9 FREE-HANGING NONUNIFORM STRING

Consider a free-hanging nonuniform string with no end mass (Wang 2011). In this case, both density and tension vary, but they are related by

$$T(x) = \int_x^1 \rho(s)\,ds \tag{2.71}$$

Let the string have a power density distribution and be pointy at the free end, i.e.,

$$\rho(x) = (1-x)^n \tag{2.72}$$

where n is any positive number. Then from Equation (2.71)

$$T(x) = \frac{(1-x)^{n+1}}{n+1} \tag{2.73}$$

Equation (2.5) becomes

$$z\frac{d^2w}{dz^2} + (n+1)\frac{dw}{dz} + (n+1)\omega^2 w = 0 \tag{2.74}$$

where $z = 1 - x$. The bounded solution is

$$w = Cz^{-n/2} J_n(2\sqrt{n+1}\,\omega\sqrt{z}) \tag{2.75}$$

If the top is fixed, the characteristic equation is

$$J_n(2\sqrt{n+1}\,\omega) = 0 \tag{2.76}$$

Table 2.9 shows the first five frequencies for various n values.

TABLE 2.9
Frequencies of the Hanging Pointy Nonuniform String

n	0	0.5	1	1.5	2
ω_1	1.2024	1.2826	1.3547	1.4209	1.4825
ω_2	2.7600	2.5651	2.4804	2.4429	2.4299
ω_3	4.3269	3.8477	3.5969	3.4482	3.3544
ω_4	5.8958	5.1302	4.7106	4.4481	4.2712
ω_5	7.4655	6.4128	5.8233	5.4457	5.1846

The $n = 0$ case represents the uniform string of Bernoulli; $n = 1$ gives a string with linear taper in one direction (wedgelike); and $n = 2$ yields a string with linear taper in two transverse directions (conelike). Note that as n increases, the fundamental frequency increases while all higher frequencies decrease.

2.10 OTHER COMBINATIONS

We did not exhaust all combinations of the product $\rho(x)T(x)$ that exactly solve Equation (2.32). Some forms may lead to Kummer functions, Mathieu functions, and hypergeometric functions (e.g., Murphy 1960). These special functions are usually not included as computer library functions, and their evaluations are very tedious and error prone, which deviates from our purpose of presenting benchmark exact solutions.

As noted in Section 2.6, there are infinitely many density-tension distributions that are isospectral to known frequencies. Most of these distributions are in unusual forms, and will not be presented here. For further research into isospectral strings, see Gottlieb (2002) and the references therein.

REFERENCES

Borg, G. 1946. Eine umkehrung der Sturm–Liouvilleschen eigenwertaufgabe: Bestimmung der differentialgleichung durch die eigenwerte. *Acta Math.* 78:1–96.

Chen, Y. 1963. Vibration of a string with attached concentrated masses. *J. Franklin Inst.* 276:191–96.

Fulcher, L. P. 1985. Study of the eigenvalues of a nonuniform string. *Am. J. Phys.* 53:730–35.

Gottlieb, H. P. W. 2002. Isospectral strings. *Inverse Prob.* 18:971–78.

Horgan, C. O., and A. M. Chan. 1999. Vibration of inhomogeneous strings, rods and membranes. *J. Sound Vibr.* 255:503–13.

Levinson, M. 1976. Vibration of stepped strings and beams. *J. Sound Vibr.* 49:287–91.

Magrab, E. B. 2004. *Vibration of elastic structural members.* New York: Springer.

Murphy, G. M. 1960. *Ordinary differential equations and their solutions.* Princeton, NJ: Van Nostrand.

Sujith, R. I., and D. H. Hodges. 1995. Exact solution for the free vibration of a hanging cord with a tip mass. *J. Sound Vibr.* 179:359–61.

Wang, C. Y. 2011. Vibration of a hanging tapered string with or without a tip mass. *Eur. J. Phys.* 32:L29–L34.

Wang, C. Y., and C. M. Wang. 2010. Exact solutions for vibration of a vertical heavy string with tip mass. *The IES Journal A: Civil and Structural Engineering* 3:278–81.

3 Vibration of Membranes

3.1 INTRODUCTION

A membrane is a two-dimensional surface whose equilibrium shape is maintained by tension. We shall restrict our study to flat membranes. There are inflated membranes, such as balloons, but there seem to be no exact vibration solutions. Membranes are used in drums, receivers, diaphragms, drapes, and tents, and they also model nets. Membrane vibration is related to the vibration of some simply supported plates (see Chapter 5) and also the TM (transverse magnetic) waves of electromagnetic waveguides.

3.2 ASSUMPTIONS AND GOVERNING EQUATIONS

We assume that the thickness of a membrane is small compared to its lateral dimensions and that the membrane has a finite area. The membrane cannot admit compressive stress and have no bending resistance. The vibrations are mainly normal to the membrane and thus do not appreciably affect the tension. By balancing an elemental area, or setting the moment terms in the plate equations to zero (Chapter 5), one can show that the membrane equation is

$$\frac{\partial}{\partial x'}\left(T'_{xx}\frac{\partial u'}{\partial x'}\right)+\frac{\partial}{\partial y'}\left(T'_{yy}\frac{\partial u'}{\partial y'}\right)+\frac{\partial}{\partial x'}\left(T'_{xy}\frac{\partial u'}{\partial y'}\right)+\frac{\partial}{\partial y'}\left(T'_{xy}\frac{\partial u'}{\partial x'}\right)=\rho'\frac{\partial^2 u'}{\partial t'^2} \quad (3.1)$$

Here T'_{xx}, T'_{yy} are the normal stresses (force per length) in the Cartesian x',y' directions, and T'_{xy} is the shear stress. Of course, the stresses are in static equilibrium, and can be derived from a plane stress function (e.g., Timoshenko and Goodier 1970). Traditionally, one considers membranes with equal and constant normal stress and zero shear stress, but there is no reason to limit to such a case (Soedel 2004). However, there seems to be no exact solution if the shear stress is not zero (Leissa and Ghamat-Rezaei 1990). Thus we shall consider normal stresses only.

We normalize all lengths by a membrane dimension L, the stresses by the maximum stress T_0, the density by the maximum density ρ_0, and the time by $L\sqrt{\rho_0/T_0}$ and drop the primes. For vibrations, set $u = w(x,y)e^{i\omega t}$, and Equation (3.1) becomes

$$\frac{\partial}{\partial x}\left(T_{xx}\frac{\partial w}{\partial x}\right)+\frac{\partial}{\partial y}\left(T_{yy}\frac{\partial w}{\partial y}\right)+\rho\omega^2 w=0 \quad (3.2)$$

Notice that if there is no stress in one direction, say $T_{yy} = 0$, then Equation (3.2) reduces to the string equation studied earlier.

The boundary conditions are usually fixed, i.e.,

$$w = 0 \tag{3.3}$$

But there are special cases where sliding, joined, or massed boundary conditions are used, which we shall develop as needed.

3.3 CONSTANT UNIFORM NORMAL STRESS AND CONSTANT DENSITY

We consider equal uniform stresses, $T_{xx} = T_{yy} = 1$ and constant density $\rho = 1$. Equation (3.2) simplifies to the Helmholtz equation

$$\frac{\partial^2 w}{\partial x^2} + \frac{\partial^2 w}{\partial y^2} + \omega^2 w = 0 \tag{3.4}$$

where ω is the frequency normalized by $\sqrt{T_0/\rho_0 L^2}$. For fixed boundaries, Equation (3.3) holds.

If the membrane is stretched by constant but unequal amounts in the x and y directions that results in different uniform stresses, say $T_{xx} = T_1$ and $T_{yy} = T_2$, then one can redefine a new independent variable

$$y = \sqrt{T_2/T_1}\,\bar{y} \tag{3.5}$$

Equation (3.2) then becomes

$$\frac{\partial^2 w}{\partial x^2} + \frac{\partial^2 w}{\partial \bar{y}^2} + \omega^2 w = 0 \tag{3.6}$$

which is the same as Equation (3.4). Thus in this section, without loss of generality, we can assume that tensions are uniform and equal.

Even for uniform stress and density, there are only a few membrane shapes that admit exact solutions.

3.3.1 Rectangular Membrane

Consider a rectangular membrane whose normalized length is 1 and normalized width a (i.e., aspect ratio $a \le 1$). Let Cartesian coordinates be placed at the lower corner as shown in Figure 3.1. The exact solution to Equations (3.3) and (3.4) is

$$w = \sin(n\pi x)\sin\left(\frac{m\pi y}{a}\right) \tag{3.7}$$

where n, m are nonzero integers that give the number of half waves in the x, y directions, respectively.

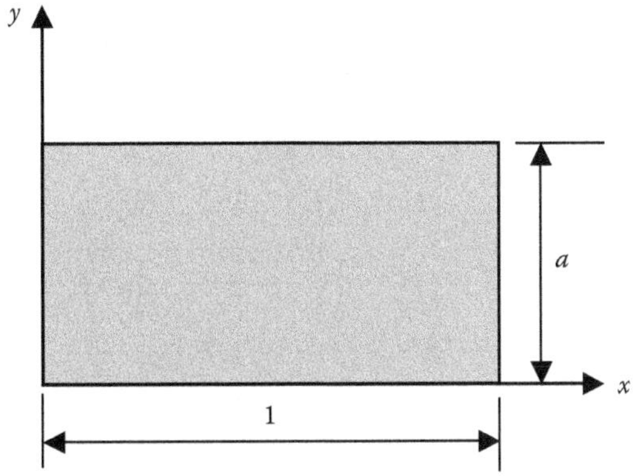

FIGURE 3.1 Rectangular membrane and coordinate axes.

The exact natural frequency of vibration of the rectangular membrane is given by

$$\omega = \pi\sqrt{n^2 + \left(\frac{m}{a}\right)^2} \tag{3.8}$$

Although the fundamental (lowest) frequency is always due to the $(m, n) = (1, 1)$ mode, the next few frequencies are not obvious, as reflected by Table 3.1. Typical mode shapes of a rectangular membrane (with $a = 0.6$) are shown in Figure 3.2.

3.3.2 Three Triangular Membranes

By combining two modes of the square membrane, such that $w = 0$ on the diagonal $x = y$, one obtains the solution to the (45°, 45°, 90°) triangular membrane as shown in Figure 3.3a

$$w = \sin(n\pi x)\sin(m\pi y) - \sin(m\pi x)\sin(n\pi y) \tag{3.9}$$

where $n \neq m$. The frequency is

$$\omega = \pi\sqrt{n^2 + m^2} \tag{3.10}$$

TABLE 3.1
Frequencies for Rectangular Membranes

a = 0.2	0.4	0.6	0.8	1
16.019 (1,1)	8.459(1,1)	6.106 (1,1)	5.029(1,1)	4.443 (1,1)
16.918 (1,2)	10.058 (1,2)	8.179 (1,2)	7.409 (1,2)	7.025 (1,2) (2,1)
18.315 (1,3)	12.268 (1,3)	10.782 (1,3)	8.459 (2,1)	8.886 (2,2)
20.116 (1,4)	14.819 (1,4)	10.933 (2,1)	10.058 (2,2)	9.935 (1,3) (3,1)
22.214 (1,5)	16.019 (2,1)	12.212 (2,2)	10.210 (1,3)	11.327 (2,3) (3,2)

Note: Mode numbers (m,n) are given in parentheses.

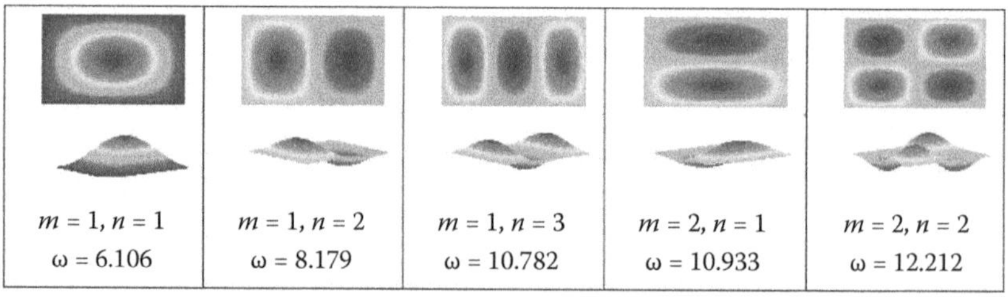

FIGURE 3.2 Mode shapes of rectangular membrane with $a = 0.6$.

Let us designate the mode by (n, m). The first five frequencies are 7.025 (1,2) (2,1), 9.935 (1,3) (3,1), 11.324 (2,3) (3,2), 12.953 (1,4) (4,1), and 14.050 (2,4) (4,2). Note that the frequencies of the 45°, 45°, 90° triangular membrane constitute only a subset of the square membrane. Thus, these shapes are not isospectral to each other.

The equilateral triangular shape (60°, 60°, 60°), as shown in Figure 3.3b, was first considered by Schelkunoff (1943), who gave an exact solution to the electromagnetic wave propagation (TM mode), which is analogous to the vibrating membrane. The length scale is the height of the triangle.

$$w = \cos\left[\frac{(m+2n)\pi y}{3\sqrt{3}}\right]\sin\left[\frac{m\pi(2+x)}{3}\right] - \cos\left[\frac{(m-n)\pi y}{3\sqrt{3}}\right]\sin\left[\frac{(m+n)\pi(2+x)}{3}\right] + \cos\left[\frac{(2m+n)\pi y}{3\sqrt{3}}\right]\sin\left[\frac{n\pi(2+x)}{3}\right] \tag{3.11}$$

Here m and n are integers that are nonzero and do not add up to zero. The frequency is

$$\omega = \frac{2\pi}{3\sqrt{3}}\sqrt{m^2 + mn + n^2} \tag{3.12}$$

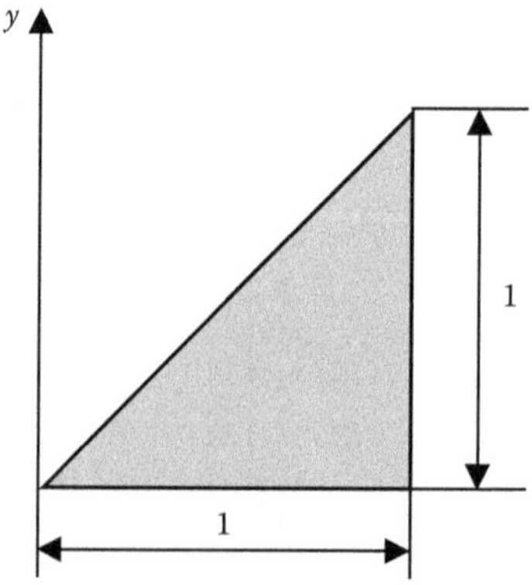

FIGURE 3.3a Right angle triangular (45°, 45°, 90°) membrane. *(continued)*

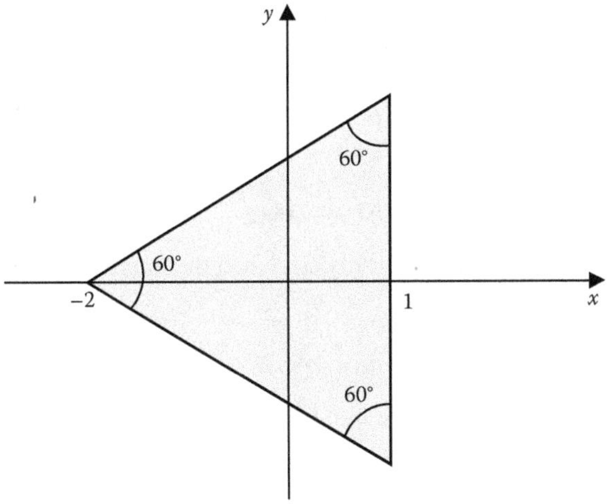

FIGURE 3.3b Equilateral triangular membrane.

Equation (3.11) is symmetric about $y = 0$ and does not include the antisymmetric modes. To complete the solution, we must add the analogous frequencies of the 30°, 60°, 90°triangular membrane (see Figure 3.3c), first given by Seth (1947).

$$w = \sin\left[\frac{(m+2n)\pi y}{3\sqrt{3}}\right]\sin\left[\frac{m\pi(2+x)}{3}\right] - \sin\left[\frac{(m-n)\pi y}{3\sqrt{3}}\right]\sin\left[\frac{(m+n)\pi(2+x)}{3}\right]$$
$$+\sin\left[\frac{(2m+n)\pi y}{3\sqrt{3}}\right]\sin\left[\frac{n\pi(2+x)}{3}\right] \tag{3.13}$$

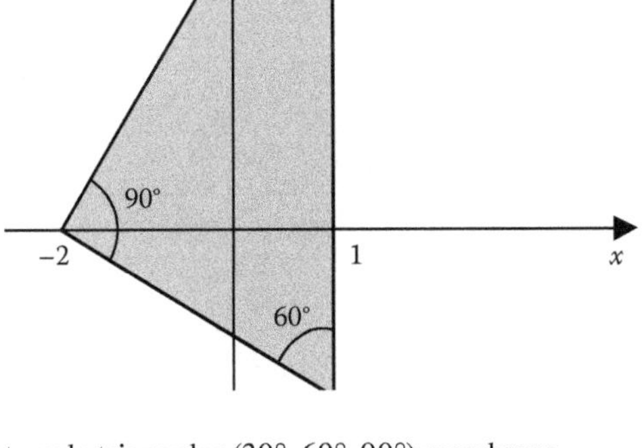

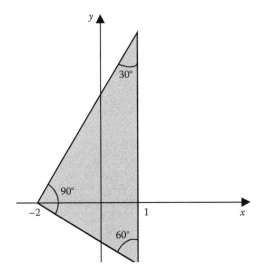

FIGURE 3.3c Right angle triangular (30°, 60°, 90°) membrane.

Note that n and m can be interchanged, and so we consider only $n \geq m$. The lowest frequencies of the equilateral triangular membrane are 2.0944 (1,1) (2,–1), 3.1992 (2,1) (3,–2) (3,–1), 4.1888 (2,2) (4,–2), 4.3598 (3,1) (4,–3) (4,–1), and 5.2708 (3,2) (5,–2) (5,–3). The lowest frequencies of the 30°, 60°, 90° membrane are 3.1992 (2,1) (3,–2) (3,–1), 4.3598 (3,1) (4,–3) (4,–1), 5.2708 (3,2) (5,–2) (5,–3), and 6.7325 (5,1) (6,–1) (6,–5).

3.3.3 Circular and Annular Membranes

In polar coordinates (r,θ), the uniformly stretched membrane satisfies

$$\frac{\partial^2 w}{\partial r^2} + \frac{1}{r}\frac{\partial w}{\partial r} + \frac{1}{r^2}\frac{\partial^2 w}{\partial \theta^2} + \omega^2 w = 0 \tag{3.14}$$

We normalize the lengths by the maximum radius. The general solution is

$$w = \cos(n\theta)[C_1 J_n(\omega r) + C_2 Y_n(\omega r)] \tag{3.15}$$

where $n \geq 0$ is an integer and J_n, Y_n are Bessel functions. For a circular membrane (Figure 3.4a), C_2 is zero, and the fixed outer boundary gives

$$J_n(\omega) = 0 \tag{3.16}$$

Let ω_p be the pth root of Equation (3.16) in ascending order. Let (n, p) denote the mode shape, i.e., n radial nodes and p concentric nodes including the outer boundary. The first five lowest frequencies are 2.4048 (0,1), 3.8317 (1,1), 5.1356 (2,1), 5.5201 (0,2), and 6.3802 (3,1).

Consider an annular membrane with an outer normalized radius 1 and inner radius $b < 1$ (Figure 3.4b). The application of the fixed boundary conditions on Equation (3.15) gives the characteristic equation

$$Y_n(\omega) J_n(\omega b) - J_n(\omega) Y_n(\omega b) = 0 \tag{3.17}$$

Table 3.2 shows the results. Notice that the fundamental frequency rises sharply from the no-core value of 2.4048. Mode shapes for a circular membrane and an annular membrane (with $b = 0.5$) are shown in Figure 3.5 and Figure 3.6, respectively.

The exact solution for an annular membrane with a solid core was given by Wang (2003).

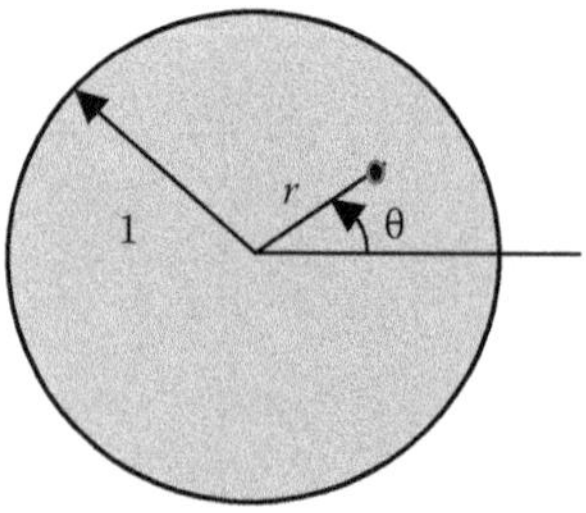

FIGURE 3.4a Circular membrane with a radius of one unit. *(continued)*

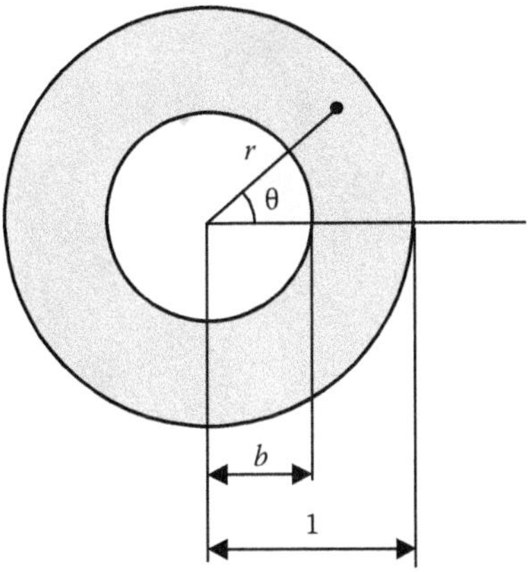

FIGURE 3.4b Annular membrane with inner to outer radius ratio of *b*.

TABLE 3.2
Frequencies for Annular Membranes

b = 0.01	0.1	0.3	0.5	0.7	0.9
2.8009 (0,1)	3.3139 (0,1)	4.4124 (0,1)	6.2461 (0,1)	10.455 (0,1)	31.412 (0,1)
3.8329 (1,1)	3.9409 (1,1)	4.7058 (1,1)	6.3932 (1,1)	10.522 (1,1)	31.429 (1,1)
5.1356 (2,1)	5.1424 (2,1)	5.4702 (2,1)	6.8138 (2,1)	10.720 (2,1)	31.482 (2,1)
6.0109 (0,2)	6.3805 (3,1)	6.4937 (3,1)	7.4577 (3,1)	11.042 (3,1)	31.570 (3,1)
6.3802 (3,1)	6.8576 (0,2)	7.6229 (4,1)	8.2667 (4,1)	11.476 (4,1)	31.693 (4,1)

Note: Mode numbers (*n*,*p*) are given in parentheses.

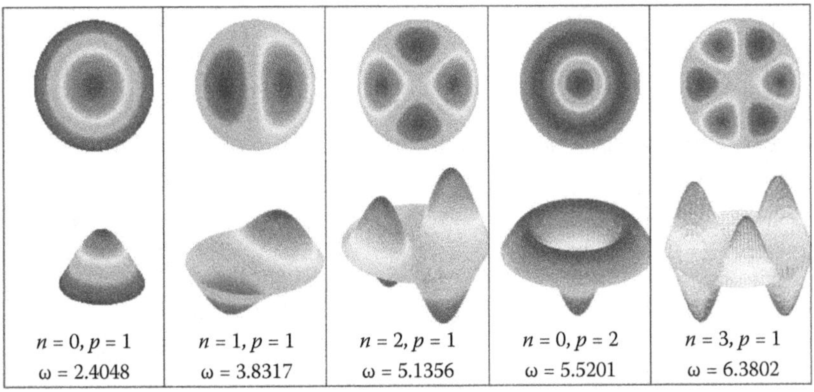

FIGURE 3.5 Mode shapes of circular membrane.

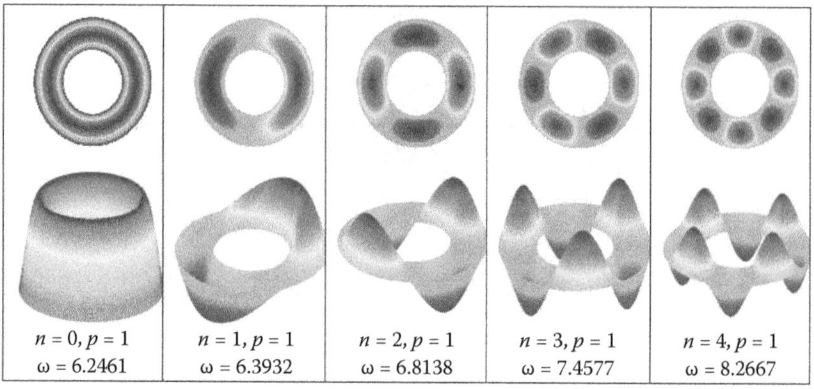

FIGURE 3.6 Mode shapes of annular membrane with $b = 0.5$.

3.3.4 Circular Sector Membrane and Annular Sector Membrane

For sector membranes (Figure 3.7), let θ_0 be the opening angle. The solution to Equation (3.14) is represented by

$$w = \sin(\beta\theta)[C_1 J_\beta(\omega r) + C_2 Y_\beta(\omega r)] \tag{3.18}$$

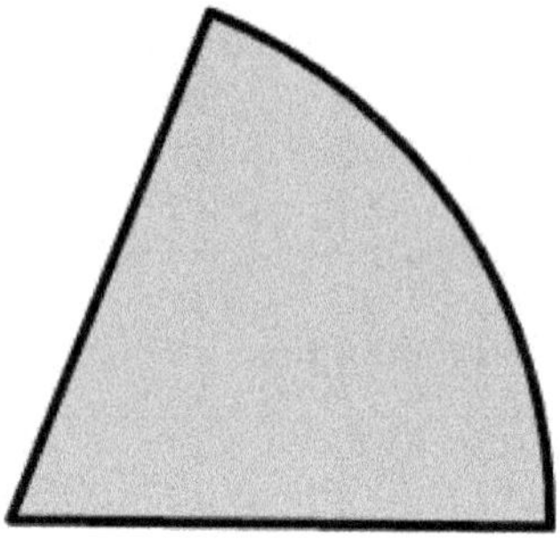

FIGURE 3.7 Circular sector membrane.

TABLE 3.3
Frequencies for Circular Sector Membranes

$\theta_0 = 30°$	45°	60°	90°	180°	360°
9.9361 (1,1)	7.5883 (1,1)	6.3802 (1,1)	5.1356 (1,1)	3.8317 (1,1)	3.1416 (1,1)
13.589 (1,2)	11.065 (1,2)	9.7610 (1,2)	7.5883 (2,1)	5.1356 (2,1)	3.8317 (2,1)
16.698 (2,1)	12.225 (2,1)	9.9361 (2,1)	8.4172 (1,2)	6.3802 (3,1)	4.4934 (3,1)
17.004 (1,3)	14.373 (1,3)	13.015 (1,3)	9.9361 (3,1)	7.0156 (1,2)	5.1356 (4,1)
20.321 (1,4)	16.038 (2,2)	13.354 (3,1)	11.065 (2,2)	7.5883 (4,1)	5.7635 (5,1)

Note: Mode numbers (n, p) are in parentheses.

where $\beta = n\pi/\theta_0$, and n is a nonzero integer. Notice that the sine function is needed for sectors.

For a circular sector membrane, $C_2 = 0$. The boundary condition requires

$$J_\beta(\omega) = 0 \tag{3.19}$$

Table 3.3 presents the natural frequencies of circular sector membranes. When $\theta_0 = 360°$, it is a circular membrane with a radial constraint.

Let the opening angle of the annular sector membrane be θ_0 and bounded by the outer radius $r = 1$ and the inner radius $r = b < 1$. The characteristic equation is

$$Y_\beta(\omega)J_\beta(\omega b) - J_\beta(\omega)Y_\beta(\omega b) = 0 \tag{3.20}$$

Table 3.4 shows the results. Note the switching of modes and the same frequencies for certain shapes. Mode shapes for a circular membrane and an annular membrane with $\theta_0 = 90°$ are shown in Figure 3.8a and Figure 3.8b, respectively.

TABLE 3.4
Frequencies for Annular Sector Membranes

b	$\theta_0 = 30°$	45°	60°	90°	180°	360°
0.1	9.9361 (1,1)	7.5884 (1,1)	6.3805 (1,1)	5.1424 (1,1)	3.9409 (1,1)	3.4907 (1,1)
	13.589 (1,2)	11.065 (1,2)	9.7641 (1,2)	7.5884 (2,1)	5.1424 (2,1)	3.9409 (2,1)
	16.698 (2,1)	12.225 (2,1)	9.9361 (2,1)	8.4574 (1,2)	6.3808 (3,1)	4.5223 (3,1)
	17.004 (1,3)	14.374 (1,3)	13.030 (1,3)	9.9361 (3,1)	7.3306 (1,2)	5.1424 (4,1)
	20.321 (1,4)	16.038 (2,2)	13.354 (3,1)	11.065 (2,2)	7.5864 (4,1)	5.7649 (5,1)
0.3	9.9386 (1,1)	7.6229 (1,1)	6.4937 (1,1)	5.4702 (1,1)	4.7056 (1,1)	4.4880 (1,1)
	13.633 (1,2)	11.348 (1,2)	9.9386 (2,1)	7.6229 (2,1)	5.4702 (2,1)	4.7058 (2,1)
	16.698 (2,1)	12.225 (2,1)	10.371 (1,2)	9.6003 (1,2)	6.4937 (3,1)	5.0427 (3,1)
	17.245 (1,3)	15.246 (1,3)	13.354 (3,1)	9.9386 (3,1)	9.1042 (1,2)	8.9760 (1,2)
	20.790 (2,2)	16.042 (2,2)	13.633 (2,2)	11.348 (2,2)	9.6003 (2,2)	9.1042 (2,2)
0.5	10.189 (1,1)	8.2667 (1,1)	7.4577 (1,1)	6.8138 (1,1)	6.3932 (1,1)	6.2832 (1,1)
	15.110 (1,2)	12.311 (2,1)	10.189 (2,1)	8.2667 (2,1)	6.8138 (1,2)	6.3932 (1,2)
	16.706 (2,1)	13.742 (1,2)	13.232 (1,2)	10.189 (3,1)	7.4577 (2,1)	6.5720 (2,1)
	20.652 (1,3)	16.706 (3,1)	13.403 (3,1)	12.311 (4,1)	8.2667 (3,1)	6.8138 (3,1)
	20.963 (2,2)	16.841 (2,2)	16.706 (4,1)	12.856 (1,2)	9.1900 (4,1)	7.1116 (4,1)
0.7	12.635 (1,1)	11.476 (1,1)	11.042 (1,1)	10.720 (1,1)	10.522 (1,1)	10.472 (1,1)
	17.581 (2,1)	14.092 (2,1)	12.635 (2,1)	11.476 (2,1)	10.720 (2,1)	10.522 (2,1)
	22.125 (1,2)	17.581 (3,1)	14.906 (3,1)	12.635 (3,1)	11.042 (3,1)	10.605 (3,1)
	23.527 (3,1)	21.472 (1,2)	17.581 (4,1)	14.092 (4,1)	11.476 (4,1)	10.720 (4,1)
	25.367 (2,2)	21.490 (4,1)	20.489 (5,1)	15.763 (5,1)	12.012 (5,1)	10.866 (5,1)
0.9	32.041 (1,1)	31.693 (1,1)	31.570 (1,1)	31.482 (1,1)	31.429 (1,1)	31.416 (1,1)
	33.859 (2,1)	32.522 (2,1)	32.041 (2,1)	31.693 (2,1)	31.482 (2,1)	31.429 (2,1)
	36.688 (3,1)	33.859 (3,1)	32.811 (3,1)	32.041 (3,1)	31.570 (3,1)	31.451 (3,1)
	40.319 (4,1)	35.646 (4,1)	33.859 (4,1)	32.522 (4,1)	31.693 (4,1)	31.482 (4,1)
	44.548 (5,1)	37.819 (5,1)	35.160 (5,1)	33.130 (5,1)	31.850 (5,1)	31.522 (5,1)

Note: Mode numbers (n, p) are in parentheses.

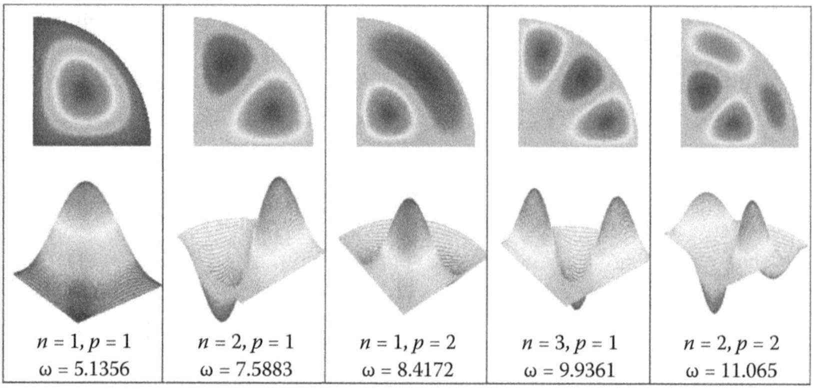

FIGURE 3.8a Mode shapes for circular sector membrane with $\theta_0 = 90°$.

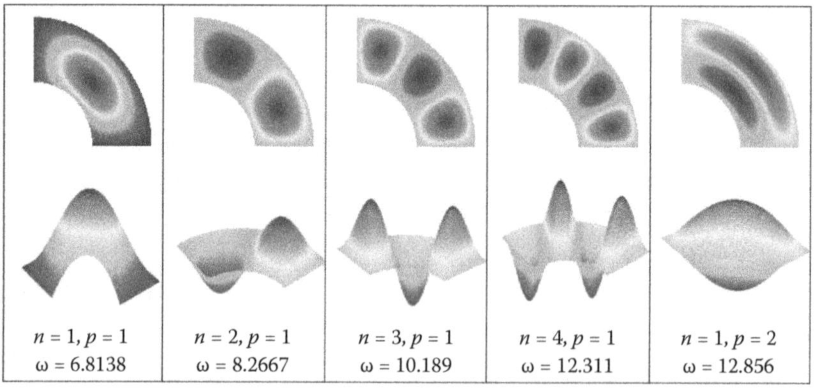

FIGURE 3.8b Mode shapes for annular sector membrane with $\theta_0 = 90°$ and $b = 0.5$.

3.4 TWO-PIECE CONSTANT-PROPERTY MEMBRANES

The two-piece membrane has constant tension, but each separate piece has different constant properties. Thus the deformation of each piece satisfies different Helmholtz equations. At the joint or common boundary, we require the deformations and the normal derivatives (force components) to match. The two-piece rectangular membrane was first studied analytically by Gottlieb (1986), and the two-piece circular membrane by Spence and Horgan (1983), but only for some extreme density ratios. The two-piece annular membrane was considered by Laura, Rossit, and La Malfa (1998).

3.4.1 Two-Piece Rectangular Membrane

Figure 3.9 shows a composite rectangular membrane with a normalized Cartesian coordinate system at one corner. Region 1 ($0 \le x \le b$, $0 \le y \le a$) has the larger density, while Region 2 ($b \le x \le 1$, $0 \le y \le a$) has the smaller density.

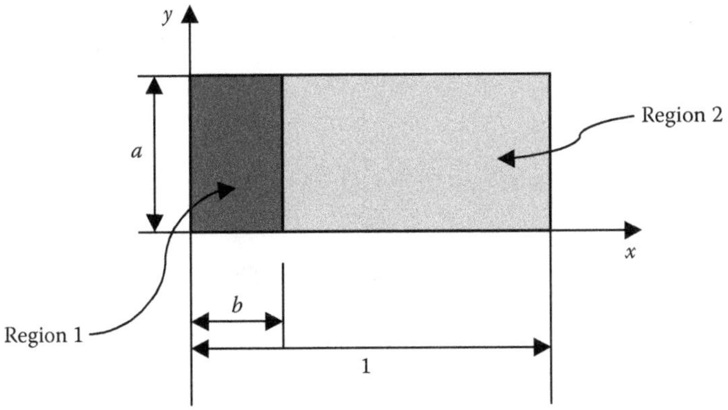

FIGURE 3.9 Two-piece rectangular membrane.

The governing equations are

$$w_{1xx} + w_{1yy} + \omega^2 w_1 = 0 \tag{3.21}$$

$$w_{2xx} + w_{2yy} + \gamma\omega^2 w_2 = 0 \tag{3.22}$$

where $\gamma = \rho_2/\rho_1 \le 1$. The boundary conditions are

$$w_1(x,0) = 0,\ \ w_1(x,a) = 0,\ w_1(0,y) = 0 \tag{3.23}$$

$$w_2(x,0) = 0,\ \ w_2(x,a) = 0,\ w_2(1,y) = 0 \tag{3.24}$$

and at the joint

$$w_1(b,y) = w_2(b,y) \tag{3.25a}$$

$$w_{1x}(b,y) = w_{2x}(b,y) \tag{3.25b}$$

The solutions that satisfy Equations (3.21) to (3.24) are

$$w_1 = C_1 \sin(\alpha y) F_1(x) \tag{3.26a}$$

$$w_2 = C_2 \sin(\alpha y) F_2(x) \tag{3.26b}$$

where $\alpha = n\pi/a$ and

$$F_1(x) = \begin{cases} \sin\left(\sqrt{\omega^2 - \alpha^2}\, x\right) & \omega > \alpha \\ x & \omega = \alpha \\ \sinh\left(\sqrt{\alpha^2 - \omega^2}\, x\right) & \omega < \alpha \end{cases} \tag{3.27}$$

$$F_2(x) = \begin{cases} \sin\left[\sqrt{\gamma\omega^2 - \alpha^2}\,(1-x)\right] & \sqrt{\gamma}\omega > \alpha \\ 1-x & \sqrt{\gamma}\omega = \alpha \\ \sinh\left[\sqrt{\alpha^2 - \gamma\omega^2}\,(1-x)\right] & \sqrt{\gamma}\omega < \alpha \end{cases} \quad (3.28)$$

For nontrivial C_1, C_2, Equations (3.25a) and (3.25b) yield the condition

$$F_1(b)F_2'(b) - F_2(b)F_1'(b) = 0 \quad (3.29)$$

Table 3.5 shows some representative results. The parentheses show the mode n at which the frequency occurs. Mode shapes of two-piece rectangular membrane for $\gamma = 0.5$, $b = 0.5$, and $a = 1$ are shown in Figure 3.10.

3.4.2 Two-Piece Circular Membrane

Figure 3.11 shows a circular membrane comprising two concentric homogeneous pieces. Let the interior circular piece be denoted by the subscript 1 and the exterior annular piece by the subscript 2. The governing equations are given by

$$\nabla^2 w_1 + \gamma_1\omega^2 w_1 = 0 \quad (3.30)$$

$$\nabla^2 w_2 + \gamma_2\omega^2 w_2 = 0 \quad (3.31)$$

TABLE 3.5a
Frequencies for the Two-Piece Rectangular Membrane ($\gamma = 0.25$)

b	$a = 0.25$	$a = 0.50$	$a = 1.00$	$a = 2.00$
0.25	12.566 (1)	6.2832 (1)	3.1416 (1)	1.5708 (1)
	15.746 (1)	10.186 (1)	6.2832 (1,2)	3.1416 (1,2)
	23.415 (1)	12.566 (1,2)	7.2447 (1)	4.7124 (3)
	25.133 (1,2)	14.986 (1)	9.4248 (3)	5.9275 (1)
	26.564 (1)	15.746 (2)	10.186 (2)	6.2832 (2,3)
0.50	12.566 (1)	6.2832 (1)	3.1416 (1)	1.5708 (1)
	13.656 (1)	7.8802 (1)	5.2491 (1)	3.1416 (1,2)
	16.606 (1)	11.778 (1)	6.2832 (1,2)	4.2378 (1)
	20.756 (1)	12.566 (1,2)	7.8802 (2)	4.7124 (3)
	25.133 (1,2)	13.656 (2)	9.4248 (3)	5.2491 (2)
0.75	12.566 (1)'	6.2832 (1)	3.1416 (1)	1.5708 (1)
	13.112 (1)	7.1860 (1)	4.5732 (1)	3.1416 (1,2)
	14.654 (1)	9.5686 (1)	6.2832 (1,2)	3.6231 (1)
	16.964 (1)	12.566 (1,2)	7.1860 (2)	4.5732 (2)
	19.805 (2)	14.654 (2)	7.6034 (1)	4.7124 (3)

Note: Mode number n is in parentheses.

TABLE 3.5b
Frequencies for the Two-Piece Rectangular Membrane ($\gamma = 0.5$)

b	*a* = 0.25	*a* = 0.50	*a* = 1.00	*a* = 2.00
0.25	12.566 (1)	6.2832 (1)	3.1416 (1)	2.2214 (1)
	15.334 (1)	8.8858 (1)	4.4429 (1)	3.1416 (2)
	17.772 (1)	9.1618 (1)	5.9443 (1)	4.4429 (2)
	18.677 (1)	11.235 (1)	6.2832 (2)	4.7124 (3)
	20.747 (1)	12.566 (2)	8.8067 (1)	4.7228 (1)
0.50	12.566 (1)	6.2832 (1)	3.1416 (1)	1.5708 (1)
	13.581 (1)	7.7005 (1)	4.4429 (1)	2.2214 (1)
	16.275 (1)	8.8858 (1)	5.0033 (1)	3.1416 (2)
	17.772 (1)	10.795 (1)	6.2832 (2)	3.9901 (1)
	19.255 (1)	12.566 (2)	7.7005 (2)	4.4429 (2)
0.75	12.566 (1)	6.2832 (1)	3.1416 (1)	1.5708 (1)
	13.089 (1)	7.1460 (1)	4.4429 (1)	2.2214 (1)
	14.560 (1)	8.8858 (1)	4.5342 (1)	3.1416 (2)
	16.750 (1)	9.3528 (1)	6.2832 (2)	3.5882 (1)
	17.772 (1)	12.262 (1)	7.1460 (2)	4.4429 (2)

Note: Mode number *n* is in parentheses.

TABLE 3.5c
Frequencies for the Two-Piece Rectangular Membrane ($\gamma = 0.75$)

b	*a* = 0.25	*a* = 0.50	*a* = 1.00	*a* = 2.00
0.25	12.566 (1)	6.2832 (1)	3.1416 (1)	1.8138 (1)
	14.462 (1)	7.2552 (1)	3.6276 (1)	3.1416 (2)
	14.510 (1)	7.9617 (1)	5.0475 (1)	3.6276 (2)
	15.577 (1)	9.8257 (1)	6.2832 (2)	3.9925 (1)
	17.498 (1)	12.537 (1)	7.2552 (2)	5.0475 (2)
0.50	12.566 (1)	6.2832 (1)	3.1416 (1)	1.8138 (1)
	13.429 (1)	7.2552 (1)	3.6276 (1)	3.1416 (2)
	14.510 (1)	7.4210 (1)	4.7260 (1)	3.6276 (2)
	15.340 (1)	9.6332 (1)	6.2832 (2)	3.7431 (1)
	16.792 (1)	12.077 (1)	7.2552 (2)	4.7124 (3)
0.75	12.566 (1)	6.2832 (1)	3.1416 (1)	1.8138 (1)
	13.047 (1)	7.0940 (1)	3.6276 (1)	3.1416 (2)
	14.387 (1)	9.1447 (1)	4.4909 (1)	3.5513 (1)
	16.331 (1)	11.797 (1)	6.2832 (2)	3.6276 (2)
	18.538 (1)	12.566 (2)	7.0940 (2)	4.4909 (2)

Note: Mode number *n* is in parentheses.

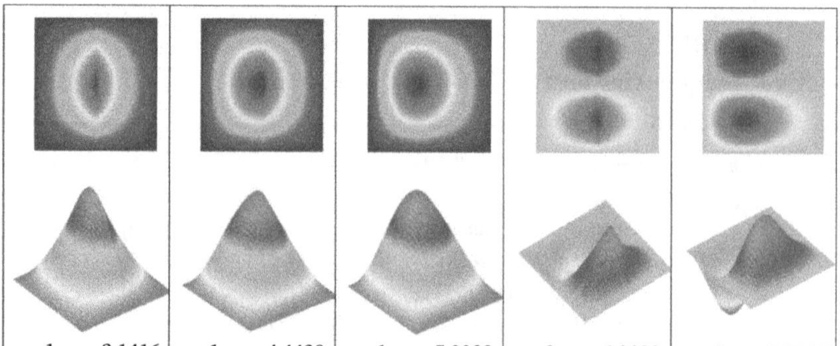

FIGURE 3.10 Mode shapes of two-piece rectangular membrane for $\gamma = 0.5$, $b = 0.5$, $a = 1$.

Here all quantities have been normalized as before, and γ represents the density ratio. If the inner piece has the higher density, then $\gamma_1 = 1, \gamma_2 = \rho_2/\rho_2 < 1$. If the outer piece has the higher density, then $\gamma_2 = 1, \gamma_1 = \rho_1/\rho_2 < 1$. The solution to Equation (3.30) satisfying the bounded condition at the center is

$$w_1 = C_1 \cos(n\theta) J_n(\sqrt{\gamma_1}\,\omega r) \tag{3.32}$$

The solution to Equation (3.31) satisfying the zero condition at the rim is

$$w_2 = C_2 \cos(n\theta)[Y_n(\sqrt{\gamma_2}\,\omega) J_n(\sqrt{\gamma_2}\,\omega r) - J_n(\sqrt{\gamma_2}\,\omega) Y_n(\sqrt{\gamma_2}\,\omega r)] \tag{3.33}$$

At the joint ($r = b$), the displacements and tension match

$$w_1(b,\theta) = w_2(b,\theta) \tag{3.34a}$$

$$w_{1r}(b,\theta) = w_{2r}(b,\theta) \tag{3.34b}$$

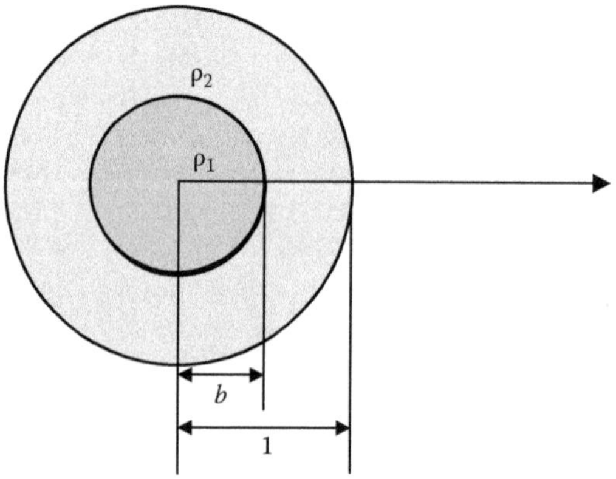

FIGURE 3.11 Two-piece circular membrane.

Equations (3.34a) and (3.34b) yield the characteristic equation

$$\sqrt{\gamma_2}J_n(\sqrt{\gamma_1}\omega b)[Y_n(\sqrt{\gamma_2}\omega)J_n'(\sqrt{\gamma_2}\omega b)-J_n(\sqrt{\gamma_2}\omega)Y_n'(\sqrt{\gamma_2}\omega b)]$$
$$-\sqrt{\gamma_1}J_n'(\sqrt{\gamma_1}\omega b)[Y_n(\sqrt{\gamma_2}\omega)J_n(\sqrt{\gamma_2}\omega b)-J_n(\sqrt{\gamma_2}\omega)Y_n(\sqrt{\gamma_2}\omega b)]=0 \quad (3.35)$$

The results are given in Tables 3.6a and 3.6b. The parentheses show the mode n at which the frequency occurs. Mode shapes for two-piece circular membrane with $\gamma_1 = 1$, $\gamma_2 = 0.5$, $b = 0.5$ are shown in Figure 3.12.

3.5 NONHOMOGENEOUS MEMBRANES

We consider nonuniform membranes with continuously varying density (or thickness). The tension is constant. Equation (3.2) becomes

$$\nabla^2 w + \omega^2 \rho w = 0 \quad (3.36)$$

TABLE 3.6a
Frequencies for the Two-Piece Circular Membrane, $\gamma_1 = 1$

b	$\gamma_2 = 0.1$	$\gamma_2 = 0.3$	$\gamma_2 = 0.5$	$\gamma_2 = 0.7$	$\gamma_2 = 0.9$
0.1	6.3364 (0)	4.1985 (0)	3.3379 (0)	2.8517(0)	2.5298(0)
	12.040 (1)	6.9862 (1)	5.4158 (1)	4.5787(1)	4.0387(1)
	12.803 (0)	9.1004 (0)	7.2628 (2)	6.1382(2)	5.4134(2)
	16.238 (2)	9.3760 (2)	7.4819(0)	6.4822(0)	5.7926(0)
	20.176 (3)	11.649 (3)	9.0229(3)	7.6258(3)	6.7253(3)
0.3	3.6770 (0)	3.2868 (0)	2.9628 (0)	2.7030 (0)	2.4942 (0)
	7.6836 (1)	6.2578 (1)	5.1968 (1)	4.5040 (1)	4.0223 (1)
	11.485 (0)	8.0435 (0)	6.7993 (0)	6.1143 (2)	5.4082 (2)
	12.086 (2)	9.0881 (2)	7.1896 (2)	6.1384 (0)	5.6972 (0)
	14.304 (1)	9.9098 (1)	8.7003 (1)	7.6192 (3)	6.7239 (3)
0.5	2.8261 (0)	2.7282 (0)	2.6314 (0)	2.5376 (0)	2.4479 (0)
	5.1438 (1)	4.8414 (1)	4.5266 (1)	4.2259 (1)	3.9547 (1)
	7.5893 (2)	7.0592 (2)	6.4362 (2)	5.8462 (2)	5.3491 (2)
	8.0728 (0)	7.3605 (0)	6.6496 (0)	6.0938 (0)	5.6839 (0)
	10.075 (3)	9.3598 (3)	8.3544 (3)	7.4164 (3)	6.6818 (3)
0.7	2.5020 (0)	2.4811 (0)	2.4598 (0)	2.4381 (0)	2.4160 (0)
	4.1616 (1)	4.0942 (1)	4.0229 (1)	3.9483 (1)	3.8711 (1)
	5.8022 (2)	5.6749 (2)	5.5333 (2)	5.3797 (2)	5.2180 (2)
	6.2649 (0)	6.1176 (0)	5.9552 (0)	5.7830 (0)	5.6072 (0)
	7.4559 (3)	7.2669 (3)	7.0429 (3)	6.7892 (3)	6.5179 (3)
0.9	2.4089 (0)	2.4080 (0)	2.4071 (0)	2.4062 (0)	2.4053 (0)
	3.8475 (1)	3.8441 (1)	3.8406 (1)	3.8371 (1)	3.8335 (1)
	5.1721 (2)	5.1644 (2)	5.1565 (2)	5.1483 (2)	5.1399 (2)
	5.5643 (0)	5.5550 (0)	5.5454 (0)	5.5355 (0)	5.5253 (0)
	6.4468 (3)	6.4331 (3)	6.4187 (3)	6.4038 (3)	6.3882 (3)

Note: Mode numbers n is in parentheses.

TABLE 3.6b
Frequencies for the Two-Piece Circular Membrane, $\gamma_2 = 1$

b	$\gamma_1 = 0.1$	$\gamma_1 = 0.3$	$\gamma_1 = 0.5$	$\gamma_1 = 0.7$	$\gamma_1 = 0.9$
0.1	2.4438 (0)	2.4353 (0)	2.4267 (0)	2.4180 (0)	2.4092 (0)
	3.8336 (1)	3.8332 (1)	3.8328 (1)	3.8323 (1)	3.8319 (1)
	5.1357 (2)	5.1357 (2)	5.1357 (2)	5.1357 (2)	5.1356 (2)
	5.7068 (0)	5.6676 (0)	5.6271 (0)	5.5852 (0)	5.5420 (0)
	6.3812 (3)	6.3802 (3)	6.3802 (3)	6.3802 (3)	6.3802 (3)
0.3	2.7464 (0)	2.6678 (0)	2.5966 (0)	2.5140 (0)	2.4405 (0)
	3.9448 (1)	3.9219 (1)	3.8978 (1)	3.8724 (1)	3.8456 (1)
	5.1698 (2)	5.1630 (2)	5.1558 (2)	5.1481 (2)	5.1399 (2)
	6.3896 (3)	6.3877 (3)	6.2036 (0)	5.8978 (0)	5.6348 (0)
	6.8517 (0)	6.5340 (0)	6.3857 (3)	6.3836 (3)	6.3813 (3)
0.5	3.4397 (0)	3.1509 (0)	2.8921 (0)	2.6714 (0)	2.4861 (0)
	4.4830 (1)	4.3472 (1)	4.2029 (1)	4.0540 (1)	3.9050 (1)
	5.5386 (2)	5.4672 (2)	5.3811 (2)	5.2892 (2)	5.1888 (2)
	6.6250 (3)	6.5821 (3)	6.5256 (0)	5.9953 (0)	5.6527 (0)
	7.7339 (4)	7.4654 (0)	6.5334 (3)	6.4780 (3)	6.4149 (3)
0.7	4.9826 (0)	3.9007 (0)	3.2386 (0)	2.8162 (0)	2.5217 (0)
	6.1148 (1)	5.4613 (1)	4.8616 (1)	4.3755 (1)	3.9925 (1)
	7.0688 (2)	6.6355 (2)	6.1697 (2)	5.7197 (2)	5.3166 (2)
	8.0037 (3)	7.6752 (0)	6.7637 (0)	6.1779 (0)	5.7184 (0)
	8.9477 (4)	7.6932 (3)	7.3354 (3)	6.9501 (3)	6.5652 (3)
0.9	7.4523 (0)	4.3703 (0)	3.3944 (0)	2.8720 (0)	2.5344 (0)
	11.359 (1)	6.9100 (1)	5.3920 (1)	4.5703 (1)	4.0368 (1)
	14.079 (2)	9.1582 (2)	7.1972 (2)	6.1154 (2)	5.4083 (2)
	14.836 (0)	9.8075 (0)	7.7258 (0)	6.5698 (0)	5.8124 (0)
	15.865 (3)	11.204 (3)	8.8949 (3)	7.5821 (3)	6.7156 (3)

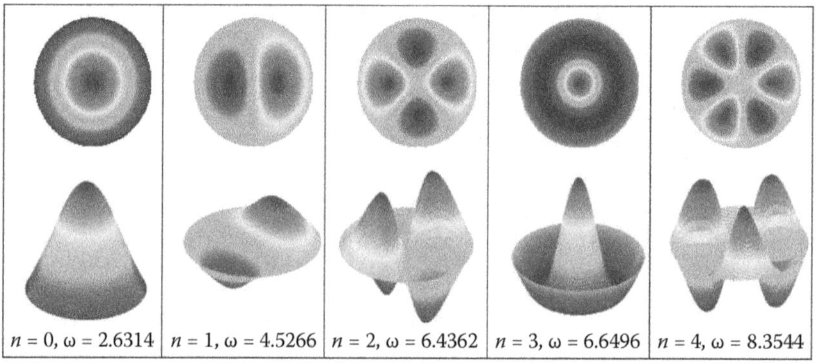

FIGURE 3.12 Mode shapes for two-piece circular membrane with $\gamma_1 = 1$, $\gamma_2 = 0.5$, $b = 0.5$.

where the normalized density ρ is a function of space. So far, exact solutions with density varying in only one direction have been found, except for the work of Gottlieb (2004), who studied isospectral membranes through conformal mapping.

3.5.1 Rectangular Membrane with Linear Density Distribution

Wang (1998) found the solution to the rectangular membrane with density varying linearly and parallel to an edge. However, that source only presented the results for the fundamental frequencies. The density is

$$\rho = 1 - bx, \quad \text{for} \quad b < 1 \tag{3.37}$$

which is a linear taper. Note that Wang (1998) used a density with constant mass, which is slightly different. We shall similarly develop the solution. Assume that

$$w = \sin(\alpha y) f(x) \tag{3.38}$$

where $\alpha = n\pi/a$. Equation (3.36) becomes

$$\frac{d^2 f}{dx^2} + [\omega^2(1 - bx) - \alpha^2] f = 0 \tag{3.39}$$

Let

$$f(x) = h(z), \quad z = [\omega^2(1 - bx) - \alpha^2](b\omega^2)^{-2/3} \tag{3.40}$$

Then Equation (3.39) becomes the Stokes equation

$$\frac{d^2 h}{dz^2} + zh = 0 \tag{3.41}$$

The solution satisfying the conditions of f being zero on the boundary is

$$f = \sqrt{z}[J_{1/3}(p_0)J_{-1/3}(p) - J_{-1/3}(p_0)J_{1/3}(p)] \tag{3.42}$$

where

$$p = \frac{2}{3} z^{3/2}, \quad p_0 = p|_{x=0}, \quad p_1 = p|_{x=1} \tag{3.43}$$

The characteristic equation is

$$J_{1/3}(p_0)J_{-1/3}(p_1) - J_{-1/3}(p_0)J_{1/3}(p_1) = 0 \tag{3.44}$$

Table 3.7 shows the frequency results. Mode shapes for rectangular membranes with a linear density distribution (i.e., $b = 0.5$ and $a = 1$) are shown in Figure 3.13.

TABLE 3.7
Normalized Frequencies ω for Rectangular Membrane with Various Aspect Ratio *a* and Taper *b* Values

a\b	0	0.1	0.25	0.5	0.75	0.9
0.25	12.953 (1)	13.276 (1)	15.052 (1)	18.312 (1)	29.303 (1)	43.361 (1)
	14.050 (1)	14.470 (1)	16.825 (1)	20.694 (1)	32.718 (1)	47.380 (1)
	15.708 (1)	16.120 (1)	19.028 (1)	23.398 (1)	36.262 (1)	51.457 (1)
	17.772 (1)	18.237 (1)	21.533 (1)	26.332 (1)	39.900 (1)	55.579 (1)
	20.116 (1)	20.642 (1)	24.247 (1)	29.426 (1)	43.610 (1)	59.737 (1)
0.5	7.0248 (1)	7.2052 (1)	7.4934 (1)	10.311 (1)	14.569 (1)	23.564 (1)
	8.8858 (1)	9.1179 (1)	9.5090 (1)	13.151 (1)	18.081 (1)	27.685 (1)
	11.327 (1)	11.623 (1)	12.122 (1)	16.309 (1)	21.770 (1)	31.874 (1)
	12.953 (2)	13.276 (2)	15.034 (1)	18.312 (2)	25.563 (1)	36.107 (1)
	14.050 (1,2)	14.417 (1)	15.052 (2)	19.636 (1)	26.001 (2)	40.371 (1)
0.75	5.2360 (1)	5.3711 (1)	5.5907 (1)	6.0068 (1)	9.6505 (1)	14.261 (1)
	7.5515 (1)	7.7484 (1)	8.0789 (1)	8.7530 (1)	13.235 (1)	18.378 (1)
	8.9473 (2)	9.1752 (2)	11.035 (1)	10.657 (2)	17.016 (1)	22.590 (1)
	10.314 (1)	10.583 (1)	11.211 (2)	11.964 (1)	19.483 (2)	26.855 (1)
	10.472 (2)	10.746 (2)	13.498 (2)	12.163 (2)	20.891 (1)	28.814 (2)
1	4.4429 (1)	4.5578 (1)	4.7454 (1)	5.1060 (1)	8.9709 (1)	13.721 (1)
	7.0248 (1,2)	7.2052 (2)	7.4934 (2)	8.1390 (1)	12.737 (1)	17.952 (1)
	8.8858 (2)	7.2080 (1)	7.5147 (1)	10.311 (2)	16.628 (1)	22.241 (1)
	9.9346 (1,3)	9.1176 (2)	9.5090 (2)	11.520 (1)	18.081 (2)	23.564 (2)
	11.327 (2,3)	10.194 (1)	10.629 (1)	13.151 (2)	20.576 (1)	26.559 (1)
1.5	3.7757 (1)	3.8735 (1)	4.0339 (1)	4.3449 (1)	4.7245 (1)	9.0599 (1)
	5.2560 (2)	5.3711 (2)	5.5907 (2)	7.6710 (1)	8.4522 (1)	13.321 (1)
	6.6231 (1)	6.7957 (1)	7.0845 (1)	8.7530 (2)	9.6505 (2)	14.261 (2)
	7.0248 (3)	7.2052 (3)	7.4934 (3)	10.311 (3)	12.370 (1)	17.642 (1)
	7.5515 (2)	7.7484 (2)	8.0789 (2)	11.193 (1)	13.235 (2)	18.378 (2)
2	3.5124 (1)	3.6034 (1)	3.7529 (1)	4.0437 (1)	4.4012 (1)	8.8572 (1)
	4.4429 (2)	4.5578 (2)	4.7455 (2)	5.1060 (2)	8.2630 (1)	13.179 (1)
	5.6636 (3)	5.8096 (3)	6.0460 (3)	7.5004 (1)	8.9709 (2)	13.721 (2)
	6.4766 (1)	6.6453 (1)	6.9276 (1)	8.1390 (2)	10.040 (3)	17.533 (1)
	7.0248 (2)	7.2080 (2)	7.5147 (2)	9.1060 (3)	12.239 (1)	17.952 (2)
3	3.3115 (1)	3.3973 (1)	3.5384 (1)	3.8137 (1)	4.1537 (1)	4.3975 (1)
	3.7757 (2)	3.8735 (2)	4.0339 (2)	4.3449 (2)	4.7245 (2)	8.7095 (1)
	4.4429 (3)	4.5578 (3)	4.7455 (3)	5.1060 (3)	8.1252 (1)	9.0599 (2)
	5.2360 (4)	5.3711 (4)	5.5907 (4)	6.0068 (4)	8.4522 (2)	9.6500 (3)
	6.1062 (5)	6.2634 (5)	6.5170 (5)	7.3762 (1)	8.9709 (3)	13.076 (1)

Note: The vertical mode number n is in parentheses.

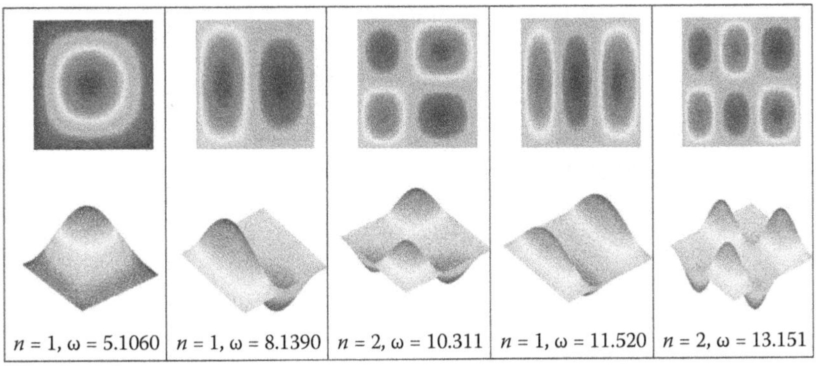

FIGURE 3.13 Mode shapes for rectangular membrane with linear density distribution ($b = 0.5$, $a = 1$).

3.5.2 Rectangular Membrane with Exponential Density Distribution

Rectangular membrane with an exponential density distribution was studied by Wang and Wang (2011a). The normalized density is given by

$$\rho = e^{-bx} \tag{3.45}$$

where $b > 0$. In view of Equation (3.38), Equation (3.36) gives

$$\frac{d^2 f}{dx^2} + (\omega^2 e^{-bx} - \alpha^2) f = 0 \tag{3.46}$$

Let

$$f(x) = h(z), \qquad z = \frac{2}{b} e^{-bx/2} \tag{3.47}$$

After some work, Equation (3.46) becomes

$$\frac{d^2 h}{dz^2} + \frac{1}{z}\frac{dh}{dz} + \left(\omega^2 - \frac{4\alpha^2}{b^2 z^2} \right) h = 0 \tag{3.48}$$

The exact solution to Equation (3.48) is

$$f = h = C_1 J_{2\alpha/b}(\omega z) + C_2 Y_{2\alpha/b}(\omega z) \tag{3.49}$$

The boundaries are at $x = 0$, 1 or

$$z_0 = \frac{2}{b}, \qquad z_1 = \frac{2}{b} e^{-b/2} \tag{3.50}$$

The boundary conditions give the exact mode shape

$$f = Y_{2\alpha/b}(\omega z_0)J_{2\alpha/b}(\omega z) - J_{2\alpha/b}(\omega z_0)Y_{2\alpha/b}(\omega z) \tag{3.51}$$

and the exact characteristic equation

$$Y_{2\alpha/b}(\omega z_0)J_{2\alpha/b}(\omega z_1) - J_{2\alpha/b}(\omega z_0)Y_{2\alpha/b}(\omega z_1) = 0 \tag{3.52}$$

The five lowest frequencies are shown in Table 3.8. The $b = 0$ column is from the frequency of the uniform membrane, Equation (3.8). Mode shapes for a rectangular membrane with $\alpha = 0.5$, $b = 0.5$ are shown in Figure 3.14.

3.5.3 Nonhomogeneous Circular or Annular Membrane

Consider a membrane with radius R, under uniform stress T_0 and maximum density T_0. The normalized membrane equation in polar coordinates (r,θ) is

$$\frac{\partial^2 w}{\partial r^2} + \frac{1}{r}\frac{\partial w}{\partial r} + \frac{1}{r^2}\frac{\partial^2 w}{\partial \theta^2} + \omega^2 \rho(r) w = 0 \tag{3.53}$$

where w is the transverse deflection, ρ is the density, a function of radius and normalized by ρ_0, and ω is the frequency normalized by $\sqrt{T_0/\rho_0 R^2}$. The boundary conditions are that $w = 0$ on the boundaries. The following solutions are mostly from Wang and Wang (2012).

3.5.3.1 Power Law Density Distribution

Let the normalized density (or thickness) be given by

$$\rho = c r^{\nu} \tag{3.54}$$

where ν is a constant exponent, and $c = 1$ if $\nu \geq 0$. If $\nu < 0$, the density is unbounded at the origin, and only the annular membrane is appropriate. In that case, let the inner radius of the annular membrane be bR, where the maximum density occurs, and set $c = b^{|\nu|}$.

For full membranes, let the solution of Equation (3.53) be

$$w = \cos(n\theta) f(r) \tag{3.55}$$

where n is an integer. Equation (3.53) becomes

$$r^2 f'' + r f' - n^2 f + \omega^2 c r^{\nu+2} f = 0 \tag{3.56}$$

TABLE 3.8
Normalized Frequencies ω for Rectangular Membrane with Various Aspect Ratio *a* and Exponential Index *b* Values

a/b	0	0.10	0.25	0.50	0.75	1.00
0.25	12.953 (1)	13.268 (1)	13.705 (1)	14.355 (1)	14.931 (1)	15.461 (1)
	14.050 (1)	14.407 (1)	14.966 (1)	15.937 (1)	16.917 (1)	17.878 (1)
	15.708 (1)	16.106 (1)	16.726 (1)	17.817 (1)	18.977 (1)	20.192 (1)
	17.772 (1)	18.221 (1)	18.916 (1)	20.132 (1)	21.421 (1)	22.785 (1)
	20.116 (1)	20.624 (1)	21.408 (1)	22.770 (1)	24.068 (1)	25.721 (1)
0.50	7.0248 (1)	7.2001 (1)	7.4615 (1)	7.8913 (1)	8.3122 (1)	8.7232 (1)
	8.8858 (1)	9.1103 (1)	9.4560 (1)	10.056 (1)	10.684 (1)	11.337 (1)
	11.327 (1)	11.613 (1)	12.053 (1)	12.813 (1)	13.610 (1)	14.442 (1)
	12.953 (2)	13.268 (2)	13.705 (2)	14.355 (2)	14.931 (2)	15.461 (2)
	14.050 (1,2)	14.404 (1)	14.948 (1)	15.888 (1)	16.869 (1)	17.878 (2)
0.75	5.2360 (1)	5.3673 (1)	5.5656 (1)	5.8990 (1)	6.2348 (1)	6.5719 (1)
	7.5515 (1)	7.7420 (1)	8.0343 (1)	8.5389 (1)	9.0651 (1)	9.6124 (1)
	8.9473 (2)	9.1690 (2)	9.4936 (2)	10.012 (2)	10.502 (2)	10.967 (2)
	10.314 (1)	10.574 (1)	10.973 (1)	11.661 (1)	12.377 (1)	13.123 (1)
	10.472 (2)	10.737 (2)	11.147 (2)	11.862 (2)	12.610 (2)	13.382 (2)
1.00	4.4429 (1)	4.5545 (1)	4.7238 (1)	5.0107 (1)	5.3026 (1)	5.5987 (1)
	7.0248 (1,2)	7.2001 (2)	7.4615 (2)	7.8913 (2)	8.3122 (2)	8.7232 (2)
	8.8858 (2)	7.2020 (1)	7.4734 (1)	7.9410 (1)	8.4276 (1)	8.9327 (1)
	9.8346 (1,3)	9.1103 (2)	9.4560 (2)	10.056 (2)	10.684 (2)	11.337 (2)
	11.327 (2,3)	10.180 (3)	10.569 (1)	11.092 (3)	11.609 (3)	12.094 (3)
1.5	3.7751 (1)	3.8707 (1)	4.1052 (1)	4.2614 (1)	4.5138 (1)	4.7718 (1)
	5.2560 (2)	5.3673 (2)	5.5656 (2)	5.8990 (2)	6.2348 (2)	6.5719 (2)
	6.6231 (1)	6.7900 (1)	7.0456 (1)	7.4852 (1)	7.9418 (1)	8.4151 (1)
	7.0248 (3)	7.2001 (3)	7.4615 (3)	7.8913 (3)	8.3122 (3)	8.7232 (3)
	7.5515 (2)	7.7420 (2)	8.0343 (2)	8.5389 (2)	9.0651 (2)	9.6124 (2)
2	3.5124 (1)	3.6008 (1)	3.7355 (1)	3.9653 (1)	4.2615 (1)	4.4436 (1)
	4.4429 (2)	4.5545 (2)	4.7238 (2)	5.0107 (2)	5.3626 (2)	5.5987 (2)
	5.6636 (3)	5.8055 (3)	6.0192 (3)	6.3768 (3)	6.7347 (3)	7.0918 (3)
	6.4766 (1)	6.6398 (1)	6.8896 (1)	7.3191 (1)	7.7649 (1)	8.2266 (1)
	7.0248 (2)	7.2001 (4)	7.4615 (4)	7.8913 (4)	8.3122 (4)	8.9327 (2)
3	3.3115 (1)	3.3949 (1)	3.5220 (1)	3.7392 (1)	3.9628 (1)	4.1925 (1)
	3.7757 (2)	3.8707 (2)	4.0152 (2)	4.2614 (2)	4.5138 (2)	4.7718 (2)
	4.4429 (3)	4.5545 (3)	4.7238 (3)	5.0107 (3)	5.3026 (3)	5.5987 (3)
	5.2360 (4)	5.3673 (4)	5.5656 (4)	5.8990 (4)	6.2348 (4)	6.5719 (4)
	6.1062 (5)	6.2590 (5)	6.4884 (5)	7.1981 (1)	7.6360 (1)	8.0894 (1)

Note: Transverse mode number *n* is in parentheses.

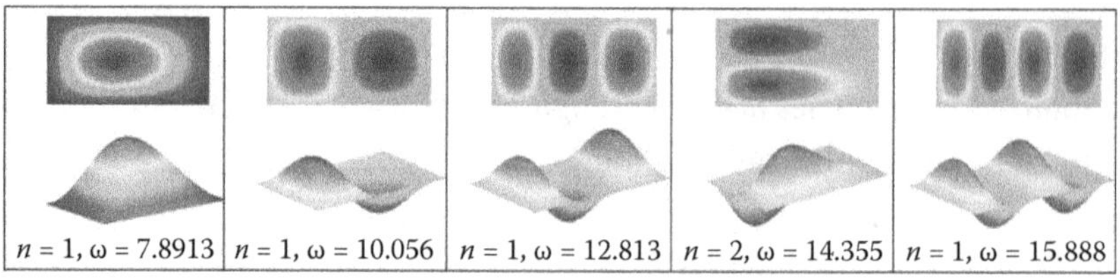

FIGURE 3.14 Mode shapes for rectangular membrane with $\alpha = 0.5$, $b = 0.5$.

Let

$$\gamma = \frac{\nu}{2} + 1 \tag{3.57}$$

The solution to Equation (3.56) is

$$f = C_1 J_{n/\gamma}\left(\frac{\omega\sqrt{c}}{\gamma} r^{\gamma}\right) + C_2 J_{-n/\gamma}\left(\frac{\omega\sqrt{c}}{\gamma} r^{\gamma}\right) \tag{3.58}$$

provided that $\gamma \neq 0$. Here J is the Bessel function of the first kind. If n/γ is an integer, the second solution is replaced by the Bessel function Y.

For a full circular membrane, $\nu \geq 0$ and $c = 1$. The boundary condition at $r = 0$ gives $C_2 = 0$. The boundary condition at $r = 1$ gives the frequency equation

$$J_{n/\gamma}\left(\frac{\omega}{\gamma}\right) = 0 \tag{3.59}$$

Thus (ω/λ) are the zeros of Equation (3.59). Table 3.9 shows the results. The $\nu = 0$ case is the uniform membrane, whose frequencies are governed by $J_n(\omega) = 0$. The $\nu = 1$ case is a membrane with linear density (or thickness). Mode shapes are shown for circular membrane with power law density at $\nu = 1.5$ in Figure 3.15.

TABLE 3.9
Frequencies for the Full Circular Membrane with Power Law Density

$\nu = 0$	0.5	1	1.5	2	3
2.4048 (0)	3.0060 (0)	3.6072 (0)	4.2084 (0)	4.8097 (0)	6.0121 (0)
3.8317 (1)	4.4497 (1)	5.0634 (1)	5.6743 (1)	6.2836 (1)	1.4971 (1)
5.1356 (2)	5.7790 (2)	6.4130 (2)	7.0406 (2)	7.6634 (2)	8.8995 (2)
5.5201 (0)	6.9001 (0)	7.7034 (3)	8.3487 (3)	8.9868 (3)	10.248 (3)
6.3802 (3)	7.0486 (3)	8.2801 (0)	9.6178 (4)	10.271 (4)	11.558 (4)

Note: The azimuthal mode n is in parentheses.

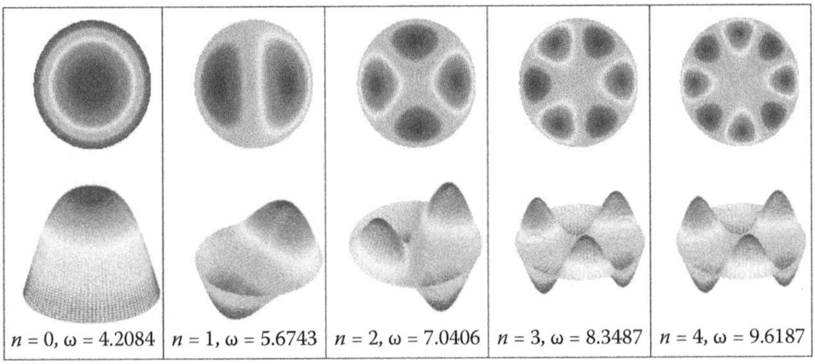

FIGURE 3.15 Mode shapes for circular membrane with power law density ν = 1.5.

For an annular membrane, ν can be either positive or negative. The boundary condition at the inner edge gives

$$f = C_1\left[J_{-n/\gamma}\left(\frac{\omega\sqrt{c}}{\gamma}b^{\gamma}\right)J_{n/\gamma}\left(\frac{\omega\sqrt{c}}{\gamma}r^{\gamma}\right) - J_{n/\gamma}\left(\frac{\omega\sqrt{c}}{\gamma}b^{\gamma}\right)J_{-n/\gamma}\left(\frac{\omega\sqrt{c}}{\gamma}r^{\gamma}\right)\right] \quad (3.60)$$

If n/γ is an integer, then $J_{-n/\gamma}$ is changed to $Y_{-n/\gamma}$.

The boundary condition at the outer edge then gives the characteristic equation

$$\left[J_{-n/\gamma}\left(\frac{\omega\sqrt{c}}{\gamma}b^{\gamma}\right)J_{n/\gamma}\left(\frac{\omega\sqrt{c}}{\gamma}\right) - J_{n/\gamma}\left(\frac{\omega\sqrt{c}}{\gamma}b^{\gamma}\right)J_{-n/\gamma}\left(\frac{\omega\sqrt{c}}{\gamma}\right)\right] = 0 \quad (3.61)$$

If $\nu = -2$, $\gamma = 0$, Equation (3.56) degenerates to

$$r^2\frac{d^2f}{dr^2} + r\frac{df}{dr} - n^2 f + \omega^2 b^2 f = 0 \quad (3.62)$$

The solution was first found by De (1971)

$$f = C\sin\left(\sqrt{\omega^2 b^2 - n^2}\,\ln r\right) \quad (3.63)$$

By setting $f = 0$ at $r = b$, the frequencies are found to be in a closed form given by

$$\omega = \frac{1}{b}\left[\left(\frac{m\pi}{\ln b}\right)^2 + n^2\right]^{1/2} \quad (3.64)$$

Here m is a positive integer. Some results for the annular membrane are given in Table 3.10. Mode shapes for annular membrane with power index $\nu = -2$ and inner

TABLE 3.10
Normalized Frequencies ω for the Full Annular Membrane with Power Index ν

$b\backslash\nu$	−2	−1	0	1	2
0.1	13.644 (0)	7.1519 (0)	3.3139 (0)	4.4410 (0)	5.6018 (0)
	16.916 (1)	8.7465 (1)	3.9409 (1)	5.1412 (1)	6.3467 (1)
	24.211 (2)	12.075 (2)	5.1424 (2)	6.4165 (2)	7.6658 (2)
	27.288 (0)	14.465 (0)	6.3805 (3)	7.7035 (3)	8.9869 (3)
	29.062 (1)	15.454 (1)	6.8576 (0)	8.9552 (4)	10.271 (4)
0.3	8.6979 (0)	6.3125 (0)	4.4124 (0)	5.4451 (0)	6.5309 (0)
	9.3147 (1)	6.7526 (1)	4.7058 (1)	5.7830 (1)	6.9046 (1)
	10.959 (2)	7.9181 (2)	5.4702 (2)	6.6464 (2)	7.8420 (2)
	13.253 (3)	9.5241 (3)	6.4937 (3)	7.7678 (3)	9.0296 (3)
	15.920 (4)	11.357 (4)	7.6229 (4)	8.9708 (4)	10.280 (4)
0.5	9.0647 (0)	7.5730 (0)	6.2461 (0)	7.7162 (0)	8.1954 (0)
	9.2827 (1)	7.7542 (1)	6.3932 (1)	7.3615 (1)	8.3776 (1)
	9.9080 (2)	8.2734 (2)	6.8138 (2)	7.8327 (2)	8.8950 (2)
	10.871 (3)	9.0715 (3)	7.4577 (3)	8.5495 (3)	9.6763 (3)
	12.090 (4)	10.080 (4)	8.2667 (4)	9.4422 (4)	10.640 (4)
0.7	12.583 (0)	11.490 (0)	10.455 (0)	11.333 (0)	12.243 (0)
	12.664 (1)	11.563 (1)	10.522 (1)	11.405 (1)	12.320 (1)
	12.903 (2)	11.782 (2)	10.720 (2)	11.618 (2)	12.548 (2)
	13.293 (3)	12.137 (3)	11.042 (3)	11.964 (3)	12.918 (3)
	13.820 (4)	12.617 (4)	11.476 (4)	12.431 (4)	13.417 (4)
0.9	33.131 (0)	32.265 (0)	31.412 (0)	32.226 (0)	33.051 (0)
	33.149 (1)	32.283 (1)	31.429 (1)	32.244 (1)	33.069 (1)
	33.205 (2)	32.337 (2)	31.482 (2)	32.298 (2)	33.125 (2)
	33.298 (3)	32.427 (3)	31.570 (3)	32.388 (3)	33.217 (3)
	33.427 (4)	32.554 (4)	31.693 (4)	32.514 (4)	33.346 (4)

Note: The azimuthal mode number n is in parentheses.

radius $b = 0.5$ are shown in Figure 3.16. The frequencies of a membrane with power $\nu = -4$ are isospectral to those of a homogeneous membrane (Gottlieb 1992).

For circular sector or annular sector membranes, let the opening angle be β. Let

$$w = \sin(\alpha\theta) f(r) \tag{3.65}$$

where $\alpha = n\pi/\beta$. Note that n is a nonzero positive integer. Then the frequency equation is similar to that for the full membrane, only with n replaced by α. For example, Equation (3.59) becomes

$$J_{n\pi/(\beta\gamma)}\left(\frac{\omega}{\gamma}\right) = 0 \tag{3.66}$$

Table 3.11 shows the frequencies for the circular sector membrane.

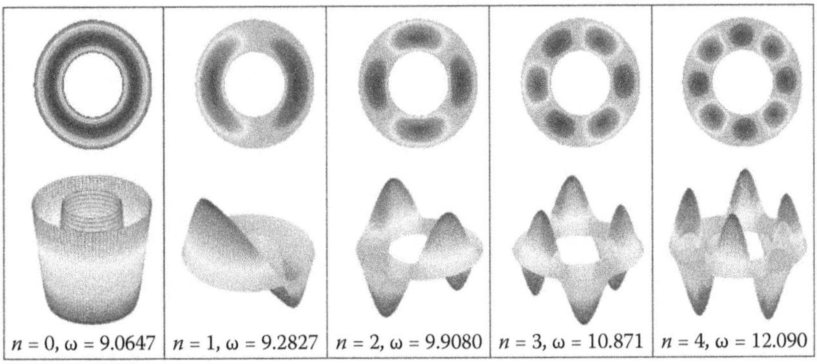

FIGURE 3.16 Mode shapes for annular membrane with power index $\nu = -2$ and inner radius $b = 0.5$.

TABLE 3.11
Frequencies for Circular Sector Membranes

$\beta \backslash \nu$	0	0.5	1	1.5	2	3
$\pi/4$	7.5883 (1)	8.2803 (1)	8.9552 (1)	9.6178 (1)	10.271 (1)	11.558 (1)
	11.065 (1)	12.531 (1)	13.742 (2)	14.467 (2)	15.177 (2)	16.561 (2)
	12.225 (2)	12.997 (2)	13.977 (1)	15.411 (1)	16.835 (1)	19.662 (1)
	14.373 (1)	16.612 (1)	18.338 (3)	19.115 (3)	19.872 (3)	21.341 (3)
	16.038 (2)	17.535 (3)	18.832 (1)	20.643 (2)	22.129 (2)	25.061 (2)
$\pi/2$	5.1356 (1)	5.7790 (1)	6.4130 (1)	7.0406 (1)	7.6634 (1)	8.8995 (1)
	7.5883 (2)	8.2803 (2)	8.9552 (2)	9.6178 (2)	10.271 (2)	11.558 (2)
	8.4172 (1)	9.8312 (1)	11.236 (1)	12.078 (3)	12.760 (3)	14.097 (3)
	9.9361 (3)	10.671 (3)	11.383 (3)	12.636 (1)	14.031 (1)	16.561 (4)
	11.065 (2)	12.531 (2)	13.742 (4)	14.467 (4)	15.177 (4)	16.814 (1)
π	3.8317 (1)	4.4497 (1)	5.0634 (1)	5.6743 (1)	6.2832 (1)	7.4971 (1)
	5.1356 (2)	5.7790 (2)	6.4130 (2)	7.0406 (2)	7.6634 (2)	8.8995 (2)
	6.3802 (3)	7.0486 (3)	7.7034 (3)	8.3487 (3)	8.9868 (3)	10.248 (3)
	7.0156 (1)	8.2803 (4)	8.9552 (4)	9.6178 (4)	10.271 (4)	11.558 (4)
	7.5883 (4)	8.4071 (1)	9.7954 (1)	10.859 (5)	11.527 (5)	12.839 (5)
2π	3.1416 (1)	3.7486 (1)	4.3539 (1)	4.9582 (1)	5.5618 (1)	6.7677 (1)
	3.8317 (2)	4.4497 (2)	5.0634 (2)	5.6743 (2)	6.2832 (2)	7.4971 (2)
	4.4934 (3)	5.1240 (3)	5.7476 (3)	6.3665 (3)	6.9820 (3)	8.2064 (3)
	5.1356 (4)	5.7790 (4)	6.4130 (4)	7.0406 (4)	7.6634 (4)	8.8995 (4)
	5.7635 (5)	6.4195 (5)	7.0640 (5)	7.7004 (5)	8.3309 (5)	9.5793 (5)

Note: The number n is in parentheses.

The substitution is similar for the annular sector. For example, for ν = −2, the formula is

$$\omega = \frac{1}{b}\left[\left(\frac{m\pi}{\ln b}\right)^2 + \left(\frac{n\pi}{\beta}\right)^2\right]^{1/2} \tag{3.67}$$

Owing to the presence of many parameters, we present only some representative cases in Table 3.12.

3.5.3.2 A Special Annular Membrane

The annulus is between $r = b < 1$ and $r = 1$. Let the density distribution be

$$\rho = \frac{b^2[1 + c\ln(r/b)]}{r^2} \tag{3.68}$$

For the density to be positive, $c > 1/\ln b$. In view of Equation (3.55), Equation (3.53) becomes

$$r^2\frac{d^2 f}{dr^2} + r\frac{df}{dr} - n^2 f + \omega^2 b^2[1 + c\ln(r/b)]f = 0 \tag{3.69}$$

The solution is

$$f = C_1\sqrt{z}[J_{1/3}(p_0)J_{-1/3}(p) - J_{-1/3}(p_0)J_{1/3}(p)] \tag{3.70}$$

TABLE 3.12a
Frequencies for Annular Sector Membranes, β = π/4

$b \backslash \nu$	−2	−1	0	1	2
0.25	18.389 (1)	12.311 (1)	7.5984 (1)	8.9592 (1)	10.273 (1)
	24.180 (1)	16.841 (1)	11.169 (1)	13.742 (2)	15.177 (2)
	31.552 (1)	21.086 (2)	12.225 (2)	14.011 (1)	16.849 (1)
	33.259 (2)	21.972 (1)	14.765 (1)	18.338 (3)	19.872 (3)
	39.632 (1)	25.443 (2)	16.038 (2)	18.957 (1)	22.129 (2)
0.50	12.090 (1)	10.080 (1)	8.2667 (1)	9.4422 (1)	10.640 (1)
	18.389 (2)	15.243 (2)	12.311 (2)	13.781 (2)	15.197 (2)
	19.816 (1)	16.583 (1)	13.742 (1)	15.939 (1)	18.289 (1)
	24.180 (2)	20.268 (2)	16.706 (3)	18.340 (3)	19.873 (3)
	25.655 (3)	21.075 (3)	16.843 (2)	19.539 (2)	22.337 (2)
0.75	15.567 (1)	14.413 (1)	13.366 (1)	14.278 (1)	15.218 (1)
	18.050 (2)	16.774 (2)	15.548 (2)	16.593 (2)	17.664 (2)
	21.634 (3)	20.694 (3)	18.616 (3)	19.836 (3)	21.075 (3)
	25.829 (4)	23.986 (4)	22.184 (4)	23.597 (4)	25.007 (4)
	29.605 (1)	27.526 (1)	25.546 (1)	27.326 (1)	29.178 (1)

Note: Mode number n is in parentheses.

TABLE 3.12b
Frequencies for Annular Sector Membranes, $\beta = \pi/2$

$b\backslash\nu$	−2	−1	0	1	2
0.25	12.090 (1)	8.2667 (1)	5.3199 (1)	6.5327 (1)	7.7518 (1)
	18.389 (2)	12.311 (2)	7.5984 (2)	8.9592 (2)	10.273 (2)
	19.816 (1)	13.742 (1)	9.1444 (1)	11.383 (3)	12.760 (3)
	24.180 (2)	16.706 (3)	9.9365 (3)	11.658 (1)	14.301 (1)
	25.655 (3)	16.841 (2)	11.169 (2)	14.011 (2)	15.177 (4)
0.50	9.9080 (1)	8.2734 (1)	6.8138 (1)	7.8327 (1)	8.8950 (1)
	12.090 (2)	10.080 (2)	8.2667 (2)	9.4422 (2)	10.640 (2)
	15.039 (3)	12.568 (3)	10.189 (3)	11.528 (3)	12.853 (3)
	18.389 (4)	15.243 (4)	12.311 (4)	13.781 (4)	15.197 (4)
	18.566 (1)	15.533 (1)	12.856 (1)	14.893 (1)	17.074 (1)
0.75	14.803 (1)	13.760 (1)	12.761 (1)	13.635 (1)	14.537 (1)
	15.507 (2)	14.413 (2)	13.366 (2)	14.278 (2)	15.218 (2)
	16.614 (3)	15.441 (3)	14.320 (3)	15.287 (3)	16.285 (3)
	18.050 (4)	16.774 (4)	15.547 (4)	16.593 (4)	17.664 (4)
	19.743 (5)	18.345 (5)	16.990 (5)	18.129 (5)	19.281 (5)

Note: Mode number n is in parentheses.

TABLE 3.12c
Frequencies for Annular Sector Membranes, $\beta = \pi$

$b\backslash\nu$	−2	−1	0	1	2
0.25	9.9080 (1)	6.8138 (1)	4.4475 (1)	5.5557 (1)	6.7021 (1)
	12.090 (2)	8.2667 (2)	5.3199 (2)	6.5327 (2)	7.7518 (2)
	15.039 (3)	10.189 (3)	6.4265 (3)	7.7270 (3)	9.0017 (3)
	18.389 (4)	12.311 (4)	7.5984 (4)	8.9592 (4)	10.273 (4)
	18.566 (1)	12.851 (1)	8.5369 (1)	10.180 (5)	11.527 (5)
0.50	9.2827 (1)	7.7542 (1)	6.3932 (1)	7.3615 (1)	8.3776 (1)
	9.9080 (2)	8.2734 (2)	6.8138 (2)	7.8327 (2)	8.8950 (2)
	10.871 (3)	9.0715 (3)	7.4577 (3)	8.5495 (3)	9.6763 (3)
	12.090 (4)	10.080 (4)	8.2667 (4)	9.4422 (4)	10.640 (4)
	13.497 (5)	11.241 (5)	9.1900 (5)	10.450 (5)	11.715 (5)
0.75	14.621 (1)	13.591 (1)	12.606 (1)	13.470 (1)	14.362 (1)
	14.803 (2)	13.760 (2)	12.790 (2)	13.635 (2)	14.537 (2)
	15.100 (3)	14.036 (3)	13.017 (3)	13.907 (3)	14.825 (3)
	15.507 (4)	14.413 (4)	13.345 (4)	14.278 (4)	15.218 (4)
	16.014 (5)	14.885 (5)	13.784 (5)	14.741 (5)	15.707 (5)

Note: Mode number n is in parentheses.

TABLE 3.12d
Frequencies for Annular Sector Membrane, $\beta = 2\pi$

$b \backslash \nu$	−2	−1	0	1	2
0.25	9.2827 (1)	6.3932 (1)	4.1888 (1)	5.2583 (1)	6.3755 (1)
	9.9080 (2)	6.8138 (2)	4.4475 (2)	5.5557 (2)	6.7021 (2)
	10.871 (3)	7.4577 (3)	4.8382 (3)	5.9979 (3)	7.1812 (3)
	12.090 (4)	8.2667 (4)	5.3199 (4)	6.5327 (4)	7.7518 (4)
	13.497 (5)	9.1900 (5)	5.8577 (5)	7.1183 (5)	8.3680 (5)
0.50	9.1197 (1)	7.6187 (1)	6.2832 (1)	7.2379 (1)	8.2414 (1)
	9.2827 (2)	7.7542 (2)	6.3932 (2)	7.3615 (2)	8.3776 (2)
	9.5483 (3)	7.9747 (3)	6.5720 (3)	7.5621 (3)	8.5982 (3)
	9.9080 (4)	8.2734 (4)	6.8138 (4)	7.8327 (4)	8.8950 (4)
	10.352 (5)	8.6419 (5)	7.1116 (5)	8.1648 (5)	9.2579 (5)
0.75	14.576 (1)	13.549 (1)	12.566 (1)	13.428 (1)	14.317 (1)
	14.621 (2)	13.591 (2)	12.634 (2)	13.470 (2)	14.362 (2)
	14.697 (3)	13.662 (3)	12.671 (3)	13.539 (3)	14.435 (3)
	14.803 (4)	13.760 (4)	12.713 (4)	13.635 (4)	14.537 (4)
	14.937 (5)	13.885 (5)	12.877 (5)	13.758 (5)	14.667 (5)

Note: Mode number n is in parentheses.

where

$$p = 2z^{3/2}/3, \quad p_0 = 2z_0^{3/2}/3, \quad p_1 = 2z_1^{3/2}/3$$

$$z = (\omega^2 b^2 c)^{-2/3}\{\omega^2 b^2[1 + c(\eta - 1 - \ln b)] - n^2\}$$

$$z_0 = (\omega^2 b^2 c)^{-2/3}[\omega^2 b^2 - n^2], \quad z_1 = (\omega^2 b^2 c)^{-2/3}[\omega^2 b^2(1 - c\ln b) - n^2] \tag{3.71}$$

$$\eta = 1 + \ln r$$

The frequency equation is

$$J_{1/3}(p_0)J_{-1/3}(p_1) - J_{-1/3}(p_0)J_{1/3}(p_1) = 0 \tag{3.72}$$

For sector membranes, we use Equation (3.65) instead of Equation (3.55). All n are replaced by $n\pi/\beta$, and for nontrivial solutions $n \geq 1$. Owing to the fact that the density distribution of Equation (3.68) is somewhat rare, we shall not tabulate the corresponding frequencies here.

3.6 HANGING MEMBRANES

The vertically hanging membrane under the action of gravity is important in the modeling of drapes, curtains, nets, and fabric panels. Pioneering work in this area was done by Soedel, Zadoks, and Alfred (1985), who studied the natural frequencies

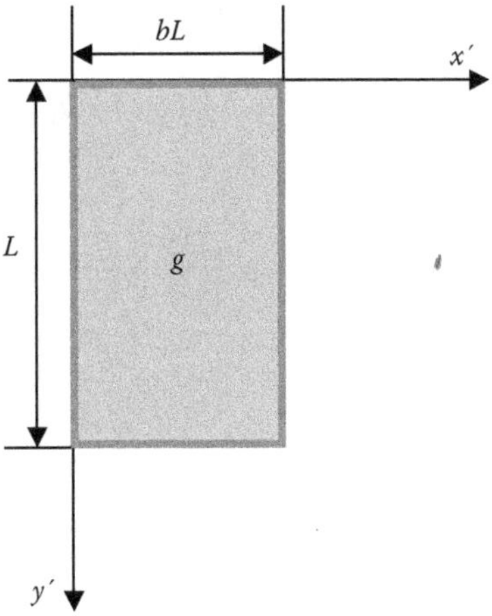

FIGURE 3.17 Hanging membrane with a weighted bottom edge.

of a vertical rectangular membrane fixed at the top and two sides, but free at the bottom edge. The following subsections are mostly from Wang and Wang (2011b).

3.6.1 Membrane with a Free, Weighted Bottom Edge

Figure 3.17 shows a vertical rectangular membrane of height L and width bL and with constant density ρ. Let (x',y') be Cartesian axes at the upper left corner as shown in the figure.

Since gravity is nonnegligible, the vertical direction tensile stress is

$$T_y = T_0 - \rho g y' \tag{3.73}$$

where T_0 is the tension at the upper edge and g is the gravitational acceleration. We assume that the horizontal tension is constant, $T_x = cT_0$.

In the case considered, the top edge and the two vertical sides are fixed, while the bottom edge is weighted down by a continuous string of masses, such that the total mass is mL (m is mass per length). Thus the tension at the bottom is

$$T_1 = T_0 - \rho g L = mg \tag{3.74}$$

A dynamic balance of vertical force at the bottom edge gives

$$T_1 \frac{\partial w'}{\partial y'} = m\bar{\omega}^2 w' \tag{3.75}$$

where $\bar{\omega}$ is the frequency. Now if there is no hanging mass, $m = 0$ and $T_1 = 0$. The boundary condition Equation (3.75) is automatically satisfied. This would be the case studied by Soedel, Zadoks, and Alfred (1985), where the only requirement at the bottom edge is that the amplitude is finite.

By normalizing all lengths by L, the tension by T_0, and dropping the primes, Equation (3.2) becomes

$$c\frac{\partial^2 w}{\partial x^2}+\frac{\partial}{\partial y}\left[(1-ay)\frac{\partial w}{\partial y}\right]+\omega^2 w=0 \tag{3.76}$$

Here

$$a=\frac{\rho g L}{T_0}\leq 1 \tag{3.77}$$

is a ratio representing the importance of gravity to tension. Since the deflections on the side edges are zero, let

$$w=\sin(\beta x)f(y) \tag{3.78}$$

where $\beta = n\pi/b$. Equation (3.76) becomes

$$\frac{d}{dy}\left[(1-ay)\frac{df}{dy}\right]+(\omega^2-c\beta^2)f=0 \tag{3.79}$$

Equation (3.79) has the general solution (Murphy 1960)

$$f=C_1 J_0\left(\frac{2k}{a}\sqrt{1-ay}\right)+C_2 Y_0\left(\frac{2k}{a}\sqrt{1-ay}\right) \tag{3.80}$$

where J_0, Y_0 are Bessel functions and

$$k=\sqrt{\omega^2-c\beta^2}=\sqrt{\omega^2-(c/b^2)n^2\pi^2} \tag{3.81}$$

The boundary condition on the top is

$$f(0)=0 \tag{3.82}$$

This gives

$$f=C\left[Y_0\left(\frac{2k}{a}\right)J_0\left(\frac{2k}{a}\sqrt{1-ay}\right)-J_0\left(\frac{2k}{a}\right)Y_0\left(\frac{2k}{a}\sqrt{1-ay}\right)\right] \tag{3.83}$$

where C is a constant. After normalization, Equation (3.75) is

$$(1-a)\frac{\partial w}{\partial y}(x,1)=\omega^2 w(x,1) \tag{3.84}$$

By using Equation (3.78), the condition is

$$(1-a)f'(1)=\omega^2 f(1) \tag{3.85}$$

or the exact characteristic frequency equation is

$$\left[Y_0\left(\frac{2k}{a}\right)J_1\left(\frac{2k}{a}\sqrt{1-a}\right)-J_0\left(\frac{2k}{a}\right)Y_1\left(\frac{2k}{a}\sqrt{1-a}\right)\right]\frac{ak}{\sqrt{1-a}}+$$
$$\omega^2\left[Y_0\left(\frac{2k}{a}\right)J_0\left(\frac{2k}{a}\sqrt{1-a}\right)-J_0\left(\frac{2k}{a}\right)Y_0\left(\frac{2k}{a}\sqrt{1-a}\right)\right]=0 \tag{3.86}$$

Notice that the normalized lateral tension c and the normalized width b come in the combination (c/b^2). If there is no added weight, $m = 0$, and $a = 1$, we have the case studied by Soedel, Zadoks, and Alfred (1985). Equation (3.86) degenerates to

$$J_0(2k)=0 \tag{3.87}$$

The results from Equation (3.86) are shown in Table 3.13. Notice that an increase in the parameter $\xi = b/\sqrt{c}$ lowers the frequencies. For small ξ (small width or large lateral tension), the mode shapes all correspond to $n = 1$, or half a sine wave in the horizontal direction. For large ξ, the first five modes correspond to $n = 1, 2, 3, 4, 5$, or multiple nodal lines in the horizontal direction. The frequencies decrease with gravity, which is reflected in the parameter a. Mode shapes for a hanging membrane with a free weighted bottom edge (i.e., $a = 0.5$, $\xi = 0.5$) are shown in Figure 3.18.

3.6.2 Vertical Membrane with All Sides Fixed

Consider a membrane with all sides fixed on a rectangular frame. In the supine state, the membrane is stretched with stress T_1 in the y direction and cT_1 in the x direction. Now, if the membrane is raised vertically as in Figure 3.19, what would be the changes in frequency?

In this case, it is more natural to normalize by the bottom stress T_1. Equations (3.74) and (3.2) give

$$T_y = T_1[1+a(1-y)] \tag{3.88}$$

$$c\frac{\partial^2 w}{\partial x^2}+\frac{\partial}{\partial y}\left[(1+a-ay)\frac{\partial w}{\partial y}\right]+\omega^2 w=0 \tag{3.89}$$

TABLE 3.13
Frequencies for Weighted, Hanging Membranes

ξ [a]	a = 0.1	0.3	0.5	0.7	0.9	1
0.1	31.565 (1)	31.548 (1)	31.530 (1)	31.509 (1)	31.482 (1)	31.439 (1)
	32.007 (1)	31.941 (1)	31.870 (1)	31.788 (1)	31.684 (1)	31.537 (1)
	32.730 (1)	32.586 (1)	32.428 (1)	32.249 (1)	32.018 (1)	31.713 (1)
	33.717 (1)	33.467 (1)	33.194 (1)	32.883 (1)	32.480 (1)	31.964 (1)
	34.945 (1)	34.568 (1)	34.154 (1)	33.681 (1)	33.065 (1)	32.291 (1)
0.2	16.003 (1)	15.970 (1)	15.934 (1)	15.892 (1)	15.837 (1)	15.754 (1)
	16.858 (1)	16.732 (1)	16.594 (1)	16.436 (1)	16.231 (1)	15.949 (1)
	18.194 (1)	17.930 (1)	17.640 (1)	17.307 (1)	16.869 (1)	16.293 (1)
	19.914 (1)	19.485 (1)	19.010 (1)	18.458 (1)	17.725 (1)	16.778 (1)
	21.929 (1)	21.319 (1)	20.639 (1)	19.842 (1)	18.770 (1)	17.392 (1)
0.5	6.9863 (1)	6.9052 (1)	6.8165 (1)	6.7147 (1)	6.5789 (1)	6.3972 (1)
	8.7668 (1)	8.5126 (1)	8.2289 (1)	7.8955 (1)	7.4409 (1)	6.8627 (1)
	11.121 (1)	10.675 (1)	10.170 (1)	9.5653 (1)	8.7197 (1)	8.6162 (1)
	12.933 (2)	12.892 (2)	12.394 (1)	11.515 (1)	10.268 (1)	9.7576 (1)
	13.756 (1)	13.120 (1)	12.846 (2)	12.795 (2)	11.980 (1)	11.005 (1)
1	4.3748 (1)	4.2295 (1)	4.0677 (1)	3.8285 (1)	3.6233 (1)	3.3638 (1)
	6.8691 (1)	6.5306 (1)	6.1414 (1)	5.6652 (1)	4.9737 (1)	4.1818 (1)
	6.9863 (2)	6.9052 (2)	6.8165 (2)	6.7147 (2)	6.5789 (2)	5.3471 (1)
	8.7668 (2)	8.5126 (2)	8.2289 (2)	7.8330 (1)	6.7392 (1)	6.3972 (2)
	9.9084 (3)	9.1756 (1)	8.5742 (1)	7.8955 (2)	7.4409 (2)	6.6805 (2)
2	3.4167 (1)	3.2060 (1)	2.9598 (1)	2.6581 (1)	2.2610 (1)	1.9782 (1)
	4.3748 (2)	4.2295 (2)	4.0677 (2)	3.8784 (2)	3.6233 (2)	3.1757 (1)
	5.6138 (3)	5.5087 (3)	5.3933 (3)	4.9295 (1)	4.0818 (1)	3.3638 (2)
	6.3050 (1)	5.9279 (1)	5.4861 (1)	5.2601 (3)	4.9737 (2)	4.1818 (2)
	6.8691 (2)	6.5306 (2)	6.1414 (2)	5.6652 (2)	5.0814 (3)	4.6032 (1)
5	3.0929 (1)	2.8421 (1)	2.5345 (1)	2.1349 (1)	1.6319 (1)	1.3567 (1)
	3.2821 (2)	3.0565 (2)	2.7888 (2)	2.4547 (2)	2.0209 (2)	1.7392 (2)
	3.5740 (3)	3.3781 (3)	3.1525 (3)	2.8799 (3)	2.5176 (3)	2.2358 (3)
	3.9450 (4)	3.7767 (4)	3.5867 (4)	3.3618 (4)	3.0595 (4)	2.7861 (4)
	4.3748 (5)	4.2295 (5)	4.0677 (5)	3.8785 (5)	3.6233 (5)	2.8307 (1)

Note: Parentheses denote lateral mode number n.

[a] Values for $\xi = b/\sqrt{c}$ and $a = 1$ are from Equation (3.87).

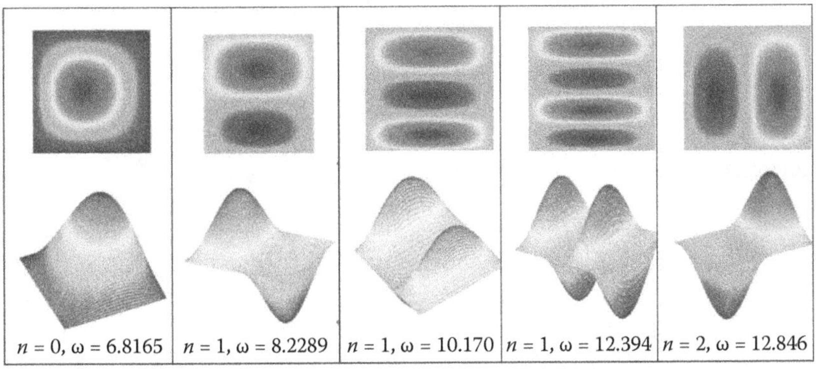

FIGURE 3.18 Mode shapes for hanging membrane with a free weighted bottom edge ($a = 0.5$, $b = 1$, $\xi = 0.5$).

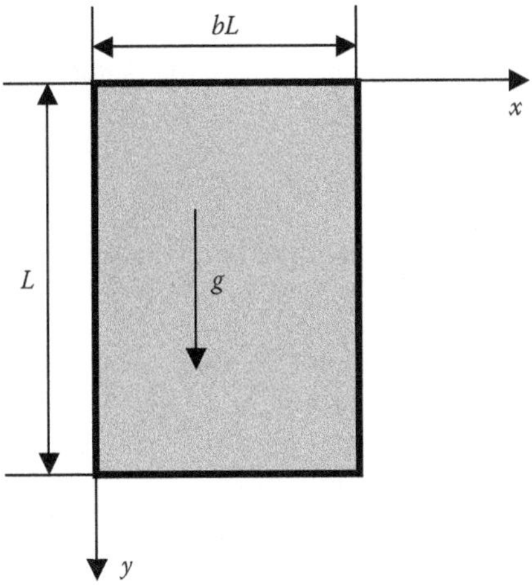

FIGURE 3.19 Vertical membrane with all sides fixed.

where

$$a = \frac{\rho g L}{T_1}, \quad \omega = \frac{\bar{\omega}}{\sqrt{T_1/(\rho L^2)}} \tag{3.90}$$

Equations (3.78) and (3.89) give

$$\frac{d}{dy}\left[(1+a-ay)\frac{df}{dy}\right] + (\omega^2 - c\beta^2)f = 0 \tag{3.91}$$

The solution that satisfies zero deflection on the top edge is

$$f = C\left[Y_0\left(\frac{2k}{a}\sqrt{1+a}\right)J_0\left(\frac{2k}{a}\sqrt{1+a-ay}\right) - J_0\left(\frac{2k}{a}\sqrt{1+a}\right)Y_0\left(\frac{2k}{a}\sqrt{1+a-ay}\right)\right] \tag{3.92}$$

The zero deflection at the bottom edge yields the frequency equation

$$Y_0\left(\frac{2k}{a}\sqrt{1+a}\right)J_0\left(\frac{2k}{a}\right) - J_0\left(\frac{2k}{a}\sqrt{1+a}\right)Y_0\left(\frac{2k}{a}\right) = 0 \tag{3.93}$$

When there is no gravity, $a = 0$. Then Equation (3.91) gives the solution

$$f = C\sin(ky), \quad \text{where } k = m\pi \tag{3.94}$$

The frequency is

$$\omega = \pi\sqrt{m^2 + (c/b^2)n^2} \tag{3.95}$$

The general results from Equation (3.93) are given in Table 3.14. Note that for an originally uniformly stressed membrane, $c = 1$. The frequencies increase with the gravity parameter a. Mode shapes for a vertical membrane with all sides fixed ($a = 0.5$, $b = 1$, $\xi = 0.5$) are shown in Figure 3.20.

3.7 DISCUSSION

Being two dimensional, a membrane always has two sets of nodes, one in each direction. An example is m and n for the uniformly stretched rectangular membrane shown in Equation (3.8). In order to be useful, an exact solution must yield the complete frequency spectrum. We did not include the exact solutions that only give partial frequency spectrums. The first kind of partial spectrum considers vibrations that are dependent on only one set of nodes, such as axisymmetric vibrations of circular or annular membranes. The second kind of partial spectrum uses the frequencies

TABLE 3.14
Frequencies when Vertical Membrane Is Fixed on All Sides

ξ	a = 0 [a]	0.25	0.5	0.75	1	2	5
	31.573 (1)	31.592 (1)	31.610 (1)	31.627 (1)	31.643 (1)	31.706 (1)	31.871 (1)
	32.038 (1)	32.113 (1)	32.184 (1)	32.252 (1)	32.318 (1)	32.565 (1)	33.223 (1)
0.1	32.799 (1)	32.963 (1)	33.119 (1)	33.268 (1)	33.412 (1)	33.950 (1)	35.363 (1)
	33.836 (1)	34.118 (1)	34.385 (1)	34.640 (1)	34.886 (1)	35.798 (1)	38.159 (1)
	35.124 (1)	35.548 (1)	35.948 (1)	36.329 (1)	36.694 (1)	38.043 (1)	41.477(1)
	16.019 (1)	16.056 (1)	16.092 (1)	16.125 (1)	16.158 (1)	16.279 (1)	16.599 (1)
	16.918 (1)	17.059 (1)	17.192 (1)	17.320 (1)	17.442 (1)	17.896 (1)	19.067 (1)
0.2	18.319 (1)	18.611 (1)	18.885 (1)	19.145 (1)	19.394 (1)	20.306 (1)	22.591 (1)
	20.116 (1)	20.587 (1)	21.027 (1)	21.441 (1)	21.836 (1)	23.265 (1)	26.756 (1)
	22.214 (1)	22.879 (1)	23.496 (1)	24.074 (1)	24.622 (1)	26.590 (1)	31.307 (1)
	7.0248 (1)	7.1094 (1)	7.1888 (1)	7.2640 (1)	7.3359 (1)	7.5998 (1)	8.2627 (1)
	8.8858 (1)	9.1515 (1)	9.3976 (1)	9.6284 (1)	9.8468 (1)	10.630 (1)	12.502 (1)
0.5	11.327 (1)	11.794 (1)	12.222 (1)	12.620 (1)	12.995 (1)	14.321 (1)	17.409 (1)
	12.953 (2)	12.992 (2)	13.043 (2)	13.084 (2)	13.125 (2)	13.274 (2)	13.664 (2)
	14.050 (1)	14.219 (2)	14.379 (2)	14.531 (2)	14.676 (1)	15.213 (2)	16.575 (2)
	4.4429 (1)	4.5755 (1)	4.6978 (1)	4.8122 (1)	4.9201 (1)	5.3055 (1)	6.2180 (1)
	7.6248 (1)	7.1694 (2)	7.1888 (2)	7.2640 (2)	7.3359 (2)	7.5998 (2)	8.2627 (2)
1	8.8858 (2)	7.3581 (1)	7.6620 (1)	7.9434 (1)	8.2067 (1)	9.1321 (1)	10.845 (3)
	9.9346 (1)	9.1515 (2)	9.3976 (2)	9.6284 (2)	9.8468 (2)	10.349 (3)	11.256 (1)
	11.327 (2)	10.464 (1)	10.944 (1)	10.105 (3)	10.157 (3)	10.630 (2)	12.502 (2)
	3.5124 (1)	3.6787 (1)	3.8298 (1)	3.9692 (1)	4.0994 (1)	4.5548 (1)	5.5912 (1)
	4.4429 (2)	4.5755 (2)	4.6978 (2)	4.8122 (2)	4.9201 (2)	5.3055 (2)	6.2180 (2)
2	5.6636 (3)	5.7682 (3)	5.8657 (3)	5.9577 (3)	6.0452 (3)	6.3628 (3)	7.1415 (3)
	6.4766 (1)	6.8366 (1)	7.1627 (1)	7.2640 (4)	7.3359 (4)	7.5998 (4)	8.2627 (4)
	7.0248 (2)	7.1094 (4)	7.1888 (4)	7.4629 (1)	7.7426 (1)	8.7174 (1)	9.5120 (5)
	3.2038 (1)	3.3853 (1)	3.5489 (1)	3.6989 (1)	3.8383 (1)	4.3213 (1)	5.4026 (1)
	3.3836 (2)	3.5559 (2)	3.7120 (2)	3.8557 (2)	3.9896 (2)	4.4563 (2)	5.5112 (2)
5	3.6637 (3)	3.8234 (3)	3.9690 (3)	4.1037 (3)	4.2297 (3)	4.6725 (3)	5.6874 (3)
	4.0232 (4)	4.1692 (4)	4.3031 (4)	4.4276 (4)	4.5447 (4)	4.9594 (4)	5.9254 (4)
	4.4429 (5)	4.5755 (5)	4.6978 (5)	4.8122 (5)	4.9201 (5)	5.3055 (5)	6.2180 (5)

Note: Lateral mode number n is in parentheses.

[a] The column for $a = 0$ is from Equation (3.95).

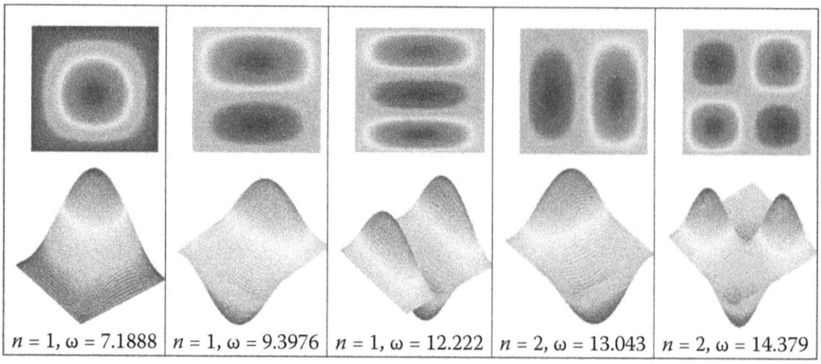

FIGURE 3.20 Mode shapes for vertical membrane with all sides fixed ($a = 0.5$, $b = 1$, $\xi = 0.5$).

of a smaller membrane to predict the frequencies of a larger membrane that is composed of several joined smaller membranes. Such an endeavor misses a great number of frequencies, especially the important fundamental frequency.

REFERENCES

De, S. 1971. Solution to the equation of a vibrating annular membrane of non-homogeneous material. *Pure Appl. Geophys.* 90:89–95.

Gottlieb, H. P. W. 1986. New types of vibration modes for stepped membranes. *J. Sound Vibr.* 110:395–411.

———. 1992. Axisymmetric isospectral annular plates and membranes. *IMA J. Appl. Math.* 49:185–92.

———. 2004. Isospectral circular membranes. *Inverse Prob.* 20:155–61.

Laura, P. A. A., C. A. Rossit, and S. La Malfa. 1998. Transverse vibrations of composite circular annular membranes: Exact solution. *J. Sound Vibr.* 216:190–93.

Leissa, A. W., and A. Ghamat-Rezaei. 1990. Vibrations of rectangular membranes subjected to shear and nonuniform tensile stress. *J. Acoust. Soc. Am.* 88:231–38.

Murphy, G. M. 1960. *Ordinary differential equations and their solutions.* Princeton, NJ: Van Nostrand.

Schelkunoff, S. A. 1943. *Electromagnetic waves.* New York: Van Nostrand, pp. 393–94.

Seth, B. R. 1947. Transverse vibrations of rectilinear plates. *Proc. Ind. Acad. Sci.* A25:25–29.

Soedel, W. 2004. *Vibrations of shells and plates.* 3rd ed. New York: Dekker.

Soedel, W., R. I. Zadoks, and J. R. Alfred. 1985. Natural frequencies and modes of hanging nets or curtains. *J. Sound Vibr.* 103:499–507.

Spence, J. P., and C. O. Horgan. 1983. Bounds on natural frequencies of composite circular membranes: Integral equation methods. *J. Sound Vibr.* 87:71–81.

Timoshenko, S. P., and J. N. Goodier. 1970. *Theory of elasticity.* 3rd ed. New York: McGraw-Hill.

Wang, C. Y. 1998. Some exact solutions of the vibration of nonhomogeneous membranes. *J. Sound Vibr.* 210:555–58.

———. 2003. Vibration of an annular membrane attached to a rigid core, *J. Sound Vibr.* 260:776–82.

Wang, C. Y., and C. M. Wang. 2011a. Exact solutions for vibrating rectangular membranes placed in a vertical plane. *Int. J. Appl. Mech.* 3:625–31.

———. 2011b. Exact solutions for vibrating nonhomogeneous rectangular membranes with exponential density distribution. *The IES J. Part A: Civil and Structural Engineering* 4:37–40.

———. 2012. Exact vibration solutions of nonhomogeneous circular, annular and sector membranes. *Adv. Appl. Math. Mech.* 4:250–58.

4 Vibration of Beams

4.1 INTRODUCTION

A beam or rod is a slender elastic body where the variations of all properties are negligible across the cross section. A beam can admit bending moment, axial force, and transverse shear. There are tens of thousands of publications on the vibrations of rods and beams. Notable treatises include Gorman (1975), Magrab (1980), and Karnovsky and Lebed (2004a, 2004b), where many exact solutions are given.

4.2 ASSUMPTIONS AND GOVERNING EQUATIONS

We consider only small transverse vibrations of originally straight beams of finite length. The effects of rotary inertia and transverse shear deformation are assumed to be negligible. Figure 4.1 shows an elemental segment of the beam.

The beam has the Euler-Bernoulli property that the local moment m' is proportional to the local curvature

$$m' = EI(x')\frac{\partial^2 y'}{\partial x'^2} \tag{4.1}$$

where EI is the flexural rigidity, x' is the longitudinal coordinate, and y' is the transverse displacement. The axial compressive force F' is constant, and the net transverse shear V' balances the transverse acceleration

$$-dV' = \rho(x')dx'\frac{\partial^2 y'}{\partial t'^2} \tag{4.2}$$

where ρ is the density (mass per length) and t' is the time. Considering a moment balance, we have

$$\frac{dm'}{dx'} + F'\frac{dy'}{dx'} - V' = 0 \tag{4.3}$$

Equations (4.1) to (4.3) combine to give

$$\frac{\partial^2}{\partial x'^2}\left(EI(x')\frac{\partial^2 y'}{\partial x'^2}\right) + \frac{\partial}{\partial x'}\left(F'\frac{\partial y'}{\partial x'}\right) + \rho(x')\frac{\partial^2 y'}{\partial t'^2} = 0 \tag{4.4}$$

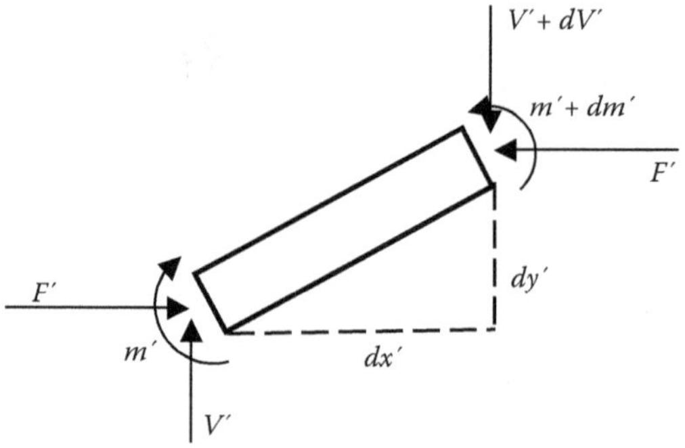

FIGURE 4.1 Elemental segment of beam.

Let

$$EI(x') = EI_0 l(x'), \quad \rho(x') = \rho_0 r(x') \tag{4.5}$$

where EI_0 is the maximum of EI, and ρ_0 is the maximum of ρ. Consider a harmonic vibration with frequency ω', i.e.,

$$y' = w'(x')e^{i\omega' t'} \tag{4.6}$$

By normalizing all length variables by the beam length L (e.g., $x = x'/L$), and the time by $L^2\sqrt{\rho_0/EI_0}$, and dropping the primes, Equation (4.4) becomes

$$\frac{d^2}{dx^2}\left[l(x)\frac{d^2w}{dx^2}\right] + a\frac{d^2w}{dx^2} - \omega^2 r(x)w = 0 \tag{4.7}$$

Here,

$$a = \frac{F'L^2}{EI_0}, \quad \omega = \omega' L^2\sqrt{\rho_0 / EI_0} \tag{4.8}$$

are nondimensional compressive force and nondimensional frequency, respectively.

At the ends of the beam, the classical boundary conditions are as follows:

For a clamped end (C),

$$w = 0, \quad \frac{dw}{dx} = 0 \tag{4.9}$$

For a pinned end (P),

$$w = 0, \quad \frac{d^2w}{dx^2} = 0 \tag{4.10}$$

For a free end (F),

$$\frac{d}{dx}\left(l\frac{d^2w}{dx^2}\right)+a\frac{dw}{dx}=0,\quad \frac{d^2w}{dx^2}=0 \tag{4.11}$$

For a sliding end (S),

$$\frac{d}{dx}\left(l\frac{d^2w}{dx^2}\right)+a\frac{dw}{dx}=0,\quad \frac{dw}{dx}=0 \tag{4.12}$$

Other nonclassical boundary conditions include the elastically supported end, where one of the boundary conditions is

$$\frac{d}{dx}\left(l\frac{d^2w}{dx^2}\right)+a\frac{dw}{dx}=\mp\left(\frac{k_1L^3}{EI_0}\right)w \tag{4.13}$$

Here, k_1 is the elastic spring constant, and the top and bottom signs refer to left or right ends, respectively. For a rotational spring,

$$l\frac{d^2w}{dx^2}=\pm\left(\frac{k_2L}{EI_0}\right)\frac{dw}{dx} \tag{4.14}$$

where k_2 is the rotational spring constant. For a mass M attached at the end

$$\frac{d}{dx}\left(l\frac{d^2w}{dx^2}\right)+a\frac{dw}{dx}=\pm\left(\frac{M}{\rho_0L}\right)\omega^2w \tag{4.15}$$

Equation (4.7) needs two boundary conditions at each end. For nontrivial solutions, the eigenvalues or the nondimensional frequencies ω are determined.

4.3 SINGLE-SPAN CONSTANT-PROPERTY BEAM

4.3.1 General Solutions

For the constant-property beam, $l = 1$ and $r = 1$. Equation (4.7) becomes

$$\frac{d^4w}{dx^4}+a\frac{d^2w}{dx^2}-\omega^2w=0 \tag{4.16}$$

The general solution is

$$w=c_1\cosh(\alpha x)+c_2\sinh(\alpha x)+c_3\cos(\beta x)+c_4\sin(\beta x) \tag{4.17}$$

where

$$\alpha = \sqrt{\frac{\sqrt{a^2 + 4\omega^2} - a}{2}}, \quad \beta = \sqrt{\frac{\sqrt{a^2 + 4\omega^2} + a}{2}} \tag{4.18}$$

The two general boundary conditions can be written as

$$\bar{s}_1^i w + \bar{s}_2^i \frac{dw}{dx} + \bar{s}_3^i \frac{d^2 w}{dx^2} + \bar{s}_4^i \frac{d^3 w}{dx^3} = 0 \tag{4.19}$$

$$\hat{s}_1^i w + \hat{s}_2^i \frac{dw}{dx} + \hat{s}_3^i \frac{d^2 w}{dx^2} + \hat{s}_4^i \frac{d^3 w}{dx^3} = 0 \tag{4.20}$$

Here, $i = 0$ for the left end and $i = 1$ for the right end. The substitution of Equation (4.17) into Equations (4.19) and (4.20) yields the exact characteristic equation

$$\begin{vmatrix} c_{11} & c_{12} & c_{13} & c_{14} \\ c_{21} & c_{22} & c_{23} & c_{24} \\ c_{31} & c_{32} & c_{33} & c_{34} \\ c_{41} & c_{42} & c_{43} & c_{44} \end{vmatrix} = 0 \tag{4.21}$$

where

$$\begin{aligned}
&c_{11} = \bar{s}_1^0 + \alpha^2 \bar{s}_3^0, c_{12} = \alpha(\bar{s}_2^0 + \alpha^2 \bar{s}_4^0),\ c_{13} = \bar{s}_1^0 - \beta^2 \bar{s}_3^0\ \ c_{14} = \beta(\bar{s}_2^0 - \beta^2 \bar{s}_4^0) \\
&c_{21} = \hat{s}_1^0 + \alpha^2 \hat{s}_3^0,\ c_{22} = \alpha(\hat{s}_2^0 + \alpha^2 \hat{s}_4^0),\ c_{23} = \hat{s}_1^0 - \beta^2 \hat{s}_3^0,\ c_{24} = \beta(\hat{s}_2^0 - \beta^2 \hat{s}_4^0) \\
&c_{31} = (\bar{s}_1^1 + \alpha^2 \bar{s}_3^1)\cosh\alpha + \alpha(\bar{s}_2^1 + \alpha^2 \bar{s}_4^1)\sinh\alpha, \\
&c_{32} = \alpha(\bar{s}_2^1 + \alpha^2 \bar{s}_4^1)\cosh\alpha + (\bar{s}_1^1 + \alpha^2 \bar{s}_3^1)\sinh\alpha, \\
&c_{33} = (\bar{s}_1^1 - \beta^2 \bar{s}_3^1)\cos\beta - \beta(\bar{s}_2^1 - \beta^2 \bar{s}_4^1)\sin\beta, \\
&c_{34} = \beta(\bar{s}_2^1 - \beta^2 \bar{s}_4^1)\cos\beta + (\bar{s}_1^1 - \beta^2 \bar{s}_3^1)\sin\beta, \\
&c_{41} = (\hat{s}_1^1 + \alpha^2 \hat{s}_3^1)\cosh\alpha + \alpha(\hat{s}_2^1 + \alpha^2 \hat{s}_4^1)\sinh\alpha, \\
&c_{42} = \alpha(\hat{s}_2^1 + \alpha^2 \hat{s}_4^1)\cosh\alpha + (\hat{s}_1^1 + \alpha^2 \hat{s}_3^1)\sinh\alpha, \\
&c_{43} = (\hat{s}_1^1 - \beta^2 \hat{s}_3^1)\cos\beta - \beta(\hat{s}_2^1 - \beta^2 \hat{s}_4^1)\sin\beta, \\
&c_{44} = \beta(\hat{s}_2^1 - \beta^2 \hat{s}_4^1)\cos\beta + (\hat{s}_1^1 - \beta^2 \hat{s}_3^1)\sin\beta
\end{aligned} \tag{4.22}$$

The frequencies are obtained by solving Equation (4.21) for ω using a root search algorithm. Since there are numerous combinations of the boundary conditions, we shall concentrate on only a few examples.

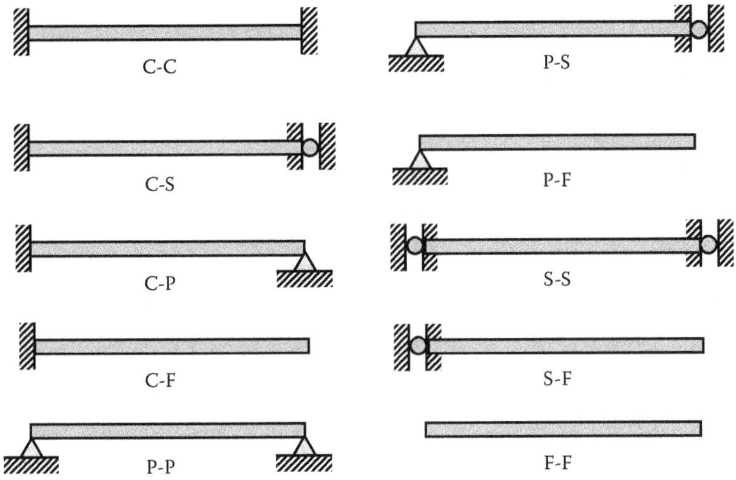

FIGURE 4.2 Ten combinations of C, P, F, S boundary conditions.

4.3.2 Classical Boundary Conditions with Axial Force

There are 10 combinations of C, P, F, S boundary conditions (Figure 4.2). We shall include the effects of axial force, which are seldom considered in the literature.

1. Both ends are clamped (C-C).
   Guided by Equation (4.9), the nonzero coefficients are

$$\bar{s}_1^0 = 1, \; \hat{s}_2^0 = 1, \; \bar{s}_1^1 = 1, \; \hat{s}_2^1 = 1 \tag{4.23}$$

   Equations (4.21) and (4.22) then give the frequencies in Table 4.1. The mode shapes for a C-C beam with $a = -20$ are shown in Figure 4.3. Positive a means that the beam is in compression, and negative a means the beam is in tension. At $a = 39.4784$, the beam buckles, showing a frequency of zero for the first mode. However, buckling does not mean that higher mode vibrations do not exist.

TABLE 4.1
Frequencies for the C-C Beam

a = −40	−20	0	20	39.478
31.347	27.274	22.373	15.848	0
75.040	68.708	61.673	53.650	44.363
136.26	128.82	120.90	112.42	103.48
216.34	208.27	199.86	191.08	182.12
315.74	307.27	298.56	289.58	280.56

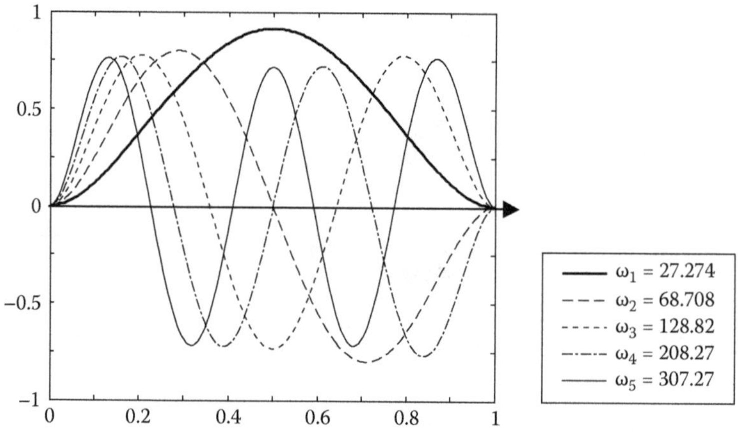

FIGURE 4.3 Mode shapes for C-C beam at $a = -20$.

2. One end is clamped and one end pinned (C-P).
   The nonzero coefficients are

$$\bar{s}_1^0 = 1, \ \hat{s}_2^0 = 1, \ \bar{s}_1^1 = 1, \ \hat{s}_3^1 = 1 \tag{4.24}$$

   The frequency results are shown in Table 4.2. Mode shapes for the C-P beam with $a = 10$ are shown in Figure 4.4. At $a = 20.190$, the beam buckles, showing a zero frequency for the first mode
3. One end is clamped and one end sliding (C-S).
   The nonzero coefficients are

$$\bar{s}_1^0 = 1, \ \hat{s}_2^0 = 1, \ \bar{s}_2^1 = 1, \ \hat{s}_4^1 = 1 \tag{4.25}$$

   The results are shown in Table 4.3. Mode shapes for a C-S beam with $a = -10$ are shown in Figure 4.5. The beam buckles at $a = 9.8696$.

TABLE 4.2
Frequencies for the C-P Beam

$a = -20$	−10	0	10	20.190
21.556	18.760	15.418	11.021	0
57.906	54.085	49.965	45.468	40.374
112.91	108.66	104.25	99.635	94.703
187.29	182.84	178.27	173.58	168.67
281.27	276.69	272.03	267.29	262.38

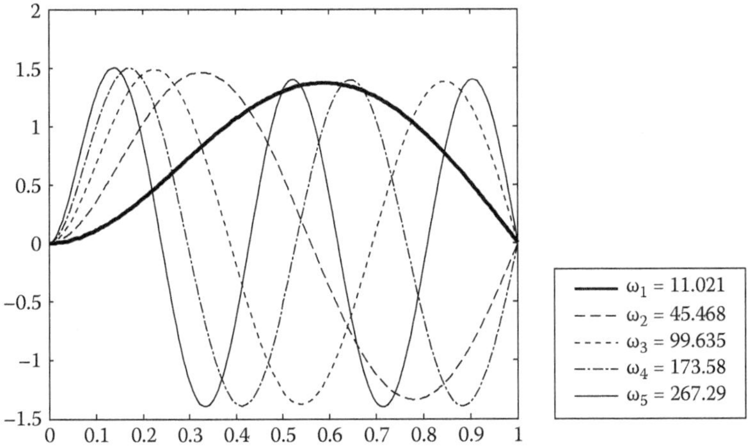

FIGURE 4.4 Mode shapes for C-P beam at $a = 10$.

TABLE 4.3
Frequencies for the C-S Beam

$a = -10$	−5	0	5	9.8696
7.8368	6.8186	5.5933	3.9619	0
34.065	32.204	30.226	28.105	25.871
78.935	76.817	74.639	72.394	70.139
143.29	141.06	138.79	136.48	134.20
227.30	225.00	222.68	220.34	218.03

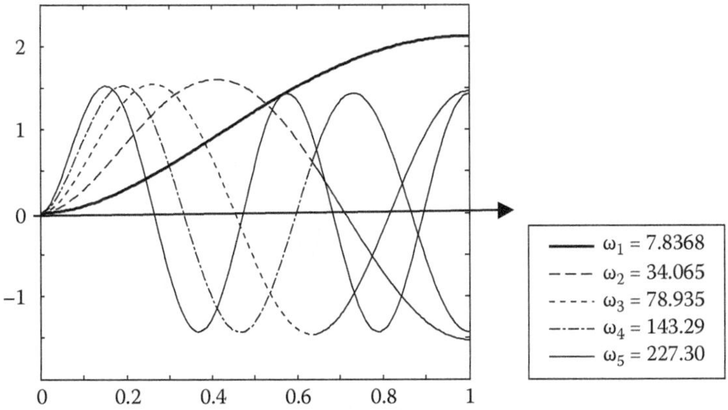

FIGURE 4.5 Mode shapes for C-S beam at $a = -10$.

TABLE 4.4
Frequencies for the C-F Beam

a = −2	−1	0	1	2.4674
4.6066	4.1102	3.5160	2.7536	0
23.453	22.757	22.035	21.285	20.129
62.937	62.321	61.697	61.068	60.132
122.08	121.49	120.90	120.31	119.44
201.00	200.43	199.86	199.29	198.45

4. One end is clamped and one end free (C-F).
   The nonzero coefficients are

$$\bar{s}_1^0 = 1,\ \hat{s}_2^0 = 1,\ \bar{s}_4^1 = 1,\ \bar{s}_2^1 = a,\ \hat{s}_3^1 = 1 \tag{4.26}$$

   Sample frequencies are presented in Table 4.4, and mode shapes for a C-F beam with $a = 0$ are shown in Figure 4.6. The beam buckles at $a = 2.4674$.
5. One end is sliding and one end free (S-F).
   The nonzero coefficients are

$$\bar{s}_2^0 = 1,\ \hat{s}_4^0 = 1,\ \bar{s}_4^1 = 1,\ \bar{s}_2^1 = a,\ \hat{s}_3^1 = 1 \tag{4.27}$$

   Sample frequencies are presented in Table 4.5, and mode shapes for an S-F beam with $a = 2.4674$ are shown in Figure 4.7.

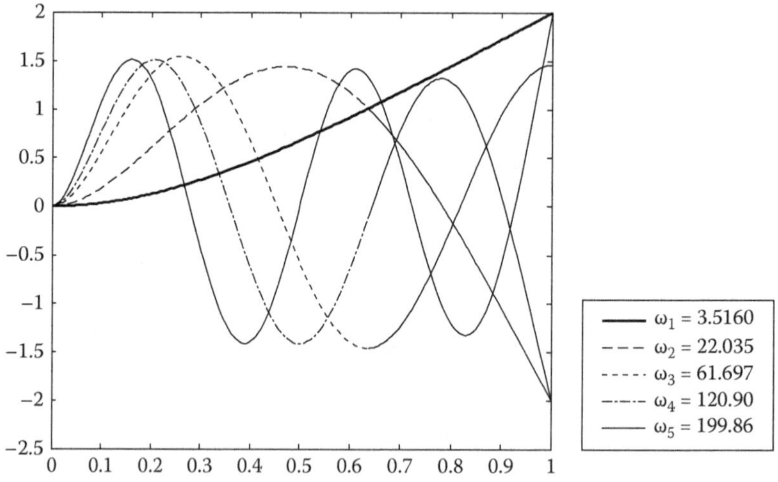

FIGURE 4.6 Mode shapes for C-F beam at $a = 0$.

TABLE 4.5
Frequencies for the S-F Beam

$a = -2$	−1	0	1	2.4674
7.4575	6.5991	5.5933	4.3357	0
31.731	30.988	30.226	29.442	28.252
75.974	75.309	74.639	73.962	72.957
140.04	139.42	138.79	138.16	137.24
223.88	223.28	222.68	222.08	221.20

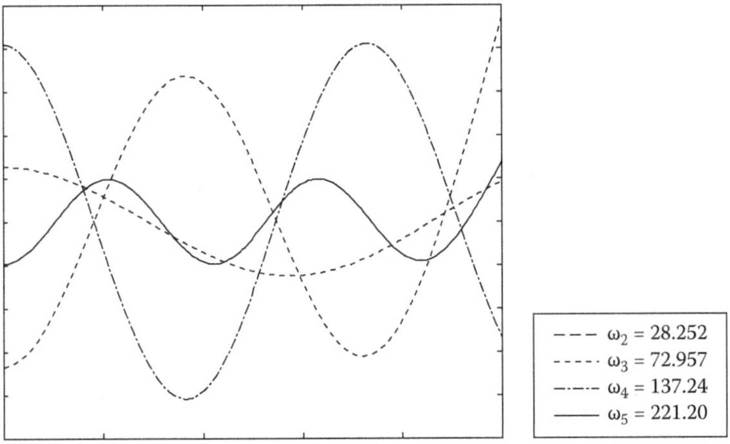

FIGURE 4.7 Mode shapes for S-F beam at $a = 2.4674$.

TABLE 4.6
Frequencies for the P-F Beam

$a = -8$	−4	−2	0
4.6948	3.3708	2.4168	0
21.219	18.582	17.084	15.418
55.351	52.733	51.369	49.965
109.30	106.80	105.53	104.25
183.10	180.70	179.49	178.27

6. One end is pinned and one end free (P-F).
   The nonzero coefficients are

$$\bar{s}_1^0 = 1,\ \hat{s}_3^0 = 1,\ \bar{s}_4^1 = 1,\ \bar{s}_2^1 = a,\ \hat{s}_3^1 = 1 \tag{4.28}$$

Sample frequency results for a P-F beam are presented in Table 4.6, and mode shapes with $a = 0$ are shown in Figure 4.8. Notice that for zero axial force, the buckling load is zero. Higher mode vibrations still exist.

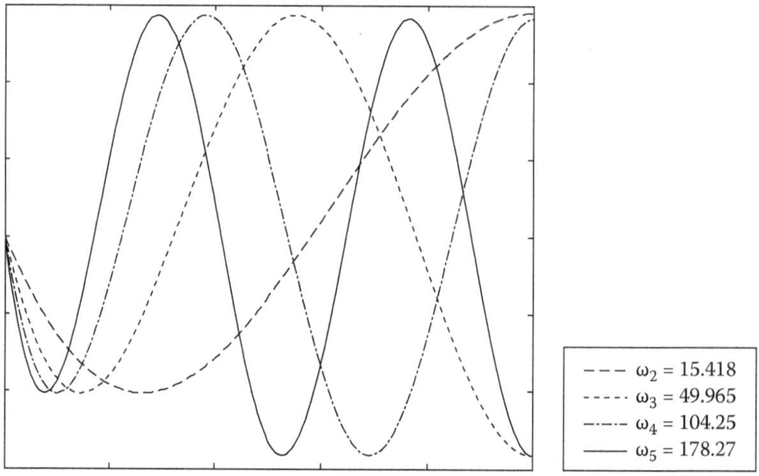

FIGURE 4.8 Mode shapes for P-F beam at $a = 0$.

7. Both ends are free (F-F).
   The nonzero coefficients are

$$\bar{s}_4^0 = 1,\ \bar{s}_2^0 = a,\ \hat{s}_3^0 = 1,\ \bar{s}_4^1 = 1,\ \bar{s}_2^1 = a,\ \hat{s}_3^1 = 1 \tag{4.29}$$

   Sample frequency results for an F-F beam are presented in Table 4.7, and mode shapes with $a = -2$ are shown in Figure 4.9. For $a = 0$, the resonance frequencies are the same as in the case of its C-C beam counterpart (cf. Table 4.1 and Table 4.7 for $a = 0$), except that in this case we have the two rigid body modes (translation and rotation at $\omega = 0$), since it is allowed by the boundary conditions.
8. Both ends are pinned (P-P).
   In this case, the frequencies are in closed form. Consider first Equation (4.16). Under P-P conditions and no axial force, the solution is

$$w = \sin(\sqrt{\omega_0}\, x), \quad \omega_0 = n^2\pi^2 \tag{4.30}$$

TABLE 4.7
Frequencies for the F-F Beam

$a = -8$	-4	-2	0
9.6433	6.8684	4.8768	0
29.830	26.396	24.476	22.373
68.335	65.100	63.412	61.673
126.93	123.95	122.44	120.90
205.47	202.69	201.28	199.86

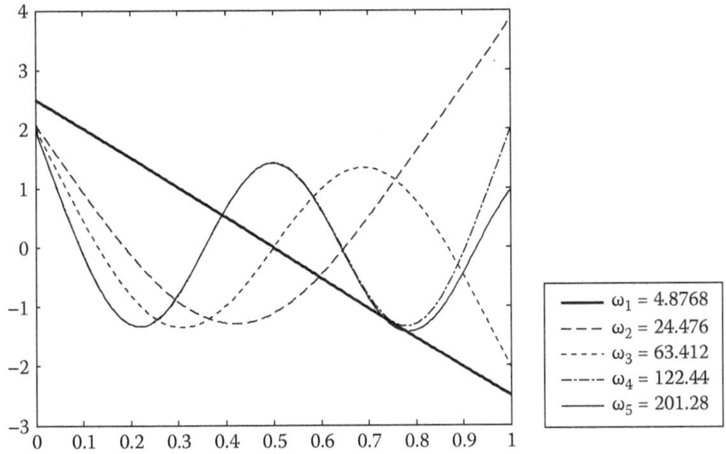

FIGURE 4.9 Mode shapes for F-F beam at $a = -2$.

Now, if axial force is present, the solution is still given by Equation (4.30). But its substitution into Equation (4.16) gives

$$\omega_0^2 - a\omega_0 - \omega^2 = 0 \tag{4.31}$$

or

$$\omega = \sqrt{\omega_0(\omega_0 - a)},\ a \le \omega_0 \tag{4.32}$$

Note that for $n > 1$, ω_0 may be a higher mode. Thus we have a relation between the frequency with axial load and that without an axial load. Equation (4.32) is somewhat simpler than the formula of Galef (1968), who used energy integrals and the buckling load.

9. Both ends are sliding (S-S).
   The formula in Equation (4.32) still holds. Here, $\omega_0 = n^2\pi^2$.
10. One end pinned and one end sliding (P-S).
   The formula in Equation (4.32) holds with $\omega_0 = (n - 1/2)^2\pi^2$.

For symmetrical end conditions, the mode shapes alternate between symmetric and antisymmetric as the frequency is increased. One can separate a symmetric mode or an antisymmetric mode into two mirror halves, with the middle end condition being sliding or pinned. The following relations can be established:

- For the C-C beam, the symmetric mode is equivalent to the C-S beam and the anti-symmetric mode equivalent to the C-P beam, with the axial force divided by 4 and the frequency divided by 4.
- For the F-F beam, the symmetric mode is equivalent to the F-S beam and the anti-symmetric mode equivalent to the F-P beam, with the axial force divided by 4 and the frequency divided by 4.

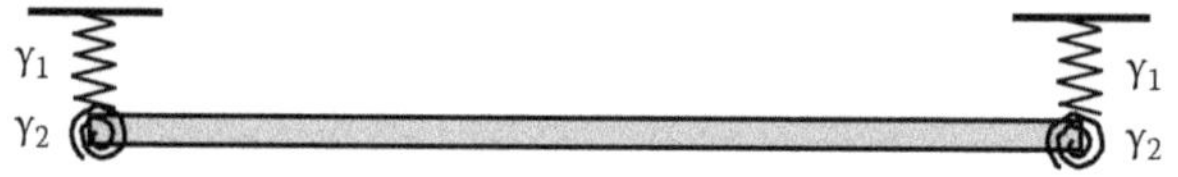

FIGURE 4.10 Beam with elastically restrained ends.

- For the P-P beam, the symmetric mode is equivalent to the P-S beam and the anti-symmetric mode equivalent to the P-P beam, with the axial force divided by 4 and the frequency divided by 4.
- For the S-S beam, the symmetric mode is equivalent to the S-S beam and the anti-symmetric mode equivalent to the P-S beam, with the axial force divided by 4 and the frequency divided by 4.

4.3.3 Elastically Supported Ends

We consider a beam with both ends constrained by elastic translational and rotational springs (Figure 4.10). In the limit of zero or infinite spring constants, some of the classical boundary conditions will be recovered. Consider a case with zero axial force where the spring constants at both ends are symmetrical. Let

$$\gamma_1 = \frac{k_1 L^3}{EI_0}, \quad \gamma_2 = \frac{k_2 L}{EI_0} \tag{4.33}$$

The boundary conditions given by Equations (4.13) and (4.14) become

$$\frac{d^3 w}{dx^3} \pm \gamma_1 w = 0, \quad \frac{d^2 w}{dx^2} \mp \gamma_2 \frac{dw}{dx} = 0 \tag{4.34}$$

and the nonzero coefficients are

$$\begin{aligned} &\bar{s}_1^0 = \gamma_1, \ \bar{s}_4^0 = 1, \ \hat{s}_2^0 = -\gamma_2, \ \hat{s}_3^0 = 1 \\ &\bar{s}_1^1 = -\gamma_1, \ \bar{s}_4^1 = 1, \ \hat{s}_2^1 = \gamma_2, \ \hat{s}_3^1 = 1 \end{aligned} \tag{4.35}$$

Equation (4.21) then furnishes the characteristic calculation. The results are given in Table 4.8, and mode shapes for a beam with elastically restrained ends ($\gamma_1 = 1$, $\gamma_2 = 10$) are shown in Figure 4.11.

Note that when both γ_1 and γ_2 are zero, the beam is an F-F beam with two zero frequencies, representing a rigid body translation and a rigid body rotation. Normally we don't accept rigid body motion as a vibration mode. The zero frequencies are included because they morph into slow vibration modes as soon as γ_1 becomes nonzero. We find the following limits. When $\gamma_1 = 0$ and $\gamma_2 \to \infty$, it's an S-S beam; when $\gamma_1 \to \infty$ and $\gamma_2 \to 0$, it's a P-P beam; and when $\gamma_1 \to \infty$ and $\gamma_2 \to \infty$, it's a C-C beam.

TABLE 4.8
Frequencies for the Symmetrically Elastically Supported Beam

$\gamma_2 \backslash \gamma_1$	0	1	10	100	1,000	∞
	0	1.4025	4.1304	8.2757	9.6787	9.8696
	0	2.4466	7.6541	21.751	36.446	36.478
0	22.373	22.552	24.141	36.920	73.369	88.826
	61.673	61.738	62.326	68.482	112.59	157.91
	120.90	120.94	121.24	124.34	158.18	246.74
	0	1.4058	4.2190	9.1789	11.251	11.552
	4.3931	4.9877	8.6041	21.753	37.637	41.309
1	25.490	25.635	26.933	38.017	73.687	90.713
	65.174	65.233	65.769	71.326	112.67	159.83
	124.54	124.57	124.86	127.79	159.40	248.66
	0	1.4106	4.3608	11.298	16.344	17.269
	8.3363	8.6046	10.699	21.757	42.604	49.959
10	34.097	34.176	34.892	41.764	75.096	101.32
	77.974	78.103	78.359	81.894	113.04	171.74
	140.51	140.54	140.74	142.83	165.67	261.51
	0	1.4121	4.4047	12.234	19.887	21.542
	9.6778	9.8859	11.580	21.760	47.051	59.447
100	38.724	38.778	39.269	44.268	76.460	116.66
	87.154	87.179	87.403	89.693	113.41	193.04
	154.99	155.00	155.13	156.44	252.35	288.64
	0	1.4122	4.4100	12.361	20.482	22.284
	9.8499	10.051	11.698	21.760	47.879	61.427
1,000	39.400	39.451	39.912	44.590	76.724	120.42
	88.650	88.673	88.878	90.984	113.48	199.06
	157.60	157.61	157.73	158.90	172.01	297.35
	0	1.4123	4.4106	12.376	20.552	22.373
	9.8696	10.070	11.712	21.760	47.980	61.673
∞	39.478	39.529	39.987	44.635	76.756	120.91
	88.826	88.849	89.052	91.137	113.49	199.86
	157.91	157.93	158.04	159.20	172.14	298.56

4.3.4 Cantilever Beam with a Mass at One End

The cantilever is an important structural member. Consider one end pinned with a rotational spring k_2 and the other end free but with an attached mass M, as shown in Figure 4.12. There is no axial force.

Define the mass ratio as

$$\nu = \frac{M}{\rho_0 L} \tag{4.36}$$

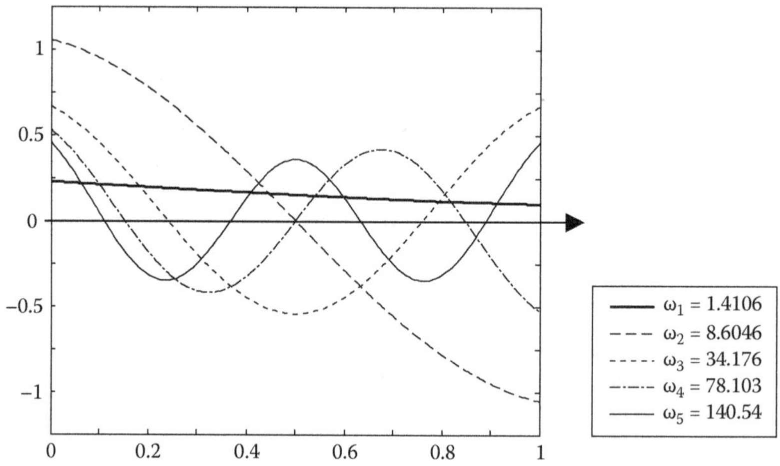

FIGURE 4.11 Mode shapes for beam with elastically restrained ends ($\gamma_1 = 1, \gamma_2 = 10$).

The nonzero coefficients are

$$\bar{s}_1^0 = 1, \ \hat{s}_2^0 = -\gamma_2, \ \hat{s}_3^0 = 1. \ \bar{s}_3^1 = 1, \ \hat{s}_1^1 = \nu\omega^2, \ \hat{s}_4^1 = 1 \tag{4.37}$$

Equation (4.21) is then solved for the frequency. The results are given in Table 4.9, and the mode shapes for the cantilever beam with a mass at one end with $\gamma_2 = 1000$ and $\nu = 10$ are shown in Figure 4.13. Notice that when $\gamma_2 = 0$ and $\nu = 0$, it is a P-F beam; when $\gamma_2 = \infty$ and $\nu = 0$, it is a C-F beam; when $\gamma_2 = 0$ and $\nu = \infty$, it is a P-P beam; and when $\gamma_2 = \infty$ and $\nu = \infty$, it is a C-P beam. Notice also that the end mass decreases frequency, especially the fundamental frequency.

4.3.5 Free Beam with Two Masses at the Ends

Figure 4.14 shows a completely free beam with masses M_1 and M_2 at the ends. This case models the vibration of a two-atom molecule. The nonzero coefficients are

$$\bar{s}_3^0 = 1, \ \hat{s}_1^0 = -\nu_1\omega^2, \ \hat{s}_4^0 = 1. \ \bar{s}_3^1 = 1, \ \hat{s}_1^1 = \nu_2\omega^2, \ \hat{s}_4^1 = 1 \tag{4.38}$$

where the ν's are defined by Equation (4.36). Without loss of generality, we can consider $\nu_1 \geq \nu_2$. Equation (4.21) gives the results presented in Table 4.10. Mode shapes for a free beam with two masses at the ends with $\nu_1 = 0.1$, $\nu_2 = 0.1$ are shown in Figure 4.15.

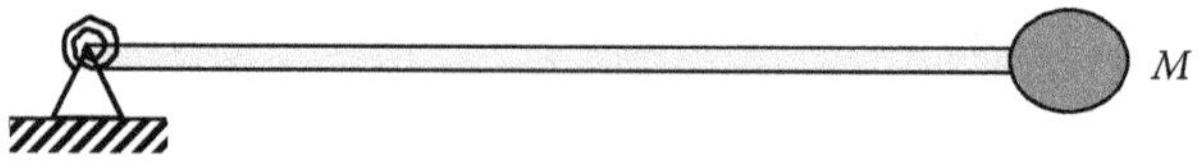

FIGURE 4.12 Cantilever beam with a mass at one end and the other end pinned with a rotational restraint.

TABLE 4.9
Frequencies for the Cantilever with Mass and Rotational Spring

$\nu \backslash \gamma_2$	0	1	10	100	1,000	∞
	0	1.5573	2.9678	3.4477	3.5090	3.5160
	15.418	16.250	19.356	21.620	21.991	22.035
0	49.965	50.896	55.518	60.570	61.575	61.697
	104.25	105.20	110.71	118.76	120.66	120.90
	178.27	179.23	185.35	196.42	199.47	199.86
	0	0.7577	1.3553	1.5330	1.5548	1.5573
	10.714	11.524	14.219	15.949	16.219	16.250
1	40.399	41.311	45.621	49.968	50.796	50.896
	89.773	90.715	96.022	103.33	104.99	105.20
	158.87	159.83	165.80	176.13	178.88	179.23
	0	0.2698	0.4744	0.5334	0.5406	0.5414
	9.9678	10.808	13.519	15.218	15.481	15.512
10	39.578	40.497	44.818	49.144	49.965	50.064
	88.926	89.872	95.191	102.48	104.14	104.35
	158.01	158.97	164.95	175.28	178.02	178.37
	0	0.0865	0.1517	0.1705	0.1727	0.1730
	9.8796	10.724	13.439	15.135	15.397	15.428
100	39.488	40.408	44.732	49.055	49.876	49.975
	88.836	89.783	95.103	102.40	104.05	104.26
	157.92	158.88	164.87	175.19	177.93	178.28
	0	0.0274	0.0480	0.0540	0.0547	0.0548
	9.8706	10.715	13.431	15.177	15.389	15.419
1,000	39.479	40.400	44.723	49.046	49.867	49.966
	88.827	89.774	95.094	102.39	104.04	104.25
	157.92	158.88	164.86	175.18	177.92	178.27
	0	2.7×10^{-5}	4.8×10^{-5}	5.4×10^{-5}	5.5×10^{-5}	5.5×10^{-5}
	9.8696	10.714	13.430	15.126	15.388	15.418
∞	36.478	40.399	44.722	49.045	49.866	49.965
	88.826	89.773	95.093	102.39	104.04	104.25
	157.91	158.87	164.86	175.18	177.92	178.27

Notice that when $v_1 = 0$ and $v_2 = 0$, it is an F-F beam; when $v_1 = \infty$ and $v_2 = 0$, it is a P-F beam; and when $v_1 = \infty$ and $v_2 = \infty$, it is a P-P beam. We have omitted the zero frequencies of rigid translation and rotation.

4.4 TWO-SEGMENT UNIFORM BEAM

There are two types of two-segment beams. In one type, the equation of motion is the same, but some condition is imposed in mid-span. In another type, the equation of motion is different for two different segments, including the stepped beam and the beam on a partial elastic foundation.

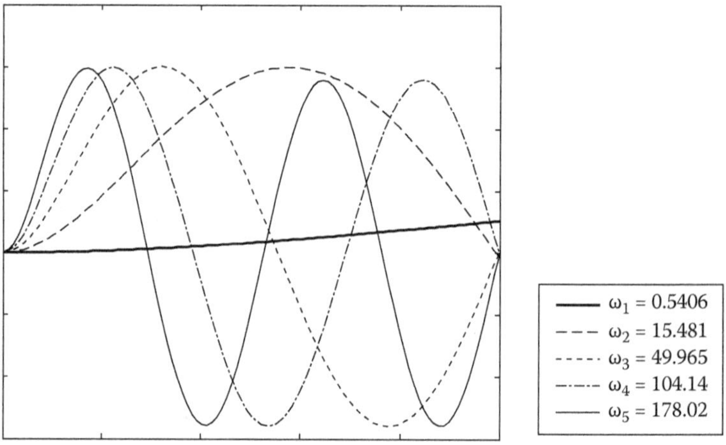

FIGURE 4.13 Mode shapes for cantilever beam with a mass at one end at $\gamma_2 = 1{,}000$ and $\nu = 10$.

FIGURE 4.14 Free beam with two different masses at the ends.

4.4.1 Beam with an Internal Elastic Support

Figure 4.16 shows a uniform beam subjected to compressive forces. On the span, there is an elastic translational spring that divides the beam into two segments. Let the subscripts 1 and 2 represent the two segments. For $0 \le x < b$, the general solution to Equation (4.16) is

$$w_1 = s_{11}\cosh(\alpha x) + s_{12}\sinh(\alpha x) + s_{13}\cos(\beta x) + s_{14}\sin(\beta x) \tag{4.39}$$

where α and β are given in Equation (4.18). For $b < x \le 1$, we write

$$w_2 = s_{21}\cosh[\alpha(x-1)] + s_{22}\sinh[\alpha(x-1)] + s_{23}\cos[\beta(x-1)]) + s_{24}\sin[\beta(x-1)] \tag{4.40}$$

At the joint, we require

$$w_1(b) = w_2(b), \quad \frac{dw_1}{dx}(b) = \frac{dw_2}{dx}(b) \tag{4.41}$$

$$\frac{d^2w_1}{dx^2}(b) = \frac{d^2w_2}{dx^2}(b), \quad \frac{d^3w_1}{dx^3}(b) - \gamma_1 w_1 = \frac{d^3w_2}{dx^3}(b) \tag{4.42}$$

TABLE 4.10
Natural Frequencies for Beams with Two Different Masses Attached at the Ends

$\nu_2 \backslash \nu_1$	0	0.1	1	10	100	∞
	22.373	19.627	16.336	15.522	15.429	15.418
	61.673	55.501	50.892	50.064	49.975	49.965
0	120.90	110.71	105.20	104.35	104.26	104.25
	199.86	185.35	179.27	178.37	178.28	178.27
	298.56	279.55	273.00	272.13	272.04	272.03
		17.270	14.293	13.528	13.440	13.430
		49.960	45.618	44.818	44.731	44.722
0.1		101.32	96.023	95.191	95.103	95.093
		171.75	16.580	164.95	164.87	164.86
		261.53	255.12	254.26	254.17	254.16
			11.552	10.811	10.724	10.714
			41.310	40.497	40.408	40.399
1			90.715	89.872	89.783	89.773
			159.83	158.97	158.88	158.87
			248.67	247.81	247.72	247.71
				10.066	9.9778	9.9678
				39.677	39.588	39.578
10				89.025	88.936	88.926
				158.11	158.02	158.01
				246.94	246.85	246.84
					9.8896	9.8796
					39.498	39.488
100					88.846	88.836
					157.93	157.92
					246.76	246.75
						9.8696
						39.478
∞						88.826
						157.91
						246.74

Here, γ_1 is as defined in Equation (4.33). There are two boundary conditions at each end and four conditions at the joint. The following 8 × 8 determinant is obtained for the eight nontrivial constants s_{ij}:

$$\begin{vmatrix} E_1 & \underline{0} \\ \underline{0} & E_2 \\ \multicolumn{2}{c}{I} \end{vmatrix} = 0 \tag{4.43}$$

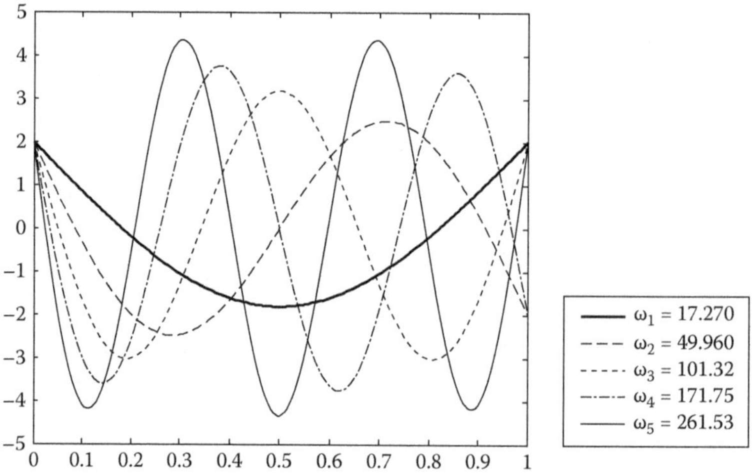

FIGURE 4.15 Mode shapes for free beam with two masses at ends at $n_1 = 0.1$, $n_2 = 0.1$.

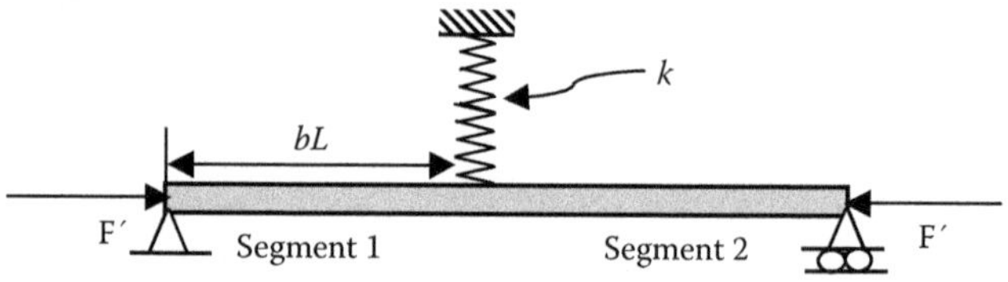

FIGURE 4.16 Beam with an internal elastic support.

Here E_1 and E_2 are 2×4 matrices from the left and right end conditions, $\underline{0}$ is the null 2×4 matrix, and I is a 4×8 matrix from the joint conditions. We find

$$
\begin{aligned}
E &= \begin{pmatrix} 1 & 0 & 1 & 0 \\ 0 & \alpha & 0 & \beta \end{pmatrix} \quad \text{for a clamped end} \\
E &= \begin{pmatrix} 1 & 0 & 0 & 0 \\ 0 & 0 & 1 & 0 \end{pmatrix} \quad \text{for a pinned end} \\
E &= \begin{pmatrix} 0 & 1 & 0 & 0 \\ 0 & 0 & 0 & 1 \end{pmatrix} \quad \text{for a sliding end} \\
E &= \begin{pmatrix} \alpha^2 & 0 & -\beta^2 & 0 \\ 0 & \alpha(\alpha^2 + a) & 0 & -\beta(\beta^2 - a) \end{pmatrix} \quad \text{for a free end}
\end{aligned} \tag{4.44}
$$

Let

$$C_{11} = \cosh(\alpha b),\ C_{12} = \sinh(\alpha b),\ C_{13} = \cos(\beta b),\ C_{14} = \sin(\beta b)$$

$$C_{21} = \cosh[\alpha(b-1)],\ C_{22} = \sinh[\alpha(b-1)],\ C_{23} = \cos[\beta(b-1)],\ C_{24} = \sin[\beta(b-1)] \tag{4.45}$$

Equations (4.41) and (4.42) give the matrix I

$$I = \begin{pmatrix} C_{11} & C_{12} & C_{13} & C_{14} & -C_{21} & -C_{22} & -C_{23} & -C_{24} \\ \alpha C_{12} & \alpha C_{11} & -\beta C_{14} & \beta C_{13} & -\alpha C_{22} & -\alpha C_{21} & \beta C_{24} & -\beta C_{23} \\ \alpha^2 C_{11} & \alpha^2 C_{12} & -\beta^2 C_{13} & -\beta^2 C_{14} & -\alpha^2 C_{21} & -\alpha^2 C_{22} & \beta^2 C_{23} & \beta^2 C_{24} \\ \alpha^3 C_{12} - \gamma_1 C_{11} & \alpha^3 C_{11} - \gamma_1 C_{12} & \beta^3 C_{14} - \gamma_1 C_{13} & -\beta^3 C_{13} - \gamma_1 C_{14} & -\alpha^3 C_{22} & -\alpha^3 C_{21} & -\beta^3 C_{24} & \beta^3 C_{23} \end{pmatrix}. \tag{4.46}$$

We shall consider only the case for equal span ($b = 0.5$) and when both ends are pinned or both ends are clamped. Table 4.11 shows the results for the P-P case. The $\gamma_1 = 0$ case is the one-span P-P beam, and the frequencies are given by Equation (4.32). For small γ_1, the fundamental mode is symmetric, the next mode antisymmetric, etc. For high γ_1, the sequence is reversed, with the fundamental mode antisymmetric. The frequencies of the antisymmetric modes are independent of the spring constant, since the spring is at a node. The zero entry means that the mode has buckled. Mode shapes for a P-P beam with a translational support at mid-span with $a = 5$ and $\gamma_1 = 10$ are shown in Figure 4.17.

The results for the clamped-clamped beam are shown in Table 4.12. Mode shapes for a C-C beam with translational support at mid-span and $a = -20$, $\gamma_1 = 10$ are shown in Figure 4.18.

4.4.2 Beam with an Internal Attached Mass

Figure 4.19 shows a beam with an internal attached mass. The analysis is the same as in the previous section, except γ_1 is replaced by $-\nu\omega^2$, where ν is defined in Equation (4.36).

Equation (4.43), the exact-frequency equation, furnishes the results for $b = 0.5$ for a P-P beam and a C-C beam in Tables 4.13 and 4.14, respectively. The results for the zero-mass case and for the infinite-mass case are identical to those for the translational spring. Except for the infinite-mass case, the fundamental mode is symmetric. The frequencies

TABLE 4.11
Frequencies for the P-P beam with Translational Elastic Support at Mid Span

$\gamma_1 \backslash a$	−10	−5	0	5	10
	14.004	12.114	9.8696	6.9526	0
	44.197	41.904	39.478	36.894	34.114
0	93.693	91.292	88.826	86.290	83.677
	162.84	160.39	157.91	155.39	152.83
	251.69	249.23	246.74	244.23	241.69
	14.699	12.911	10.833	8.2461	4.3181
	44.197	41.904	39.478	36.894	34.114
10	93.800	91.402	88.939	86.406	83.797
	162.84	160.39	157.91	155.39	152.83
	251.73	249.27	246.78	244.27	241.73
	19.763	18.466	17.070	15.548	13.859
	44.197	41.904	39.478	36.894	34.114
100	94.773	92.402	89.968	87.466	84.891
	162.84	160.39	157.91	155.39	152.83
	252.09	249.63	247.15	244.64	242.10
	41.231	40.400	39.478	36.894	34.114
	44.197	41.904	39.531	38.640	37.663
1,000	105.25	103.20	101.11	98.982	96.819
	167.84	160.39	157.91	155.39	152.83
	255.83	253.41	250.97	248.50	246.01
	44.197	41.904	39.478	36.894	34.114
	65.292	63.510	61.673	59.775	57.810
∞	162.84	160.39	157.91	155.39	152.83
	204.11	201.99	199.86	197.70	195.52
	360.27	357.80	355.30	352.80	350.27

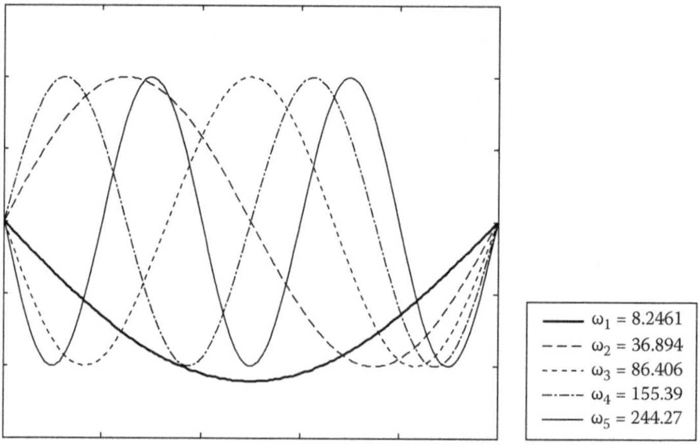

FIGURE 4.17 Mode shapes for P-P beam with translational support at mid span at $a = 5$, $\gamma_1 = 10$.

TABLE 4.12
Frequencies for the C-C Beam with Translational Support at Mid Span

$\gamma_1 \backslash a$	−40	−20	0	20	40
	31.347	27.274	22.373	15.848	0
	75.040	68.708	61.673	53.650	44.086
0	136.26	128.82	120.90	112.42	103.23
	216.34	208.27	199.86	191.08	181.87
	315.74	307.27	298.56	289.58	280.31
	31.733	27.723	22.929	16.642	4.4461
	75.040	68.708	61.673	53.650	44.086
10	136.33	128.90	120.99	122.51	103.32
	216.34	208.27	199.86	191.08	181.87
	315.77	307.30	298.59	289.61	280.35
	34.964	31.421	27.357	22.466	15.950
	75.040	68.708	61.673	53.650	44.086
100	137.01	129.61	121.73	113.29	104.15
	216.34	208.27	199.86	191.08	181.87
	316.06	307.60	298.89	289.92	280.67
	55.444	53.291	51.020	48.605	44.086
	75.040	68.708	61.673	53.650	46.014
1,000	144.20	137.20	129.81	121.94	113.51
	216.34	208.27	199.86	191.08	181.87
	319.03	310.64	302.02	293.14	283.99
	75.040	68.708	61.673	53.650	44.086
	99.830	94.817	89.493	83.796	77.639
∞	216.34	208.27	199.86	191.08	181.87
	261.17	254.04	246.69	239.10	231.24
	434.66	425.92	416.99	407.87	398.54

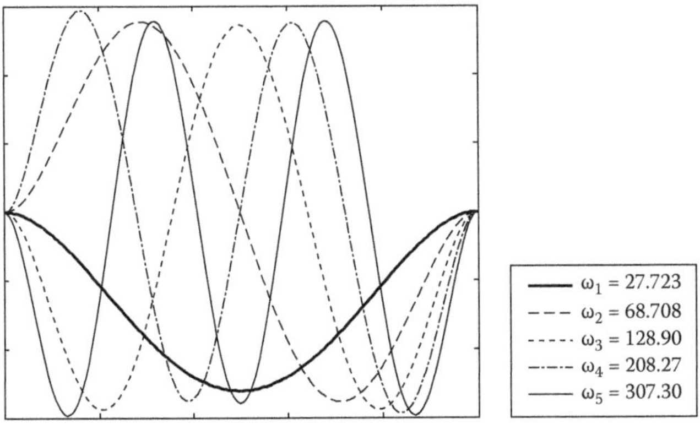

FIGURE 4.18 Mode shapes for C-C beam with translational support at mid span at $a = -20$, $\gamma_1 = 10$.

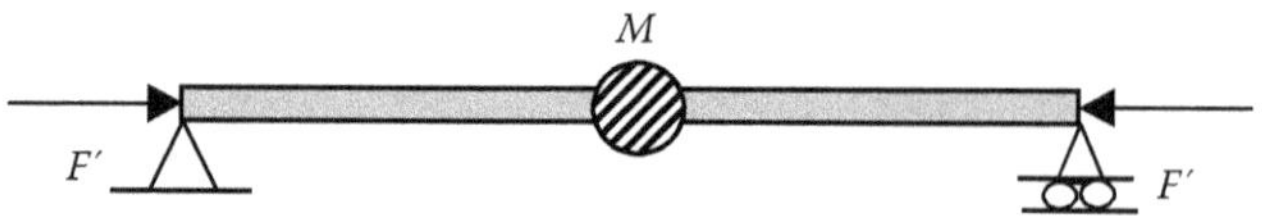

FIGURE 4.19 Beam with an internal attached mass.

for symmetrical modes decrease with increased mass. The fundamental (symmetric) mode becomes a very slow oscillation for large ν. Mode shapes for a P-P beam with a mass at mid-span when $a = 5$, $\nu = 10$ are shown in Figure 4.20, and mode shapes for a C-C beam with a mass at mid-span when $a = -40$, $\nu = 1$ are shown in Figure 4.21.

TABLE 4.13
Frequencies for the P-P Beam with a Mass at Mid Span

$\nu \backslash a$	−10	−5	0	5	10
	14.004	12.114	9.8696	6.9326	0
	44.197	41.904	39.478	36.894	34.114
0	93.693	91.292	88.826	86.290	83.677
	162.84	160.39	157.91	155.39	152.83
	251.69	249.23	246.74	244.23	241.69
	8.0366	6.9614	5.6796	3.9958	0
	44.197	41.904	39.478	36.894	34.114
1	71.694	69.819	67.888	65.899	63.844
	162.84	160.39	157.91	155.39	152.83
	211.08	208.95	206.79	204.61	202.41
	3.0192	2.6186	2.1395	1.5076	0
	44.197	41.904	39.478	36.894	34.114
10	66.088	64.297	62.452	60.546	58.574
	162.84	160.39	157.91	155.39	152.83
	204.90	202.78	200.05	198.49	196.30
	0.9749	0.8457	0.6911	0.4872	0
	44.197	41.904	39.478	36.894	34.114
100	65.374	63.591	61.753	59.854	57.889
	162.84	160.39	157.91	155.39	152.83
	204.19	202.07	199.94	197.78	195.60
	0.3089	0.2680	0.2190	0.1544	0
	44.197	41.904	39.478	38.640	34.114
1,000	65.300	63.518	61.681	59.783	57.818
	162.84	160.39	157.91	155.39	152.83
	204.11	202.00	199.87	197.71	195.53
	44.197	41.904	39.478	36.894	34.114
	65.292	63.510	61.673	59.775	57.810
∞	162.84	160.39	157.91	155.39	152.83
	204.11	201.99	199.86	197.70	195.52
	360.27	357.80	355.30	352.80	350.27

TABLE 4.14
Frequencies for the C-C Beam with a Mass at Mid Span

$\nu \backslash a$	−40	−20	0	20	40
	31.347	27.274	22.373	15.848	0
	75.040	68.708	61.673	53.650	44.086
0	136.26	128.82	120.90	112.42	103.23
	216.34	208.27	199.86	191.08	181.87
	315.74	307.27	298.56	289.58	280.31
	16.673	14.463	11.818	8.3296	0
	75.040	68.708	61.673	53.650	44.086
1	106.95	101.52	95.757	89.592	82.932
	216.34	208.27	199.86	191.08	181.87
	268.40	261.18	253.73	246.05	238.10
	6.0691	5.2649	4.3025	3.0325	0
	75.040	68.708	61.673	53.650	44.086
	100.68	95.620	90.428	84.499	78.288
10	216.34	208.27	199.86	191.08	181.87
	261.98	254.84	247.48	239.88	232.02
	1.9507	1.6923	1.3831	0.9749	0
	75.040	68.708	61.673	53.650	44.086
100	99.916	94.899	89.570	83.868	77.706
	216.34	208.27	199.86	191.08	181.87
	261.25	254.12	246.27	239.18	231.32
	0.6179	0.5361	0.4381	0.3088	0
	75.040	68.708	61.673	53.650	44.086
1,000	99.838	94.825	89.301	83.803	77.646
	216.34	208.27	199.86	191.08	181.87
	261.18	254.05	246.70	239.11	231.25
	75.040	68.708	61.673	53.650	44.086
	99.830	94.817	89.493	83.796	77.639
∞	216.34	208.27	199.86	191.08	181.87
	261.17	254.04	246.69	239.10	231.24
	434.66	425.92	416.99	407.87	398.54

4.4.3 Beam with an Internal Rotational Spring

Figure 4.22 shows a beam with an internal rotational spring, which models a partial crack. The condition at the spring is

$$w_1(b) = w_2(b), \quad \frac{d^2 w_1}{dx^2}(b) = \gamma_2\left(\frac{dw_2}{dx}(b) - \frac{dw_1}{dx}(b)\right) \tag{4.47}$$

$$\frac{d^2 w_1}{dx^2}(b) = \frac{d^2 w_2}{dx^2}(b), \quad \frac{d^3 w_1}{dx^3}(b) + a\frac{dw_1}{dx}(b) = \frac{d^3 w_2}{dx^3}(b) + a\frac{dw_2}{dx}(b) \tag{4.48}$$

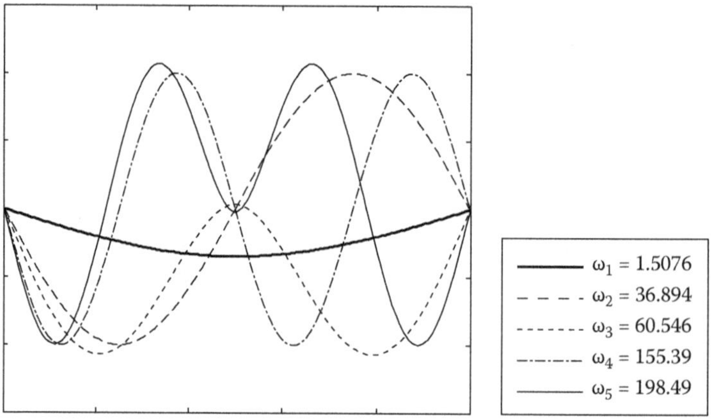

FIGURE 4.20 Mode shapes for P-P beam with mass at mid span at $a = 5$, $v = 10$.

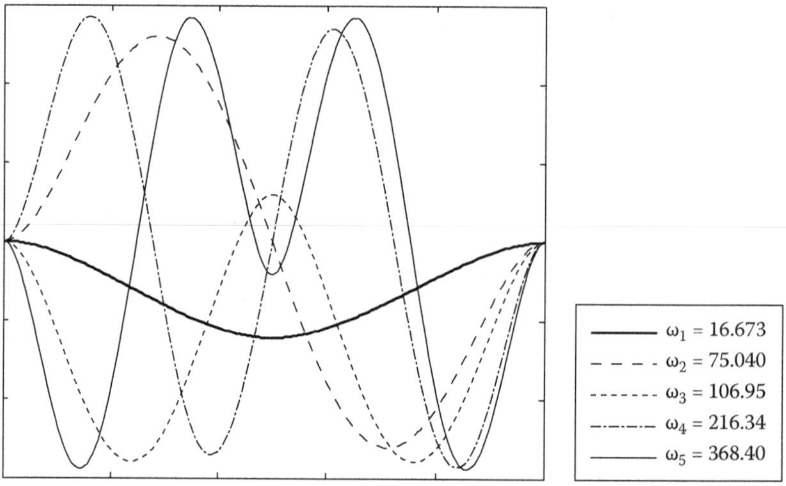

FIGURE 4.21 Mode shapes for C-C beam with mass at mid span at $a = -40$, $v = 1$.

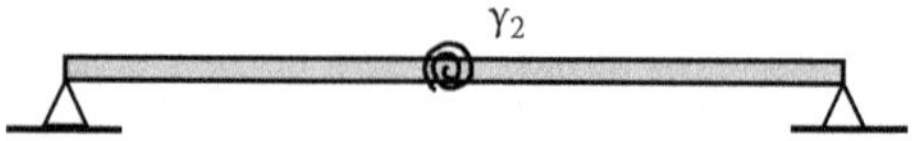

FIGURE 4.22 Beam with an internal rotational spring at mid span.

The matrix I becomes

$$I = \begin{pmatrix} C_{11} & C_{12} & C_{13} & C_{14} & -C_{21} & -C_{22} & -C_{23} & -C_{24} \\ \alpha^2 C_{11} + \gamma_2 \alpha C_{12} & \alpha^2 C_{12} + \gamma_2 \alpha C_{11} & -\beta^2 C_{13} - \gamma_2 \beta C_{14} & -\beta^2 C_{14} + \gamma_2 \beta C_{13} & -\gamma_2 \alpha C_{22} & -\gamma_2 \alpha C_{21} & \gamma_2 \beta C_{24} & -\gamma_2 \beta C_{23} \\ \alpha^2 C_{11} & \alpha^2 C_{12} & -\beta^2 C_{13} & -\beta^2 C_{14} & -\alpha^2 C_{21} & -\alpha^2 C_{22} & \beta^2 C_{23} & \beta^2 C_{24} \\ \alpha(\alpha^2 + a) C_{12} & \alpha(\alpha^2 + a) C_{11} & \beta(\beta^2 - a) C_{14} & -\beta(\beta^2 - a) C_{13} & -\alpha(\alpha^2 + a) C_{22} & -\alpha(\alpha^2 + a) C_{21} & -\beta(\beta^2 - a) C_{24} & \beta(\beta^2 - a) C_{23} \end{pmatrix} \tag{4.49}$$

The results for $b = 0.5$ are given for a P-P beam and a C-C beam in Tables 4.15 and 4.16, respectively. The fundamental mode is always symmetrical. As before, the antisymmetrical mode is independent of the spring constant. When $\gamma_2 = 0$, the joint is a free hinge (Wang and Wang 2001), and when $\gamma_2 = \infty$, the beam becomes entirely continuous. Mode shapes for a P-P beam and a C-C beam with an internal rotational spring at the mid-span at $a = 0$, $\gamma_2 = 1$ are shown in Figure 4.23 and Figure 4.24, respectively.

4.4.4 Stepped Beam

Figure 4.25 shows a stepped beam, where the density and the rigidity are different for the two segments. We normalize with the properties of the (larger) segment 1, and let

$$\rho_2 / \rho_1 = \mu \le 1, \quad EI_2 / EI_1 = \lambda \le 1 \tag{4.50}$$

If the two segments are made of the same material, $\lambda = \mu^n$, where $n = 1$ if the width of segment 2 is smaller (with the same height), $n = 2$ if cross sections of the two segments are similar, and $n = 3$ if the height of segment 2 is smaller (with the same width). The governing equations, without axial force, are

$$\frac{d^4 w_1}{dx^4} - \omega^2 w_1 = 0 \tag{4.51}$$

$$\frac{d^4 w_2}{dx^4} - \frac{\mu}{\lambda} \omega^2 w_2 = 0 \tag{4.52}$$

The solutions are

$$w_1 = s_{11} \cosh(px) + s_{12} \sinh(px) + s_{13} \cos(px) + s_{14} \sin(px) \tag{4.53}$$

$$w_2 = s_{21} \cosh[q(x-1)] + s_{22} \sinh[q(x-1)] + s_{23} \cos[q(x-1)]) + s_{24} \sin[q(x-1)] \tag{4.54}$$

TABLE 4.15
Frequencies for the P-P Beam with a Rotational Spring at Mid Span

γ_2\a	−10	−5	0	5	10
	10.747	7.6643	0	0	0
	44.197	41.904	39.478	36.894	34.114
0	69.889	65.925	61.673	57.071	52.036
	162.84	160.39	157.91	155.39	152.83
	206.85	203.39	199.86	196.27	192.60
	11.789	9.2730	5.6796	0	0
	44.197	41.904	39.478	36.894	34.114
1	75.020	71.553	67.888	63.991	59.817
	162.84	160.39	157.91	155.39	152.83
	213.38	210.11	206.79	203.41	199.96
	13.431	11.436	9.0078	5.6101	0
	44.197	41.904	39.478	36.894	34.114
10	87.421	84.791	82.075	79.265	76.349
	162.84	160.39	157.91	155.39	152.83
	235.61	232.91	230.17	227.90	224.59
	13.936	12.035	9.7723	6.7931	0
	44.197	41.904	39.478	36.894	34.114
100	92.879	90.455	87.965	85.402	82.760
	162.84	160.39	157.91	155.39	152.83
	249.38	246.90	244.38	241.85	239.28
	14.004	12.114	9.8696	6.9326	0
	44.197	41.904	39.478	36.894	34.114
∞	93.693	91.292	88.826	86.290	83.677
	162.84	160.39	157.91	155.39	152.83
	251.69	249.23	246.74	244.23	241.69

Note: Entries with a zero indicate that the beam has buckled.

where

$$p = \sqrt{\omega}, \quad q = (\mu / \lambda)^{1/4} \sqrt{\omega} \tag{4.55}$$

At the joint, we require

$$w_1(b) = w_2(b), \quad \frac{dw_1}{dx}(b) = \frac{dw_2}{dx}(b) \tag{4.56}$$

$$\frac{d^2 w_1}{dx^2}(b) = \lambda \frac{d^2 w_2}{dx^2}(b), \quad \frac{d^3 w_1}{dx^3}(b) = \lambda \frac{d^3 w_2}{dx^3}(b) \tag{4.57}$$

TABLE 4.16
Frequencies for the C-C Beam with a Rotational Spring at Mid Span

γ_2\a	−40	−20	0	20	40
	28.670	23.073	14.064	0	0
	75.040	68.708	61.673	53.650	44.086
0	113.18	101.62	88.138	71.852	52.005
	216.34	208.27	199.86	191.08	181.87
	270.63	259.01	246.79	233.93	220.37
	29.261	24.161	16.875	0	0
	75.040	68.708	61.673	53.650	44.086
1	117.24	106.75	94.813	80.821	63.880
	216.34	208.27	199.86	191.08	181.87
	276.49	265.41	253.81	241.61	228.77
	30.649	26.341	20.998	13.308	0
	75.040	68.708	61.673	53.650	44.086
10	128.99	120.76	111.88	102.18	91.400
	216.34	208.27	199.86	191.08	181.87
	298.09	288.76	279.10	269.07	258.66
	31.256	27.158	22.211	15.578	0
	75.040	68.708	61.673	53.650	44.086
100	135.26	127.73	119.72	111.12	101.78
	216.34	208.27	199.86	191.08	181.87
	313.09	304.53	295.72	286.63	277.25
	31.347	27.274	22.373	15.848	0
	75.040	68.708	61.673	53.650	44.086
∞	136.26	128.82	120.90	112.42	103.23
	216.34	208.27	199.86	191.08	181.87
	315.74	307.27	298.56	289.58	280.31

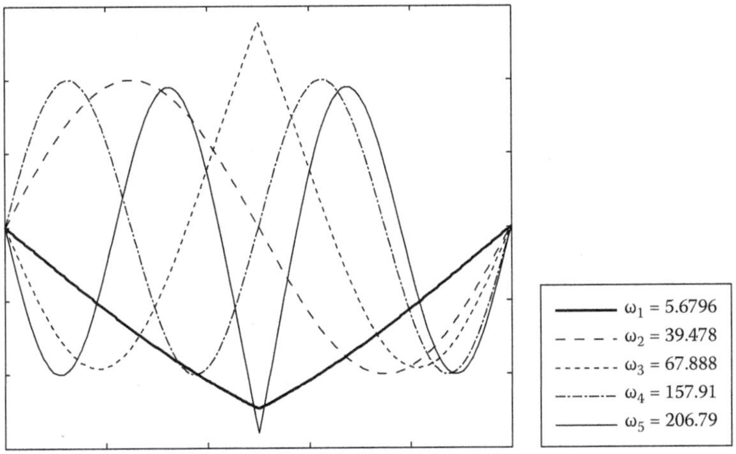

FIGURE 4.23 Mode shapes for P-P beam with a rotational spring at mid span at $a = 0, \gamma_2 = 1$.

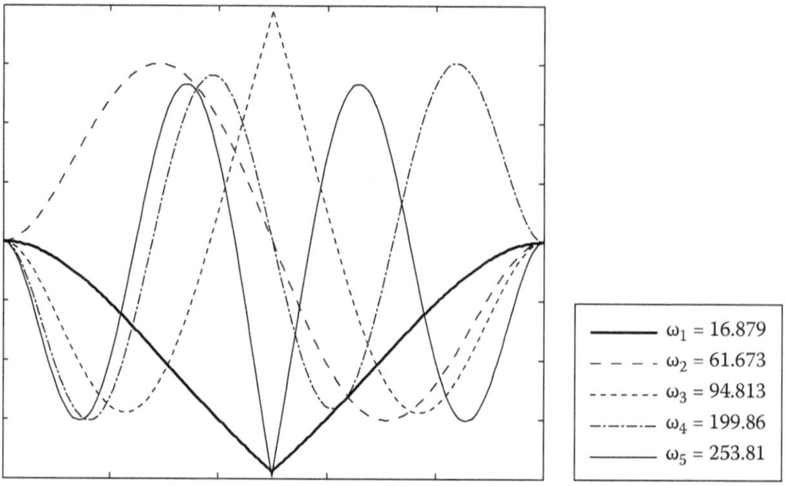

FIGURE 4.24 Mode shapes for C-C beam with a rotational spring at the mid span at $a = 0$, $\gamma_2 = 1$.

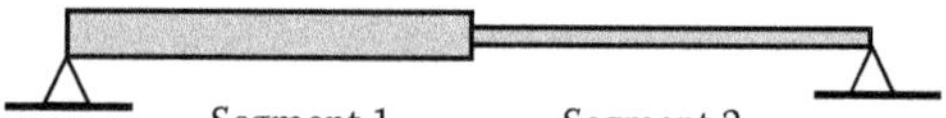

FIGURE 4.25 Beam with two segments of different densities and rigidities.

The characteristic equation is still Equation (4.43), with the end conditions

$$E = \begin{pmatrix} 1 & 0 & 1 & 0 \\ 0 & 1 & 0 & 1 \end{pmatrix} \quad \text{for a clamped end}$$

$$E = \begin{pmatrix} 1 & 0 & 0 & 0 \\ 0 & 0 & 1 & 0 \end{pmatrix} \quad \text{for a pinned end}$$

$$E = \begin{pmatrix} 0 & 1 & 0 & 0 \\ 0 & 0 & 0 & 1 \end{pmatrix} \quad \text{for a sliding end} \tag{4.58}$$

$$E = \begin{pmatrix} 1 & 0 & -1 & 0 \\ 0 & 1 & 0 & -1 \end{pmatrix} \quad \text{for a free end}$$

The I matrix is

$$I = \begin{pmatrix} C_{11} & C_{12} & C_{13} & C_{14} & -C_{21} & -C_{22} & -C_{23} & -C_{24} \\ pC_{12} & pC_{11} & -pC_{14} & pC_{13} & -qC_{22} & -qC_{21} & qC_{24} & -qC_{23} \\ p^2C_{11} & p^2C_{12} & -p^2C_{13} & -p^2C_{14} & -\lambda q^2C_{21} & -\lambda q^2C_{22} & \lambda q^2C_{23} & \lambda q^2C_{24} \\ p^3C_{12} & p^3C_{11} & p^3C_{14} & -p^3C_{13} & -\lambda q^3C_{22} & -\lambda q^3C_{21} & -\lambda q^3C_{24} & \lambda q^3C_{23} \end{pmatrix} \tag{4.59}$$

where

$$C_{11} = \cosh(pb),\ C_{12} = \sinh(pb),\ C_{13} = \cos(pb),\ C_{14} = \sin(pb)$$
$$C_{21} = \cosh[q(b-1)],\ C_{22} = \sinh[q(b-1)],\ C_{23} = \cos[q(b-1)],\ C_{24} = \sin[q(b-1)] \tag{4.60}$$

The following results are for the same material and similar cross sections where $\lambda = \mu^2$.

Table 4.17 is for the P-P case. When $b = 1$, it is a uniform P-P beam, and the frequencies are given by Equation (4.30), which also describes the $\mu = 1$ entries. When $b = 0$, it is again a uniform beam but with smaller cross section. The frequencies are decreased by a factor of $\sqrt{\lambda / \mu}$, or in our case $\sqrt{\mu}$.

Table 4.18 gives the clamped-free case. Mode shapes for a C-F beam with $\mu = 0.5$, $b = 0.5$ are shown in Figure 4.26. When $b = 1$, it is a uniform C-F beam with frequencies given by Table 4.4 ($a = 0$). Similarly, $b = 0$ frequencies are those of $b = 1$ multiplied by $\sqrt{\mu}$.

Table 4.19 shows the clamped-pinned case. Mode shapes for a C-P beam with $\mu = 0.5$, $b = 0.5$ are shown in Figure 4.27. The values for the uniform beam are given in Table 4.2 with $a = 0$.

Table 4.20 gives the results for the clamped-clamped case. The limits for $b = 0$, $b = 1$, and $\mu = 1$ can be obtained from Table 4.1

As we can see from these tables, the frequency variation with increased b is not quite monotonically increasing. This property is peculiar to stepped beams.

4.4.5 Beam with a Partial Elastic Foundation

We consider the Winkler foundation, where resistance is linearly proportional to the deflection. Let c be the force per width of foundation per displacement. For a uniform beam with no axial force, the equation governing the beam on the foundation is

$$\frac{d^4 w}{dx^4} - \omega^2 w + \xi w = 0 \tag{4.61}$$

TABLE 4.17
Frequencies for a Stepped Beam with P-P Ends

μ\b	0.1	0.3	0.5	0.7	0.9
	3.0475	2.3714	2.1686	2.9177	7.6245
	11.614	11.669	20.010	29.702	24.757
0.1	25.505	33.355	54.329	56.081	64.063
	46.038	67.565	70.684	101.40	128.93
	74.223	113.17	131.96	173.20	217.78
	5.3834	5.1532	5.3510	6.8157	9.5905
	21.362	22.597	31.122	31.059	35.966
0.3	47.884	57.617	62.093	78.214	77.535
	85.509	107.65	114.10	122.47	138.89
	135.09	159.04	182.61	203.99	223.41
	6.9743	6.9850	7.4268	8.6532	9.7934
	27.880	29.709	34.527	33.988	38.438
0.5	62.831	70.786	71.894	81.764	84.762
	112.22	125.04	134.45	141.06	149.06
	176.54	189.23	203.27	217.72	233.17
	8.2600	8.3553	8.7573	9.4319	9.8467
	33.080	34.609	36.566	36.698	39.153
0.7	74.608	79.193	80.234	84.452	87.458
	133.09	139.02	145.37	150.32	154.55
	208.78	215.97	223.68	231.90	240.75
	9.3656	9.4236	9.5873	9.7779	9.8654
	37.490	38.095	38.497	38.719	39.416
0.9	84.448	85.800	86.416	87.350	88.549
	150.33	151.92	153.91	155.75	157.18
	235.17	237.61	240.11	242.67	245.30

where

$$\xi = \frac{cL^4}{EI} \tag{4.62}$$

is a nondimensional parameter describing the foundation. For a full foundation, the frequency is simply

$$\omega = \sqrt{\omega_0^2 + \xi} \tag{4.63}$$

where ω_0 is the frequency of the same beam without the foundation. For partial foundations as shown in Figure 4.28, let the subscript 1 represent the left segment that has the foundation, and the subscript 2 represent the right segment that has no foundation.

TABLE 4.18
Frequencies for a Stepped Beam with C-F Ends

μ\b	0.1	0.3	0.5	0.7	0.9
	1.3691	2.2365	4.1786	6.1240	4.2326
	8.5715	13.643	13.205	13.433	26.261
0.1	23.974	32.768	29.211	43.286	71.494
	46.915	45.296	72.898	126.01	111.64
	77.421	79.635	92.724	211.87	155.10
	2.3241	3.4869	5.0254	5.1228	4.0376
	14.449	18.309	15.509	20.705	24.842
0.3	40.169	39.548	46.865	47.429	67.919
	77.722	75.658	85.944	105.16	127.10
	126.53	132.85	142.82	157.86	193.58
	2.8879	3.8155	4.5371	4.4650	3.8663
	17.805	19.276	18.342	22.217	25.779
0.5	49.038	47.663	53.139	53.547	65.326
	94.267	94.679	98.269	109.65	124.92
	152.55	158.86	169.52	180.31	200.33
	3.2416	3.7743	4.0548	4.0034	3.7145
	19.943	20.322	20.357	22.344	22.959
0.7	54.943	54.358	56.809	58.047	63.529
	106.02	107.33	109.27	114.13	122.98
	172.99	176.70	186.86	190.41	200.78
	3.4513	3.6105	3.6742	3.6585	3.5787
	21.429	21.471	21.619	22.159	22.310
0.9	59.592	59.567	60.084	60.800	62.224
	116.16	116.78	117.61	118.74	121.51
	191.33	192.45	194.79	196.96	200.23

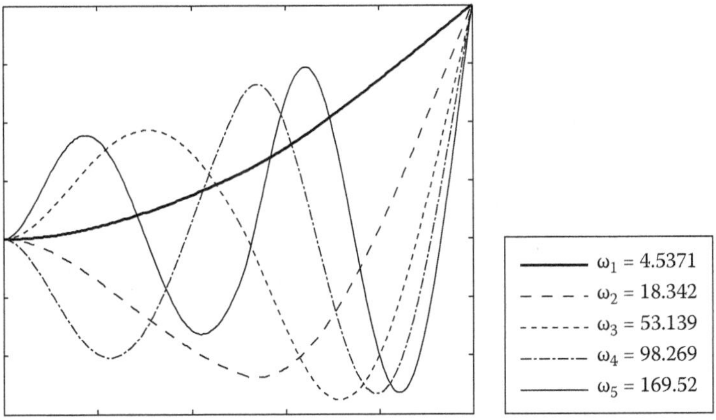

FIGURE 4.26 Mode shapes for a Stepped C-F beam at $\mu = 0.5$, $b = 0.5$.

TABLE 4.19
Frequencies for a Stepped Beam with C-P Ends

μ\b	0.1	0.3	0.5	0.7	0.9
	5.9994	9.6374	12.083	7.7981	11.117
	19.421	28.642	21.692	40.550	31.953
0.1	40.468	40.858	61.455	59.305	77.974
	69.090	69.333	88.252	123.01	148.94
	105.20	115.52	134.47	177.90	243.57
	10.132	13.583	12.154	11.142	14.812
	32.559	33.774	36.311	41.395	44.810
0.3	67.224	64.387	77.576	89.264	90.808
	113.35	117.06	123.59	138.21	157.82
	233.40	182.51	205.56	229.35	248.17
	12.512	14.231	13.165	13.476	15.253
	39.888	39.178	42.818	43.800	48.380
0.5	81.691	80.978	85.750	91.098	99.095
	136.81	141.79	149.31	157.47	168.16
	204.35	212.40	227.50	241.99	257.48
	14.019	14.635	14.231	14.680	15.369
	44.676	44.167	46.266	46.573	49.462
0.7	91.750	92.345	94.118	99.501	102.47
	154.76	158.06	163.93	169.01	174.28
	233.72	238.44	246.99	255.74	265.36
	15.027	15.139	15.085	15.260	15.410
	48.337	48.2580	48.802	48.973	49.867
0.9	100.27	100.71	101.26	102.65	103.88
	170.79	171.81	173.93	175.75	177.38
	259.98	261.68	264.51	267.39	270.38

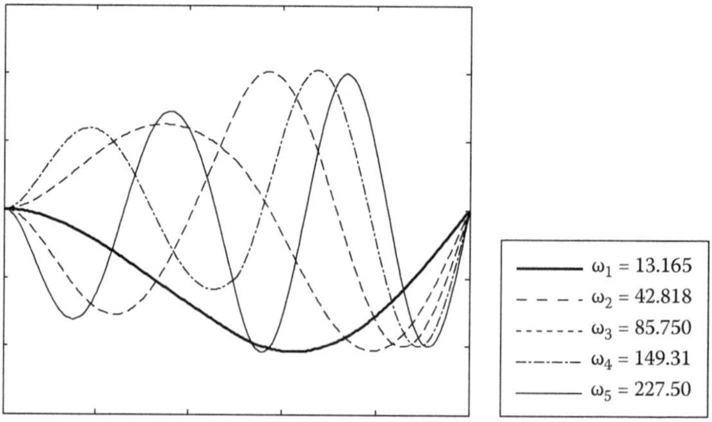

FIGURE 4.27 Mode shapes for a C-P beam at $\mu = 0.5$, $b = 0.5$.

TABLE 4.20
Frequencies for a Stepped Beam with C-C Ends

μ\b	0.1	0.3	0.5	0.7	0.9
	8.7037	13.867	13.305	9.2518	15.015
	23.964	32.754	29.530	43.453	39.427
0.1	46.916	45.292	72.872	79.855	82.543
	77.421	79.636	92.721	126.01	152.21
	115.38	129.00	154.19	211.81	247.07
	14.676	18.670	15.341	15.544	17.649
	40.092	39.548	47.257	47.098	53.246
0.3	77.723	75.649	85.937	105.88	105.51
	126.53	132.85	142.80	158.34	174.29
	184.84	198.55	222.34	242.76	263.74
	18.090	19.652	18.034	18.779	19.073
	49.016	47.677	53.447	52.673	56.271
0.5	94.268	94.670	98.298	110.13	112.98
	152.55	158.86	169.50	180.78	188.00
	223.16	231.37	246.72	262.87	281.20
	20.263	20.675	20.195	20.533	20.505
	54.919	54.366	56.991	57.261	58.490
0.7	106.02	107.32	109.31	114.37	116.60
	172.99	176.70	183.84	190.72	194.20
	256.00	261.37	270.81	281.36	290.86
	21.765	21.808	21.765	21.803	21.802
	59.568	59.546	60.127	60.475	60.631
0.9	116.16	116.78	117.63	118.79	119.51
	191.33	192.45	194.78	197.07	198.16
	285.21	287.29	290.54	293.95	296.44

The governing equations are

$$\frac{d^4 w_1}{dx^4} - (\omega^2 - \xi) w_1 = 0, \quad 0 \le x < b \tag{4.64}$$

$$\frac{d^4 w_2}{dx^4} - \omega^2 w_2 = 0, \quad b < x \le 1 \tag{4.65}$$

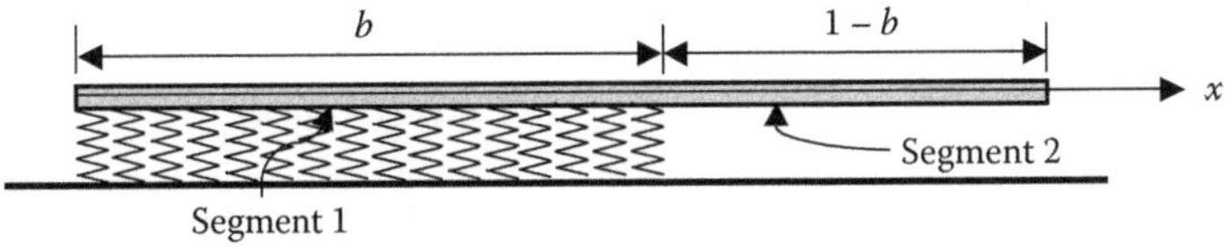

FIGURE 4.28 Beam with partial foundation on left segment.

If $\omega^2 > \xi$, let

$$p = (\omega^2 - \xi)^{1/4}, \quad q = \sqrt{\omega} \tag{4.66}$$

The general solutions are

$$w_1 = s_{11}\cosh(px) + s_{12}\sinh(px) + s_{13}\cos(px) + s_{14}\sin(px) \tag{4.67}$$

$$w_2 = s_{21}\cosh[q(x-1)] + s_{22}\sinh[q(x-1)] + s_{23}\cos[q(x-1)] + s_{24}\sin[q(x-1)] \tag{4.68}$$

The condition at the joint at $x = b$ is that w is completely continuous. The characteristic equation is the 8×8 determinant as Equation (4.43), where E_1 and E_2 are given by Equation (4.58) and

$$I = \begin{pmatrix} C_{11} & C_{12} & C_{13} & C_{14} & -C_{21} & -C_{22} & -C_{23} & -C_{24} \\ pC_{12} & pC_{11} & -pC_{14} & pC_{13} & -qC_{22} & -qC_{21} & qC_{24} & -qC_{23} \\ p^2C_{11} & p^2C_{12} & -p^2C_{13} & -p^2C_{14} & -q^2C_{21} & -q^2C_{22} & q^2C_{23} & q^2C_{24} \\ p^3C_{12} & p^3C_{11} & p^3C_{14} & -p^3C_{13} & -q^3C_{22} & -q^3C_{21} & -q^3C_{24} & q^3C_{23} \end{pmatrix} \tag{4.69}$$

Here,

$$\begin{aligned} &C_{11} = \cosh(pb),\ C_{12} = \sinh(pb),\ C_{13} = \cos(pb),\ C_{14} = \sin(pb) \\ &C_{21} = \cosh[q(b-1)],\ C_{22} = \sinh[q(b-1)],\ C_{23} = \cos[q(b-1)],\ C_{24} = \sin[q(b-1)] \end{aligned} \tag{4.70}$$

The forms are different for the foundation segment when $\omega^2 < \xi$. Let

$$r = \frac{(\xi - \omega^2)^{1/4}}{\sqrt{2}} \tag{4.71}$$

The general solution is

$$\begin{aligned} w_2 = {} & s_{11}\cosh(rx)\cos(rx) + s_{12}\cosh(rx)\sin(rx) \\ & + s_{13}\sinh(rx)\sin(rx) + s_{14}\sinh(rx)\cos(rx) \end{aligned} \tag{4.72}$$

We find for the left end

$$E_C = \begin{pmatrix} 1 & 0 & 1 & 0 \\ 0 & 1 & 0 & 1 \end{pmatrix} \quad \text{for a clamped end}$$

$$E_P = \begin{pmatrix} 1 & 0 & 0 & 0 \\ 0 & 0 & 1 & 0 \end{pmatrix} \quad \text{for a pinned end} \tag{4.73}$$

$$E_S = \begin{pmatrix} 0 & 1 & 0 & 0 \\ 0 & 0 & 0 & 1 \end{pmatrix} \quad \text{for a sliding end}$$

$$E_F = \begin{pmatrix} 1 & 0 & -1 & 0 \\ 0 & 1 & 0 & -1 \end{pmatrix} \quad \text{for a free end}$$

The I matrix is

$$I = \begin{pmatrix} t_1t_3 & t_1t_4 & t_2t_4 & t_2t_3 & -C_{21} & -C_{22} & -C_{23} & -C_{24} \\ r(t_2t_3 - t_1t_4) & r(t_2t_4 + t_1t_3) & r(t_1t_4 + t_2t_3) & r(t_1t_3 - t_2t_4) & -qC_{22} & -qC_{21} & qC_{24} & -qC_{23} \\ -2r^2t_2t_4 & 2r^2t_2t_3 & 2r^2t_1t_3 & -2r^2t_1t_4 & -q^2C_{21} & -q^2C_{22} & q^2C_{23} & q^2C_{24} \\ -2r^3(t_1t_4 + t_2t_3) & 2r^3(t_1t_3 - t_2t_4) & 2r^3(t_2t_3 - t_1t_4) & -2r^3(t_2t_4 + t_1t_3) & -q^3C_{22} & -q^3C_{21} & -q^3C_{24} & q^3C_{23} \end{pmatrix} \tag{4.74}$$

where

$$t_1 = \cosh(rb), \quad t_2 = \sinh(rb), \quad t_3 = \cos(rb), \quad t_4 = \sin(rb) \tag{4.75}$$

We shall present only the results for the C-C, C-P, C-F, P-P, P-F, and F-F cases. Table 4.21 shows the C-C case with a partial support in between $0 \le x \le b$. Let ω_0 represent the frequencies for the C-C beam with neither axial force nor elastic support (Table 4.1). Then ω_0 represents the frequencies for either $b = 0$ or $\xi = 0$ in Table 4.21. If $b = 1$, the beam is fully supported, and Equation (4.63) holds. If $\xi = \infty$, the beam is equivalent to a shorter C-C beam, with frequencies $\omega_0/(1 - b)^2$. Mode shapes for a partially supported C-C beam with $\xi = 10{,}000$, $b = 0.5$ are shown in Figure 4.29.

TABLE 4.21
Frequencies for a Partially Supported Beam, C-C case

$\xi \backslash b$	0.1	0.3	0.5	0.7	0.9
	22.374	22.394	22.485	22.575	22.595
	61.673	61.696	61.713	61.731	61.754
10	120.90	120.92	120.92	120.93	120.94
	199.86	199.87	199.87	199.88	199.88
	298.56	298.56	298.56	298.57	298.57
	22.375	22.580	23.453	24.312	24.505
	61.677	61.901	62.081	62.252	61.475
100	120.91	121.05	121.11	121.18	121.31
	199.87	199.93	199.99	200.04	200.10
	298.56	298.60	298.64	298.68	298.71
	22.390	24.191	30.797	37.360	38.727
	61.711	63.896	65.940	67.239	69.274
1,000	120.96	122.35	122.96	123.61	124.91
	199.94	200.57	201.12	201.66	202.27
	298.64	299.03	299.39	299.75	300.14
	22.535	30.636	50.185	87.252	102.43
	62.036	78.050	104.58	108.49	117.28
10,000	121.48	136.28	140.94	148.04	156.43
	200.60	207.62	212.95	217.72	222.79
	299.40	303.33	306.86	310.37	314.03

Table 4.22 gives the results for the C-P case. Again, if ω_0 represents the frequencies for no axial force in Table 4.2, then the limits are similarly obtained as in the C-C case in Table 4.21. Mode shapes for a partially supported C-P beam with ξ = 10,000, $b = 0.5$ are shown in Figure 4.30.

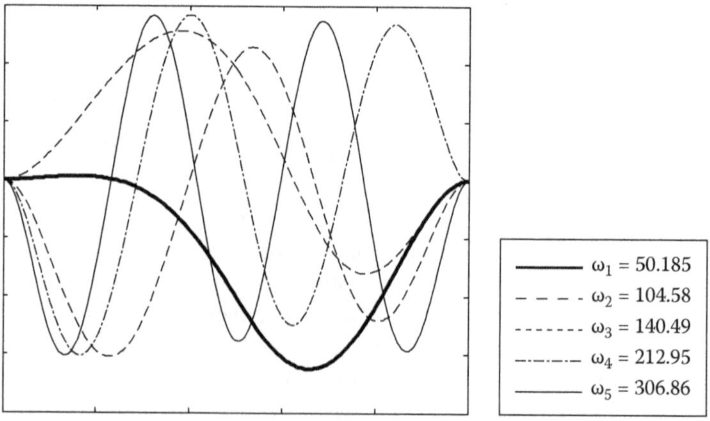

FIGURE 4.29 Mode shapes for a partially supported C-C beam at $\xi = 10{,}000$, $b = 0.5$.

TABLE 4.22
Frequencies for a Partially Supported Beam, C-P Case

$\xi \backslash b$	0.1	0.3	0.5	0.7	0.9
	15.418	15.435	15.528	15.667	15.736
	49.965	49.989	50.020	50.029	50.062
10	104.25	104.26	104.27	104.28	104.29
	178.27	178.28	178.28	178.29	178.30
	272.03	272.04	272.04	272.04	272.05
	15.419	15.587	16.465	17.741	18.348
	49.968	50.202	50.521	50.603	50.926
100	104.25	104.42	104.47	104.58	104.70
	178.28	178.35	178.42	178.48	178.53
	272.04	272.08	272.12	272.16	272.20
	15.430	16.859	22.736	31.122	35.027
	49.997	52.234	55.678	56.231	58.874
1,000	104.30	105.93	106.48	107.58	108.68
	178.34	179.11	179.75	180.33	180.82
	272.12	272.54	272.91	273.29	273.67
	15.534	21.371	35.465	66.255	100.50
	50.275	64.758	94.532	105.64	110.33
10,000	104.77	121.05	127.92	136.95	142.45
	178.98	187.67	193.48	198.22	202.36
	272.86	277.16	280.82	284.44	288.03

The results for the C-F case are given in Table 4.23. Similar limiting properties apply using ω_0 from Table 4.4. Mode shapes for a partially supported C-F beam with $\xi = 10$, $b = 0.7$ are shown in Figure 4.31.

The results for the P-P case are given in Table 4.24. For the limiting cases, we use $\omega_0 = n^2\pi^2$ from Equation (4.30), except when $\xi = \infty$, where the shorter C-P frequencies apply.

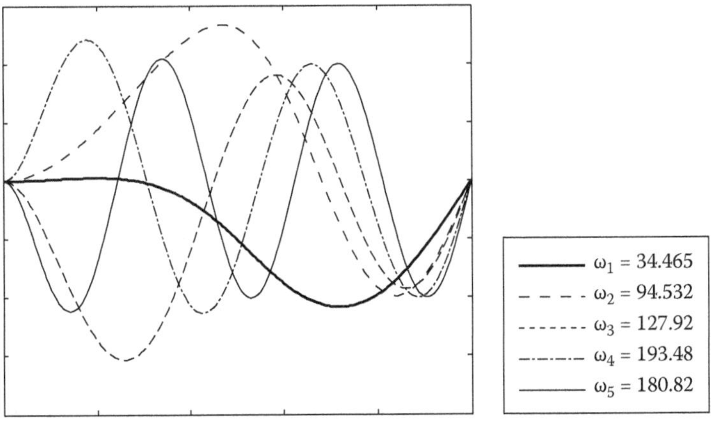

FIGURE 4.30 Mode shapes for a partially supported C-P beam at $\xi = 10{,}000$, $b = 0.5$.

TABLE 4.23
Frequencies for a Partially Supported Beam, C-F Case

$\xi \backslash b$	0.1	0.3	0.5	0.7	0.9
	3.5161	3.5227	3.5868	3.8207	4.3438
	22.035	22.054	22.134	22.195	22.207
10	61.698	61.720	61.738	61.753	61.765
	120.90	120.92	120.92	120.93	120.94
	199.86	199.87	199.87	199.88	199.88
	3.5163	3.5806	4.1223	5.7127	8.7035
	22.036	22.231	23.012	23.631	23.731
100	61.701	61.926	62.110	62.250	62.369
	120.91	121.04	121.11	121.17	121.27
	199.87	199.93	199.99	200.04	200.09
	3.5193	4.0071	6.1982	11.006	22.871
	22.051	23.777	29.998	35.930	36.875
1,000	61.735	63.930	66.013	67.066	68.256
	120.96	122.35	122.96	123.62	124.50
	199.94	200.57	201.12	201.67	202.11
	3.5465	5.0288	8.4995	17.669	53.176
	22.192	30.105	49.198	84.860	101.30
10,000	62.061	78.103	104.70	108.59	114.61
	121.48	136.27	140.95	148.18	154.09
	200.60	207.62	212.95	217.79	221.51

The P-F case is given in Table 4.25. Notice that the ω_0 from Table 4.6 starts with a zero frequency, which is a rigid body rotation. With the elastic support, the fundamental frequency is also very low, especially for small b or small ξ. The limiting cases are similarly obtained as in the C-F case.

The results for the F-F case are given in Table 4.26.

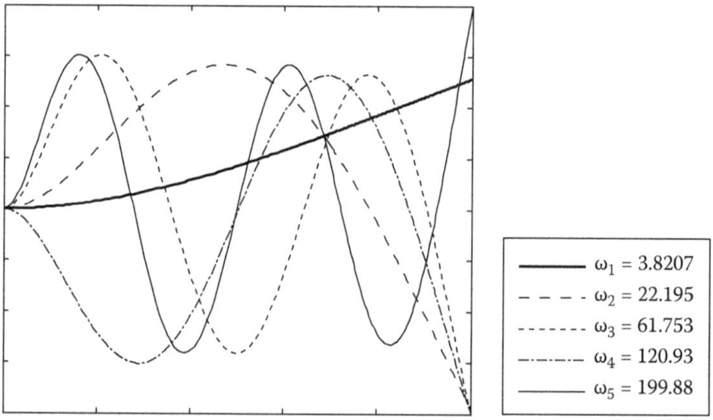

FIGURE 4.31 Mode shapes for a partially supported C-F beam at $\xi = 10$, $b = 0.7$.

TABLE 4.24
Frequencies for a Partially Supported Beam, P-P Case

$\xi \backslash b$	0.1	0.3	0.5	0.7	0.9
	9.8729	9.9444	10.119	10.292	10.361
	39.482	39.522	39.542	39.561	39.602
10	88.829	88.845	88.855	88.864	88.880
	157.92	157.92	157.93	157.94	157.94
	246.74	246.75	246.75	246.75	246.76
	9.9021	10.577	12.090	13.496	14.027
	39.509	39.917	40.121	40.299	40.695
100	88.854	89.014	89.108	89.203	89.360
	157.94	158.00	158.07	158.15	158.21
	246.76	246.80	246.84	246.88	246.92
	10.182	14.592	21.940	30.015	33.027
	39.782	43.760	46.508	47.293	50.339
1,000	89.104	90.769	91.610	92.609	94.024
	158.16	158.75	159.51	160.24	160.81
	246.94	247.35	247.75	248.16	248.56
	12.123	21.369	35.416	66.238	100.04
	42.159	64.100	93.877	102.64	106.26
10,000	91.465	112.00	116.54	126.01	131.83
	160.30	167.47	174.84	180.49	184.84
	248.77	252.81	256.75	260.60	264.36

4.5 NONUNIFORM BEAM

Nonuniform beams are those where the density and flexural rigidity vary along the beam axis. The first solution of the vibration of a nonuniform beam was due to Kirchhoff (1882). He studied beams with linear taper in one or both cross-sectional directions, and expressed the solutions in what are now known as Bessel functions. In this section, we shall study three types of exact solutions: Bessel-type solutions, polynomial solutions, and exponential solutions. We exclude solutions that are in terms of hypergeometric functions or Jacobi functions, since they are uncommon and very tedious to evaluate numerically.

We assume that the left end of the beam has the higher density and flexural rigidity. Equation (4.7) without axial force is

$$\frac{d^2}{dx^2}\left[I \frac{d^2 w}{dx^2} \right] - \omega^2 r w = 0 \tag{4.76}$$

Since I and r vary with x, this fourth-order differential equation has very few exact solutions. The boundary conditions could be clamped, pinned, free, or sliding.

TABLE 4.25
Frequencies for a Partially Supported Beam, P-F Case

$\xi \backslash b$	0.1	0.3	0.5	0.7	0.9
	0.1000	0.5171	1.1060	1.8336	2.6941
	15.421	15.481	15.589	15.632	15.656
10	49.968	50.001	50.010	50.035	50.046
	104.25	104.26	104.07	104.28	104.29
	178.27	178.28	178.28	178.29	178.30
	0.3152	1.5680	3.1958	5.3215	8.3494
	15.449	16.034	17.085	17.584	17.723
100	49.995	50.331	50.422	50.666	50.778
	104.28	104.40	104.50	104.56	104.66
	178.29	178.34	178.40	178.47	178.52
	0.9101	3.6628	6.1934	10.930	22.805
	15.714	20.276	27.832	33.080	33.791
1,000	50.264	53.657	54.909	56.667	57.825
	104.52	105.78	106.80	107.36	108.37
	178.50	179.01	179.61	180.26	180.78
	2.4733	5.0038	8.4815	17.669	53.176
	17.751	30.100	49.151	84.846	100.50
10,000	52.670	76.541	101.66	104.13	109.22
	106.84	123.65	128.24	135.07	141.48
	180.57	186.22	192.37	198.08	202.24

4.5.1 Bessel-Type Solutions

Many papers have extended Kirchhoff's work, notably Cranch and Adler (1956) and Sanger (1968). Let

$$z = 1 - cx, \text{ for } 0 \le c \le 1 \tag{4.77}$$

Equation (4.76) becomes

$$c^4 \frac{d^2}{dz^2}\left[l(z)\frac{d^2 w}{dz^2} \right] - \omega^2 r(z) w = 0 \tag{4.78}$$

Assume

$$l(z) = z^m, \quad r(z) = z^n \tag{4.79}$$

If

$$m = n + 2 \tag{4.80}$$

TABLE 4.26
Frequencies for a Partially Supported Beam, F-F case

$\xi \backslash b$	0.1	0.3	0.5	0.7	0.9
	0.0539	0.3265	0.8145	1.5494	2.5583
	1.8491	2.7471	3.0535	3.1453	3.1618
10	22.428	22.440	22.485	22.530	22.542
	61.687	61.699	61.713	61.728	61.740
	120.91	120.92	120.92	120.93	120.94
	0.1704	1.0167	2.4702	4.6746	7.9903
	5.7055	8.5346	9.6222	9.9445	9.9985
100	22.943	23.087	23.481	23.940	24.036
	61.811	61.930	62.081	62.224	62.345
	120.95	121.05	121.11	121.08	121.27
	0.5347	2.8143	5.8044	10.832	22.735
	13.791	20.247	27.746	31.366	31.618
1,000	29.392	33.592	33.839	36.267	36.996
	63.232	64.251	65.940	67.030	68.236
	121.43	122.34	122.96	123.63	124.50
	1.5729	4.8749	8.4806	17.666	53.175
	17.430	29.514	49.114	84.801	99.982
10,000	51.921	76.270	99.721	100.18	101.33
	88.518	101.52	105.11	108.45	114.60
	128.97	136.52	140.86	148.18	154.09

Equation (4.78) can be factored into

$$\left[z^{-n}\frac{d}{dz}\left(z^{n+1}\frac{d}{dz}\right)+\frac{\omega}{c^2}\right]\left[z^{-n}\frac{d}{dz}\left(z^{n+1}\frac{d}{dz}\right)-\frac{\omega}{c^2}\right]w=0 \tag{4.81}$$

Each one of the brackets in Equation (4.81) is a Bessel operator. When n is an integer, the solution is

$$w=z^{-n/2}[C_1J_n(u)+C_2Y_n(u)+C_3I_n(u)+C_4K_n(u)],\quad u=2\sqrt{\omega z}\,/\,c \tag{4.82}$$

Here, J and Y are Bessel functions, and I and K are modified Bessel functions. If an end is clamped, the boundary condition is

$$w=0,\quad \frac{dw}{dz}=0 \tag{4.83}$$

Using the properties of Bessel functions and simplifying, we obtain the coefficient matrix

$$E_C(u)=\begin{pmatrix} J_n(u) & Y_n(u) & I_n(u) & K_n(u) \\ J_{n+1}(u) & Y_{n+1}(u) & -I_{n+1}(u) & K_{n+1}(u) \end{pmatrix} \tag{4.84}$$

If an end is pinned,

$$w = 0, \quad \frac{d^2w}{dz^2} = 0 \tag{4.85}$$

This gives

$$E_P(u) = \begin{pmatrix} J_n(u) & Y_n(u) & I_n(u) & K_n(u) \\ J_{n+2}(u) & Y_{n+2}(u) & I_{n+2}(u) & K_{n+2}(u) \end{pmatrix} \tag{4.86}$$

If an end is free,

$$\frac{d^2w}{dz^2} = 0, \quad \frac{d}{dz}\left(z^{n+2} \frac{d^2w}{dz^2} \right) = 0 \tag{4.87}$$

This gives

$$E_F(u) = \begin{pmatrix} J_{n+1}(u) & Y_{n+1}(u) & I_{n+1}(u) & -K_{n+1}(u) \\ J_{n+2}(u) & Y_{n+2}(u) & I_{n+2}(u) & K_{n+2}(u) \end{pmatrix} \tag{4.88}$$

For the sliding end

$$\frac{dw}{dz} = 0, \quad \frac{d}{dz}\left(z^{n+2} \frac{d^2w}{dz^2} \right) = 0 \tag{4.89}$$

whose corresponding matrix is simplified to

$$E_S(u) = \begin{pmatrix} J_{n+1}(u) & Y_{n+1}(u) & 0 & 0 \\ 0 & 0 & I_{n+1}(u) & -K_{n+1}(u) \end{pmatrix} \tag{4.90}$$

For the single-span beam, there are 16 different combinations of clamped, pinned, sliding, or free end conditions. Let the end values be

$$u_0 = 2\sqrt{\omega} / c, \qquad u_1 = 2\sqrt{\omega(1-c)} / c \tag{4.91}$$

The exact characteristic equations are obtained by setting the following determinant to zero, i.e.,

$$\begin{vmatrix} E_s(u_0) \\ E_t(u_1) \end{vmatrix} = 0 \tag{4.92}$$

where (s,t) could indicate any of the C, P, S, F conditions.

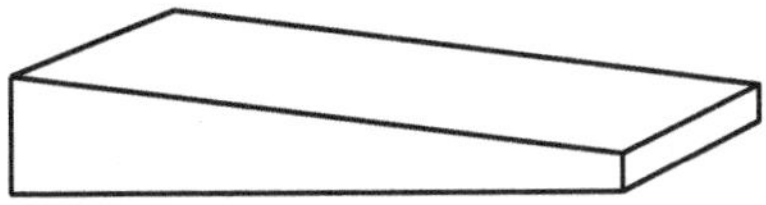

FIGURE 4.32a A beam with tapered height and constant width ($m = 3$, $n = 1$).

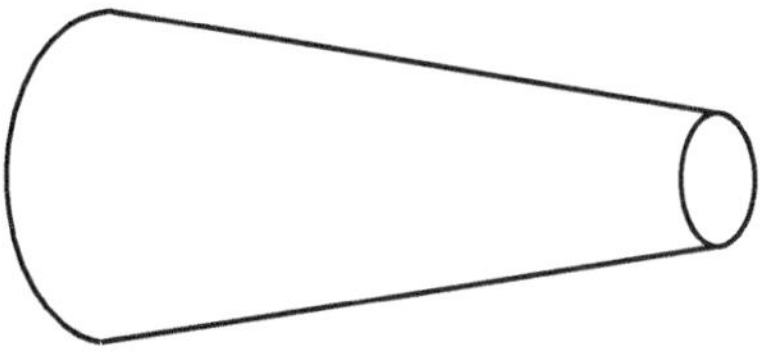

FIGURE 4.32b A beam with both tapered height and width ($m = 4$, $n = 2$).

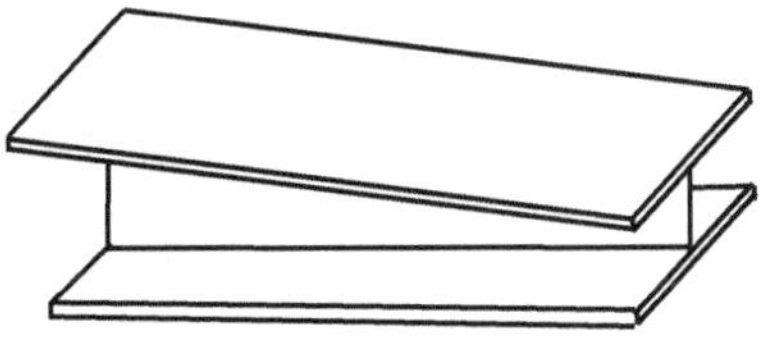

FIGURE 4.32c A beam composed of two constant-width plates ($m = 2$, $n = 0$).

4.5.1.1 The Beam with Linear Taper

We shall consider the three important cases with linear taper. Figure 4.32a shows a beam with tapered height and constant width, in which case $m = 3$, $n = 1$. If both width and height are linearly tapered with similar cross sections, as shown in Figure 4.32b, then $m = 4$, $n = 2$. For open-web or constant-width I-beams with linear height and negligible bracing, as shown in Figure 4.32c, then $m = 2$, $n = 0$.

We present frequency results for the more common C-C, C-P, C-F, P-P, P-F, and F-F cases. Table 4.27 shows the first five frequencies for $m = 3$, $n = 1$. The $c = 0$ case is the uniform beam, whose frequencies can be obtained from Section 4.3.2, with no axial force. The $c = 1$ case is a beam that tapers to a point at the right end, and is not too practical. Notice that when taper increases, the frequencies decrease, except for the fundamental frequency of the C-F beam.

Table 4.28 shows the results for the linearly tapered beam with similar cross sections, including the rectangular and circular cross sections. For this case, $m = 4$, $n = 2$. Note that the frequencies for $m = 4$, $n = 2$ are slightly larger than those of $m = 3$, $n = 1$. Also, the frequencies decrease with increased taper, except for the first two modes of the C-F or P-F beams. Table 4.29 gives the frequencies for $m = 2$, $n = 0$.

Mode shapes for a linear taper height ($m = 3$, $n = 1$) C-F beam with $c = 0.5$ are shown in Figure 4.33. Mode shapes for a beam with linear taper in both height and width with similar cross section ($m = 4$, $n = 2$) at $c = 0.5$ are shown in Figure 4.34. Mode shapes for a composition beam with linear taper in height and constant cross-sectional area ($m = 2$, $n = 0$) at $c = 0.5$ are shown in Figure 4.35.

TABLE 4.27
Frequencies for a Beam with Linear Taper in Height Only ($m = 3, n = 1$)

c	0.1	0.3	0.5	0.7	0.9
	21.241	18.879	16.336	13.483	9.8846
	58.550	52.026	44.981	37.053	27.008
C-C	114.78	101.98	88.138	72.537	52.708
	189.74	168.57	145.67	119.83	86.933
	283.43	251.80	217.57	178.94	129.70
	14.849	13.640	12.300	10.737	8.6301
	47.637	42.774	37.527	31.633	24.204
C-P	99.172	88.567	77.122	64.265	48.099
	169.44	151.00	131.07	108.66	80.402
	258.46	230.07	199.40	164.86	121.18
	3.5587	3.6668	3.8238	4.0817	4.6307
	21.338	19.881	18.317	16.625	14.931
C-F	58.980	53.322	47.265	40.588	32.833
	115.19	103.27	90.451	76.182	58.917
	190.15	169.86	148.00	123.54	93.388
	9.3675	8.3019	7.1215	5.7454	3.8895
	37.484	33.352	28.952	24.094	18.123
P-P	84.335	74.992	64.979	53.834	40.011
	149.92	133.27	115.35	95.308	70.241
	234.25	208.18	180.09	148.57	108.92
	14.854	13.693	12.494	11.291	10.357
	47.645	42.855	37.801	32.349	26.316
P-F	99.180	88.655	77.419	65.035	50.346
	169.45	151.09	131.39	109.46	82.723
	258.47	230.17	199.72	165.68	123.55
	21.253	18.997	16.724	14.461	12.491
	58.566	52.186	45.503	38.344	30.399
F-F	114.80	102.15	88.711	73.959	56.490
	189.76	168.75	146.27	121.33	90.964
	283.45	251.99	218.19	180.49	133.90

4.5.1.2 Two-Segment Symmetric Beams with Linear Taper

It is possible to join two segments of different taper into one beam. At the joint, the displacement, slope, moment, and shear are equal. A special case is the stepped uniform beam (zero taper) of Section 4.4.3. We shall present only some symmetric beams of linear taper shown in Figure 4.36. Since there is geometrical symmetry, the beam can only vibrate either symmetrically, where the midpoint is equivalent to a sliding end condition, or antisymmetrically, where the midpoint is equivalent to a pinned end condition.

Consider a C-C symmetrically tapered beam with the largest cross section at the ends, as shown in Figure 4.36a. Let the half-length of the beam be L. Then

TABLE 4.28
Frequencies for a Beam with Linear Taper in Both Height and Width and with Similar Cross Sections ($m = 4$, $n = 2$)

c	0.1	0.3	0.5	0.7	0.9
	21.245	18.923	16.479	13.835	10.764
	58.556	52.086	45.176	37.533	28.235
C-C	114.79	102.04	88.353	73.068	54.101
	189.74	168.63	145.89	120.39	88.429
	283.44	251.87	217.81	179.52	131.26
	14.955	13.962	12.851	11.557	9.9086
	47.742	43.128	38.199	32.744	26.109
C-P	99.279	88.938	77.851	65.523	50.384
	169.55	151.38	131.84	110.01	82.947
	258.57	230.46	200.18	166.26	123.91
	3.6737	4.0669	4.6252	5.5093	7.2049
	21.550	20.556	19.548	18.641	18.680
C-F	58.189	54.015	48.579	42.810	37.124
	115.40	103.98	91.813	78.521	63.505
	190.36	170.58	149.39	125.95	98.166
	9.3624	8.2562	6.9566	5.3589	3.0513
	37.489	33.401	29.110	24.480	19.094
P-P	84.342	75.069	65.228	54.438	41.494
	149.93	133.36	115.65	96.030	72.044
	234.26	208.28	180.41	149.36	110.94
	14.962	14.049	13.189	12.594	13.246
	47.754	43.253	38.634	33.932	29.879
P-F	99.292	89.074	78.316	66.759	54.240
	169.57	151.52	132.32	111.27	86.824
	258.58	230.61	200.67	167.54	127.79
	21.263	19.102	17.079	15.412	15.206
	58.580	52.326	45.963	39.502	33.621
F-F	114.81	102.31	89.214	75.216	59.958
	189.77	168.91	146.80	122.65	94.595
	283.47	252.16	218.74	181.85	137.65

the frequencies for the antisymmetrical modes can be found from the single-taper C-P beam in the previous section. The frequencies for the symmetrical mode can be found from the single-taper C-S beam. Tables 4.30, 4.31, and 4.32 are for the single-taper C-S, S-C, P-S, S-P, S-F, and P-C cases for ($m = 3$, $n = 1$), ($m = 4$, $n = 2$), and ($m = 2$, $n = 0$), respectively. By themselves, these results have few applications. However, in conjunction with a companion end condition, they represent the complete spectrum of a symmetric tapered beam.

For the C-C symmetrically tapered beam of Figure 4.36a, for the $m = 3$, $n = 1$ beam, the frequencies are the union of the C-S case in Table 4.30 and the C-P case in Table 4.27. For the $m = 4$, $n = 2$ beam, we take the C-S case in Table 4.31 and the

TABLE 4.29
Frequencies for a Composite Beam with Linear Taper in Height and Constant Cross-Sectional Area ($m = 2, n = 0$)

c	0.1	0.3	0.5	0.7	0.9
	21.240	18.864	16.288	13.371	9.6280
	58.548	52.006	44.916	36.894	26.613
C-C	114.78	101.96	88.067	72.360	52.251
	189.73	168.54	145.59	119.64	86.439
	283.43	251.78	217.49	178.75	129.18
	14.744	13.326	11.784	10.031	7.7891
	47.535	42.450	36.949	30.749	22.910
C-P	99.068	88.232	76.509	63.289	46.534
	169.34	150.66	130.44	107.64	78.660
	258.35	229.73	198.76	163.80	119.31
	3.4466	3.2984	3.1336	2.9442	2.7100
	21.128	19.228	17.169	14.856	12.010
C-F	58.775	52.666	46.072	38.677	29.497
	114.98	102.60	89.225	74.188	55.339
	189.94	169.19	146.76	121.50	89.658
	9.3706	8.3330	7.2219	5.9871	4.4640
	37.481	33.323	28.859	23.877	17.674
P-P	84.330	74.946	64.830	53.479	39.196
	149.92	133.21	115.17	94.878	69.204
	234.25	208.12	179.89	148.09	107.73
	14.746	13.350	11.863	10.229	8.2965
	47.539	42.489	37.076	31.062	23.715
P-F	99.073	88.275	76.650	63.641	47.462
	169.35	150.70	130.59	108.01	79.663
	258.367	229.78	198.91	164.18	120.37
	21.245	18.923	16.479	13.835	10.764
	58.556	52.086	45.176	37.533	28.235
F-F	114.79	102.04	88.353	73.018	54.101
	189.74	168.63	145.89	120.39	88.429
	283.44	251.87	217.81	179.52	131.26

C-P case in Table 4.28. For the $m = 2$, $n = 0$ beam, we take the C-S case in Table 4.32 and the C-P case in Table 4.29.

Other symmetrically tapered beams with P-P, S-S, or F-F end conditions may be similarly constructed using single-taper results.

4.5.1.3 Linearly Tapered Cantilever with an End Mass

If a mass is attached to an end, Equation (4.15) gives

$$\frac{d}{dz}\left(z^{n+2}\frac{d^2w}{dz^2}\right) = \mp\frac{\nu\omega^2}{c^3}w \tag{4.93}$$

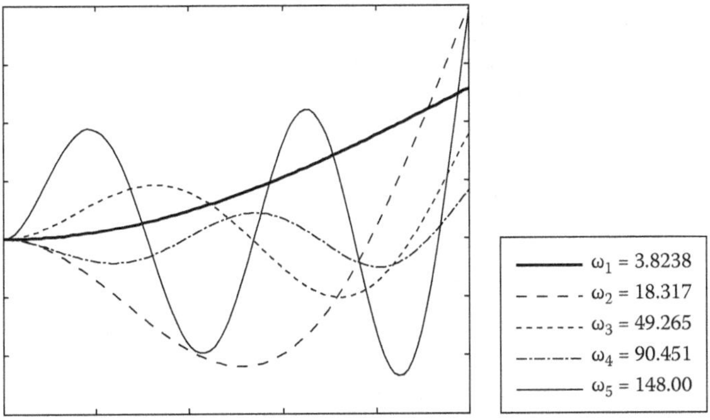

FIGURE 4.33 Mode shapes for C-F beam with linear tapered height ($m = 3, n = 1$) at $c = 0.5$.

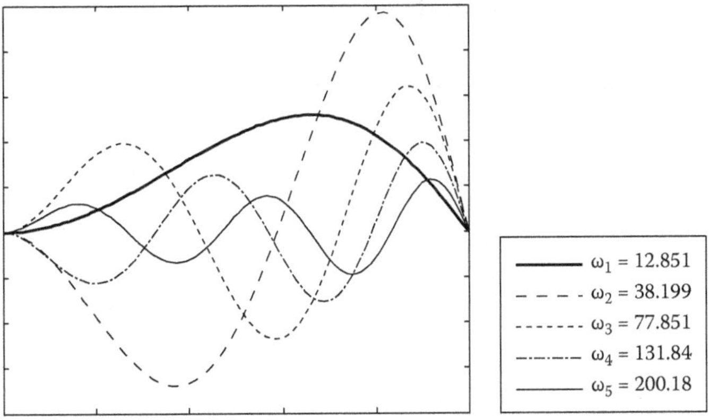

FIGURE 4.34 Mode shapes for C-P beam with linear taper in both height and width with similar cross section ($m = 4, n = 2$) at $c = 0.5$.

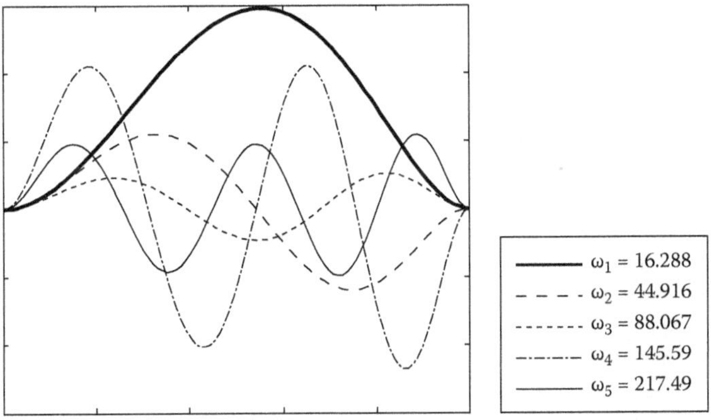

FIGURE 4.35 Mode shapes for C-C beam with linear taper in height and constant cross-sectional area ($m = 2, n = 0$) at $c = 0.5$.

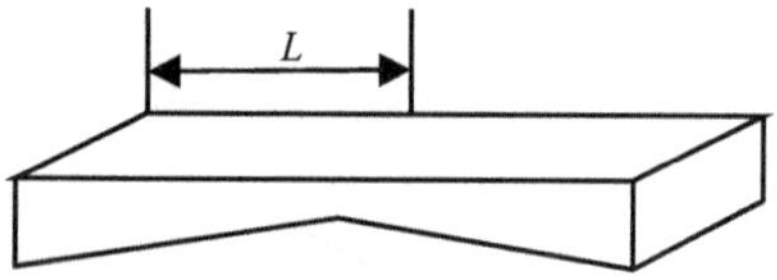

FIGURE 4.36a Symmetrically tapered beam with the largest cross section at the ends.

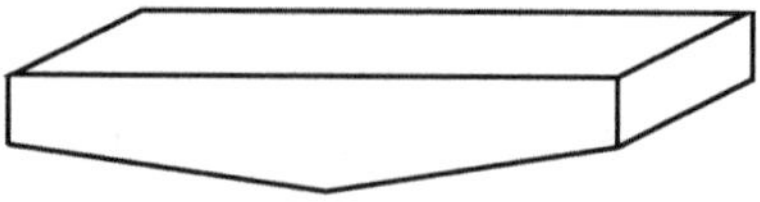

FIGURE 4.36b Symmetrically tapered beam with the smallest cross section at the ends.

TABLE 4.30
Frequencies for a Beam with Linear Taper in Height Only ($m = 3$, $n = 1$)

c	0.1	0.3	0.5	0.7	0.9
	5.4591	5.1976	4.9573	4.7766	4.8275
	28.850	25.999	22.973	19.694	16.092
C-S	71.014	63.456	55.340	46.321	35.482
	131.92	117.56	102.08	84.751	63.324
	211.56	188.31	163.20	135.00	99.686
	5.1693	4.3227	3.4717	2.5969	1.6126
	28.550	25.092	21.427	17.399	12.458
S-C	70.713	62.549	53.793	44.018	31.766
	131.62	116.65	100.53	82.446	59.588
	211.26	187.40	161.66	132.69	95.945
	2.2910	1.9169	1.5033	1.0251	0.4170
	21.039	18.651	16.182	13.627	11.141
P-S	58.519	51.960	44.992	37.372	28.518
	114.74	101.91	88.155	72.876	54.271
	189.70	168.50	145.69	120.18	88.521
	2.3910	2.2170	2.0033	1.7231	1.2841
	21.138	18.941	16.620	14.068	10.897
S-P	58.619	52.254	45.454	37.904	28.548
	114.84	102.20	88.630	73.463	54.522
	189.80	168.80	146.17	120.80	88.933
	5.5559	5.4911	5.4565	5.5057	5.8656
	28.956	26.359	23.657	20.809	17.891
S-F	71.121	63.835	56.088	47.607	37.757
	132.03	117.95	102.86	86.128	65.904
	211.67	188.70	164.00	136.43	102.47
	14.426	12.371	10.183	7.7681	4.7875
	47.238	41.576	35.524	28.807	20.482
P-C	98.772	87.363	75.100	61.379	44.136
	169.04	149.79	129.05	105.74	76.271
	258.06	228.87	197.36	161.91	116.93

TABLE 4.31
Frequencies for a Beam with Linear Taper in Both Height and Width ($m = 4$, $n = 2$)

c	0.1	0.3	0.5	0.7	0.9
	5.5616	5.5527	5.6728	6.0781	7.2980
	28.959	26.392	23.788	21.221	19.394
C-S	71.123	63.856	56.178	47.908	39.045
	132.03	117.96	102.93	86.362	66.983
	211.67	188.71	164.06	136.62	103.39
	5.0783	4.0838	3.1310	2.1942	1.1483
	28.459	24.877	21.195	17.323	12.984
S-C	70.623	62.342	53.587	44.014	32.528
	131.53	116.45	100.34	82.476	60.477
	211.17	187.20	161.47	132.74	96.915
	2.2888	1.8951	1.4367	0.8874	0.2450
	21.046	18.736	16.486	14.478	13.661
P-S	58.529	52.066	45.352	38.327	31.389
	114.75	103.02	88.539	73.879	57.258
	189.71	168.62	146.09	121.21	91.563
	2.3888	2.1945	1.9317	1.5572	0.9333
	21.146	19.011	16.830	14.534	11.914
S-P	58.628	52.349	45.748	37.578	30.067
	114.85	102.31	88.963	74.247	56.364
	189.81	168.91	146.53	121.66	90.997
	5.6586	5.8490	6.1859	6.8556	8.4742
	29.068	26.778	24.541	22.466	21.421
S-F	71.235	61.270	57.028	49.402	41.689
	132.14	118.39	103.83	87.998	70.058
	211.78	189.15	164.99	138.35	106.77
	14.320	12.054	9.6496	6.9976	3.6988
	47.144	41.333	35.212	28.578	20.816
P-C	98.679	87.135	74.837	61.265	44.756
	168.95	149.57	128.81	105.69	77.045
	257.96	228.65	197.14	161.89	117.80

where $v = M/\rho_0 L$ and the top and bottom signs are for the left or right ends, respectively. If the mass is at a free end, after some work, the boundary condition is

$$E_M(u) = \begin{pmatrix} J_{n+1}(u) \pm \phi J_n(u) & Y_{n+1}(u) \pm \phi Y_n(u) & I_{n+1}(u) \pm \phi I_n(u) & -K_{n+1}(u) \pm \phi K_n(u) \\ J_{n+2}(u) & Y_{n+2}(u) & I_{n+2}(u) & K_{n+2}(u) \end{pmatrix} \tag{4.94}$$

where

$$\phi = v\sqrt{\omega}\left(\frac{2\sqrt{\omega}}{cu}\right)^{2n+1} \tag{4.95}$$

TABLE 4.32
Frequencies for a Beam Composed of Two Constant-Width Plates ($m = 2$, $n = 0$)

c	0.1	0.3	0.5	0.7	0.9
	5.3591	4.8707	4.3460	3.7632	3.0585
	28.745	25.649	22.302	18.539	13.831
C-S	70.909	63.104	54.659	45.138	33.088
	131.81	117.21	101.40	83.557	60.881
	211.45	187.95	162.52	133.80	97.219
	5.2625	4.5804	3.8591	3.0700	2.1148
	28.645	25.347	21.791	17.791	12.708
S-C	70.809	62.802	54.146	44.383	31.922
	131.71	116.90	100.88	82.798	59.695
	211.35	187.65	162.00	133.04	96.022
	2.2924	1.9301	1.5448	1.1187	0.5889
	21.034	18.600	16.002	13.134	9.6579
P-S	58.513	51.897	44.777	36.803	26.825
	114.73	101.84	87.925	72.277	52.499
	189.69	168.43	145.45	119.57	86.709
	2.3924	2.2305	2.0473	1.8290	1.5329
	21.134	18.899	16.497	13.811	10.452
S-P	58.613	52.198	45.278	37.507	27.725
	114.83	102.14	88.430	72.996	53.469
	189.79	168.73	145.96	120.30	87.726
	5.4558	5.1620	4.8355	4.4595	3.9909
	28.849	25.981	22.906	19.489	15.334
S-F	71.013	63.445	55.295	46.177	34.873
	131.92	117.55	102.05	84.640	62.828
	211.56	188.30	163.18	134.91	99.271
	14.532	12.691	10.722	8.5333	5.8115
	47.335	41.850	35.939	29.299	20.881
P-C	98.868	87.629	75.491	61.815	44.398
	169.14	150.06	129.42	106.15	76.460
	258.15	229.13	197.73	162.30	117.07

We shall consider two relevant cases. Figure 4.37 shows a tapered cantilever beam with a mass at the smaller free end. The characteristic equation is

$$\begin{vmatrix} E_C(u_0) \\ E_M(u_1) \end{vmatrix} = 0 \tag{4.96}$$

Table 4.33 shows the results for $m = 3$, $n = 1$. Notice that the $\nu = 0$ (no end mass) case is the C-F case of Table 4.27. For larger mass, the results quickly approach those of the C-P case. Notice also that the effect of a small end mass on the fundamental frequency is not monotonic as the taper is increased.

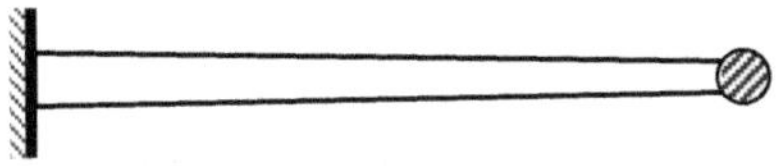

FIGURE 4.37 Tapered cantilever beam with a mass at the smaller free end.

Table 4.34 shows the results when both height and width are tapered. The frequencies are in general higher than those with tapered height only.

The results for the composite beam with constant area ($m = 2$, $n = 0$) are shown in Table 4.32.

Mode shapes for a C-F beam tapered in height only ($m = 3$, $n = 1$) with an end mass at $c = 0.5$, $\nu = 1$ are shown in Figure 4.38. Mode shapes for a C-F beam tapered in both height and width ($m = 4$, $n = 2$) with an end mass $c = 0.5$, $\nu = 10$ are shown in Figure 4.39. Mode shapes for a composite beam with linear taper in height and constant cross-sectional area ($m = 2$, $n = 0$) at $c = 0.5$, $\nu = 0.1$ are shown in Figure 4.40.

The second case studied is the F-F linearly tapered beam with a mass at the larger end shown in Figure 4.41. The characteristic equation is

$$\begin{vmatrix} E_M(u_0) \\ E_F(u_1) \end{vmatrix} = 0 \tag{4.97}$$

TABLE 4.33
Frequencies for the C-F Beam Tapered in Height Only ($m = 3$, $n = 1$) with an End Mass

$\nu \backslash c$	0.1	0.3	0.5	0.7	0.9
	3.5587	3.6668	3.8238	4.0817	4.6307
	21.338	19.881	18.317	16.625	14.931
0	58.980	53.322	47.265	40.588	32.833
	115.19	103.27	90.451	76.182	58.917
	190.15	169.86	148.00	123.54	93.388
	2.9591	2.9304	2.8737	2.7494	2.3864
	18.533	16.766	14.776	12.432	9.3434
0.1	52.660	46.696	40.299	33.229	24.688
	104.91	92.849	79.976	65.778	48.493
	175.65	155.51	133.55	110.13	80.750
	1.5121	1.4084	1.2804	1.1112	0.8435
	15.587	14.195	12.681	10.957	8.7053
1	48.441	43.342	37.887	31.817	24.254
	99.983	89.126	77.464	64.432	48.139
	170.26	151.55	131.90	108.82	80.437
	0.5212	0.4770	0.4258	0.3626	0.2700
	14.931	13.700	12.340	10.760	8.6377
10	47.722	42.834	37.564	31.652	24.209
	99.256	88.624	77.156	64.282	48.103
	169.53	151.06	131.11	108.68	80.405

TABLE 4.34
Frequencies for the C-F Beam Tapered in Both Height and Width ($m = 4$, $n = 2$) with an End Mass

ν\c	0.1	0.3	0.5	0.7	0.9
	3.6737	4.0669	4.6252	5.5093	7.2049
	21.550	20.556	19.548	18.641	18.680
0	58.189	54.015	48.579	42.810	37.124
	115.40	103.98	91.813	78.521	63.505
	190.36	170.58	149.39	125.95	98.166
	3.0114	3.0612	3.0000	2.6732	1.7036
	18.542	16.728	14.649	12.364	10.018
0.1	52.525	46.340	39.949	33.378	26.176
	104.67	92.308	79.540	66.080	50.436
	175.34	154.84	133.49	110.52	82.992
	1.5035	1.3685	1.1869	0.9370	0.5468
	15.643	14.394	13.080	11.644	9.9194
1	48.480	43.550	38.400	32.810	26.115
	100.02	89.347	78.037	65.580	50.389
	170.30	115.78	132.01	110.06	82.952
	0.5147	0.4555	0.3861	0.2996	0.1732
	15.030	14.007	12.874	11.566	9.9097
10	47.820	43.171	38.219	32.750	26.109
	99.356	88.980	77.870	65.528	50.385
	169.63	151.42	131.85	110.01	82.948

The frequencies are given in Tables 4.36 to 4.38. The changes in the fundamental frequency for this mode are not monotonic with increased taper.

Mode shapes for an F-F beam tapered in height only ($m = 3$, $n = 1$) with an end mass at the larger end at $c = 0.5$, $v = 1$ are shown in Figure 4.42. Mode shapes for an F-F beam tapered in both height and width ($m = 4$, $n = 2$) with a mass at the larger end at $c = 0.5$, $v = 1$ are shown in Figure 4.43.

4.5.1.4 Other Bessel-Type Solutions

Cranch and Adler (1956) gave some power relations such as $m = -n$, $m = n + 6$, $m = (n + 8)/3$, which also lead to exact Bessel-type solutions. However, these solutions are physically less relevant and shall not be presented here.

4.5.2 Power-Type Solutions

Cranch and Adler (1956) noted that Equation (4.78) is homogeneous when

$$I = z^{n+4}, \qquad r = z^n \tag{4.98}$$

TABLE 4.35
Frequencies for a C–F Beam with Linear Taper in Height and Constant Cross-Sectional Area and with End Mass ($m = 2, n = 0$)

$\nu \backslash c$	0.1	0.3	0.5	0.7	0.9
	3.4466	3.2984	3.1336	2.9442	2.7100
	21.128	19.228	17.169	14.856	12.010
0	58.775	52.666	46.072	38.677	29.497
	114.98	102.60	89.225	74.188	55.339
	189.94	169.19	146.76	121.50	89.658
	2.9047	2.7692	2.6171	2.4390	2.2088
	18.518	16.757	14.842	12.668	9.9055
0.1	52.796	47.101	40.939	33.992	25.205
	105.15	93.536	80.967	66.796	48.827
	175.98	156.40	135.20	111.30	80.934
	1.5194	1.4375	1.3445	1.2333	1.0824
	15.534	14.030	12.393	10.533	8.1556
1	48.409	43.207	37.579	31.239	23.226
	99.957	88.992	77.132	63.761	46.819
	170.24	151.42	131.06	108.10	78.929
	0.5276	0.4978	0.4639	0.4234	0.3679
	14.832	13.405	11.851	10.086	7.8284
10	47.628	42.531	37.016	30.800	22.943
	99.161	88.311	76.573	62.338	46.563
	169.43	150.74	130.51	107.68	78.688

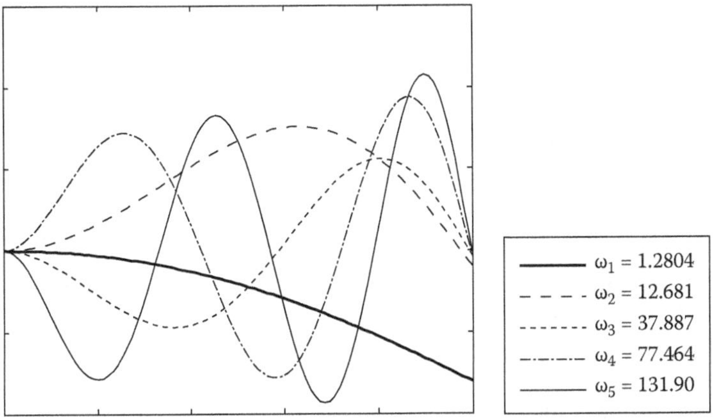

FIGURE 4.38 Mode shapes for C-F beam tapered in height only ($m = 3$, $n = 1$) with an end mass at $c = 0.5$, $\nu = 1$.

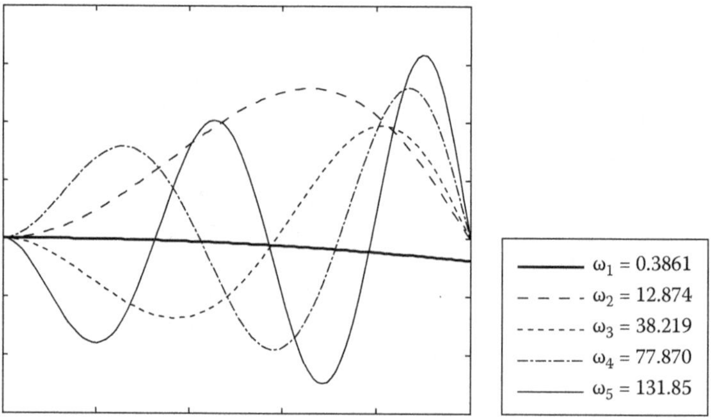

FIGURE 4.39 Mode shapes for C-F beam tapered in both height and width ($m = 4$, $n = 2$) with an end mass at $c = 0.5$, $v = 10$.

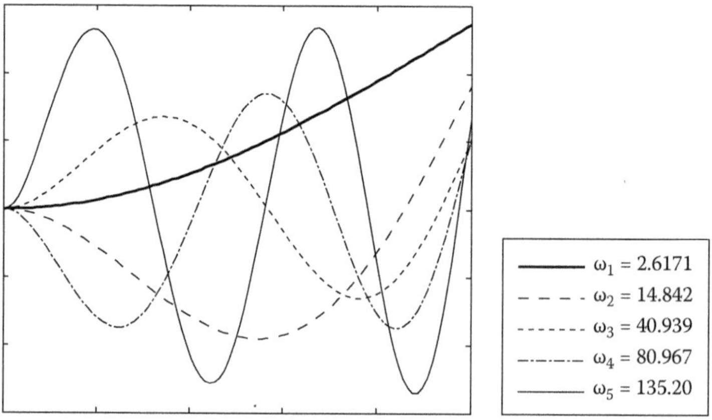

FIGURE 4.40 Mode shapes for C–F beam with linear taper in height and constant cross-sectional area ($m = 2$, $n = 0$) at $c = 0.5$, $\nu = 0.1$.

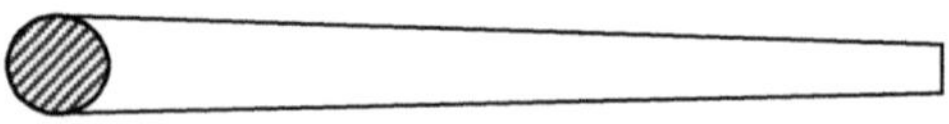

FIGURE 4.41 Free tapered beam with a mass at the larger (left) end.

but they did not give any details. Here we shall give explicit polynomial-type solutions and the results for several applications (Wang and Wang 2013). Let

$$y = z^{\lambda} \tag{4.99}$$

and one obtains the indicial equation

$$c^4\lambda(\lambda - 1)(\lambda + n + 2)(\lambda + n + 1) - \omega^2 = 0 \tag{4.100}$$

TABLE 4.36
Frequencies for an F-F Beam Tapered in Height Only (m = 3, n = 1) with an End Mass at the Larger End

$\nu \backslash c$	0.1	0.3	0.5	0.7	0.9
	21.253	18.997	16.724	14.461	12.491
	58.566	52.186	45.503	38.344	30.399
0	114.80	102.15	88.711	73.959	56.490
	189.76	168.75	146.27	121.33	90.964
	283.45	251.99	218.19	180.49	133.90
	18.719	16.885	15.030	13.184	11.628
	52.837	47.349	41.569	35.344	28.418
0.1	105.30	94.064	82.069	68.843	53.111
	176.19	157.13	136.65	113.86	86.001
	265.65	236.66	205.44	170.52	127.23
	15.695	14.384	13.041	11.697	10.629
	48.519	43.621	38.453	32.878	26.697
1	100.09	89.468	78.130	65.631	50.795
	170.38	151.93	132.13	110.10	83.215
	259.40	231.02	200.48	166.34	124.07
	14.949	13.772	12.556	11.336	10.388
	47.738	42.931	37.871	32.406	26.357
10	99.275	88.741	77.494	65.098	50.393
	169.55	151.18	131.46	109.53	82.774
	258.56	230.26	199.80	165.74	123.60

The formal solution of this fourth-order algebraic equation is extremely tedious. However, we are fortunate to factor Equation (4.100) as follows:

$$\frac{1}{16}\left[(2\lambda+n+1)^2-(n^2+4n+5)\right]^2-\left(\frac{n}{2}+1\right)2-\omega^2/c^4=0 \tag{4.101}$$

The solutions are

$$\lambda_{1,2}=\frac{1}{2}\left[-(n+1)\pm\sqrt{(n^2+4n+5)+4\sqrt{(n/2+1)^2+\omega^2/c^4}}\right] \tag{4.102}$$

$$\lambda_{3,4}=\frac{1}{2}\left[-(n+1)\pm\sqrt{(n^2+4n+5)-4\sqrt{(n/2+1)^2+\omega^2/c^4}}\right] \tag{4.103}$$

The general solution is thus

$$y=C_1z^{\lambda_1}+C_2z^{\lambda_2}+C_3z^{\lambda_3}+C_4z^{\lambda_4} \tag{4.104}$$

TABLE 4.37
Frequencies for an F-F Beam Tapered in Both Height and Width ($m = 4$, $n = 2$) with a Mass at the Larger End

$\nu \backslash c$	0.1	0.3	0.5	0.7	0.9
	21.263	19.102	17.079	15.412	15.206
	58.580	52.326	45.963	39.502	33.621
0	114.81	102.31	89.214	75.216	59.958
	189.77	168.91	146.80	122.65	94.595
	283.47	252.16	218.74	181.85	137.65
	18.745	17.030	15.441	14.188	14.323
	52.879	47.563	42.132	36.608	31.687
0.1	105.35	94.321	82.718	70.259	56.687
	176.26	157.41	137.36	115.38	89.790
	265.72	236.97	206.20	172.12	131.18
	15.779	14.680	13.656	12.916	13.457
	48.615	43.982	39.230	34.392	30.192
1	100.19	89.859	78.983	67.299	54.626
	170.48	152.34	133.03	111.86	87.261
	259.50	231.44	201.41	168.16	128.27
	15.055	14.120	13.242	12.630	13.269
	47.846	43.331	38.698	33.981	29.912
10	99.386	89.156	78.386	66.816	54.281
	169.66	151.61	132.39	111.33	86.870
	258.67	230.69	200.75	167.60	127.84

The quantity under the outer square root of Equation (4.103) may be negative. In such a case, we set

$$a = -\frac{n+1}{2}, \qquad b = \sqrt{\sqrt{(n/2+1)^2 + \omega^2/c^4} - (n^2+4n+5)/4} \tag{4.105}$$

and Equation (4.104) is replaced by

$$y = C_1 z^{\lambda_1} + C_2 z^{\lambda_2} + C_3 z^a \cos(b \ln z) + C_4 z^a \sin(b \ln z) \tag{4.106}$$

We shall consider the end conditions C-C, C-P, C-F, P-P, P-F, and F-F. For nontrivial solutions, the determinant of the coefficients of C_1 to C_4 is set to zero, giving the frequencies ω. Since n is arbitrary, there are infinite solutions. We shall present three cases that have parabolic axial variation.

TABLE 4.38
Frequencies for an F-F Beam Tapered in Height and Constant Cross Sectional Area ($m = 2, n = 0$) with an End Mass at the Larger End

$\nu \backslash c$	0.1	0.3	0.5	0.7	0.9
	21.245	18.923	16.479	13.835	10.764
	58.556	52.086	45.176	37.533	28.235
0	114.79	102.04	88.353	73.018	54.101
	189.74	168.63	145.89	120.39	88.429
	283.44	251.87	217.81	179.52	131.26
	18.696	16.769	14.728	12.501	9.8885
	52.799	47.173	41.136	34.419	26.187
0.1	105.25	93.849	81.560	67.784	50.596
	176.14	156.88	136.09	112.71	83.289
	265.59	236.39	204.84	169.30	124.37
	15.612	14.108	12.507	10.749	8.6728
	48.426	43.294	37.789	31.670	24.186
1	99.988	89.115	77.406	64.298	47.985
	170.27	151.56	131.37	108.69	80.216
	259.30	230.65	199.70	164.88	120.94
	14.844	13.437	11.936	10.289	8.3400
	47.634	42.575	37.153	31.128	23.766
10	99.168	88.363	76.730	63.710	47.518
	169.44	150.79	130.67	108.08	79.721
	258.45	229.87	198.99	164.26	120.43

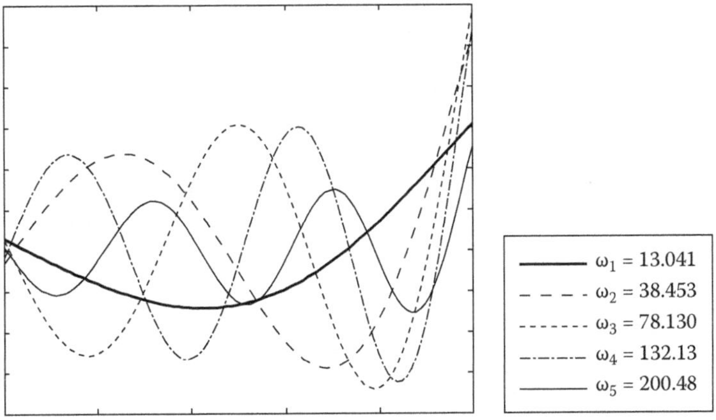

FIGURE 4.42 Mode shapes for F-F beam tapered in height only ($m = 3, n = 1$) with an end mass at the larger end at $c = 0.5$, $\nu = 1$.

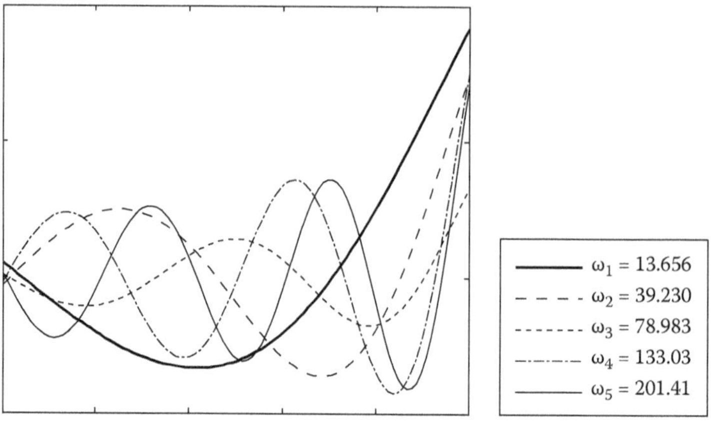

FIGURE 4.43 Mode shapes for F-F beam tapered in both height and width ($m = 4$, $n = 2$) with a mass at the larger end at $c = 0.5$, $v = 1$.

4.5.2.1 Results for $m = 6, n = 2$

Figure 4.44 shows the variable-thickness beam with constant width. Let the height vary parabolically as z^2. Then

$$l = z^6, \quad r = z^2 \tag{4.107}$$

The exponents are

$$\lambda_k = \frac{1}{2}\left(-3 \pm \sqrt{17 \pm 4\sqrt{4 + \omega^2 / c^4}}\right) \tag{4.108}$$

The results are given in Table 4.39. Mode shapes are shown in Figure 4.45a for a beam of which the cross section is solid rectangular with constant width and height varying parabolically as z^2 for $c = 0.5$ for the C-C condition, while the mode shapes for the F-F condition are shown in Figure 4.45b. Notice that they have the same frequencies but have different mode shapes.

4.5.2.2 Results for $m = 8, n = 4$

Figure 4.46 shows the beam's cross section is similar, where both width and height vary parabolically. The density is proportional to h^2 and the rigidity proportional to h^4. Thus

$$l = z^8, \quad r = z^4 \tag{4.109}$$

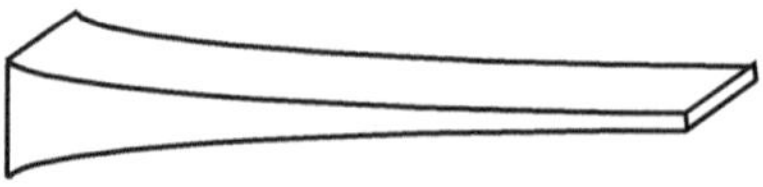

FIGURE 4.44 Variable-thickness beam with constant width.

TABLE 4.39
Frequencies for Variable Thickness Beam with Constant Width, $m = 6$, $n = 2$ Case

c	0	0.1	0.3	0.5	0.7	0.9
C-C F-F	22.373	20.178	16.040	12.247	8.8189	5.7931
	61.673	55.589	43.917	32.887	22.420	12.084
	120.90	108.95	85.846	63.781	42.578	21.313
	199.86	180.08	141.72	104.91	69.349	33.494
	298.56	268.99	211.55	155.29	102.76	48.655
C-P P-F	15.418	14.297	12.048	9.8026	7.5956	5.4611
	49.965	45.413	36.576	28.059	19.756	11.212
	104.25	94.320	75.031	56.440	38.348	19.816
	178.27	160.00	127.42	95.030	63.508	31.325
	272.03	245.47	193.77	143.86	95.290	45.785
C-F	3.5160	3.5978	3.7754	3.9706	4.1717	4.3017
	22.035	20.653	17.832	14.905	11.787	8.1260
	61.697	56.366	45.925	35.711	25.534	14.628
	120.90	109.71	87.851	66.626	45.745	23.951
	199.86	180.84	143.73	107.77	72.546	36.186
P-P	9.8696	8.8778	6.8658	4.8183	2.7474	0.7304
	39.478	35.581	28.084	20.982	14.251	7.7893
	88.826	80.043	63.071	46.862	31.301	15.767
	157.91	142.28	111.98	82.915	54.849	26.591
	246.74	222.30	174.85	129.20	84.995	40.338

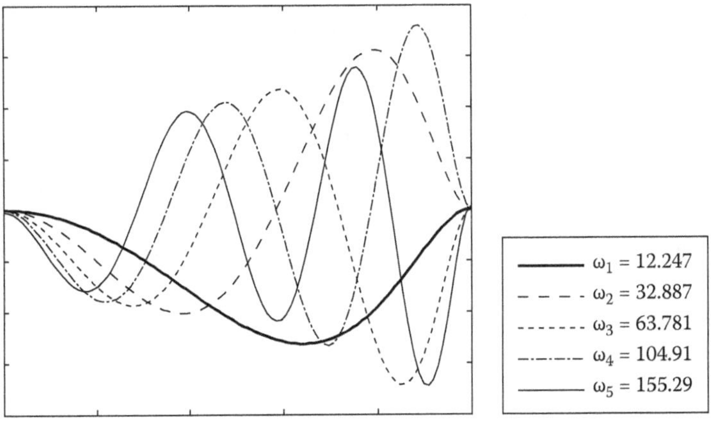

FIGURE 4.45a Vibration modes for C-C beam ($m = 6$, $n = 2$) at $c = 0.5$. (*continued*)

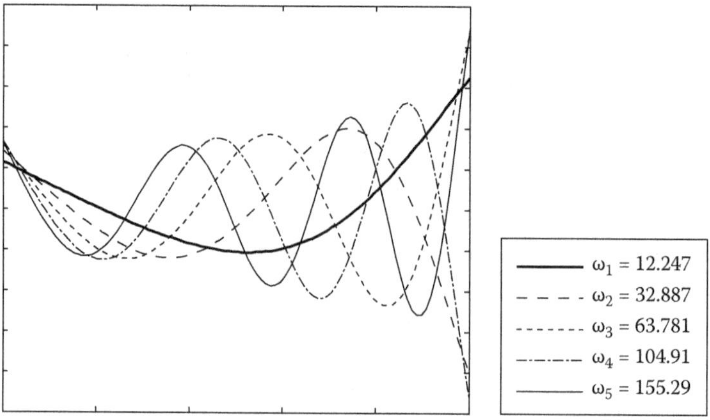

FIGURE 4.45b Vibration modes for F-F beam ($m = 6$, $n = 2$) at $c = 0.5$.

The exponents are

$$\lambda_k = \frac{1}{2}\left(-5 \pm \sqrt{37 \pm 4\sqrt{9 + \omega^2 / c^4}}\right) \tag{4.110}$$

The results are given in Table 4.40.

Mode shapes are shown in Figure 4.47a for a beam whose cross section is similar, where both width and height vary parabolically and the density is proportional to h^2 and the rigidity proportional to h^4 for $c = 0.5$ for the C-C condition, while the mode shapes for the F-F condition are shown in Figure 4.47b. Notice that they have the same frequencies but have different mode shapes.

We find the frequencies of the C-C and F-F cases, and also the C-P and P-F cases, are identical. This is due to the fact that the characteristic determinants are identical, even for our nonuniform beams. For uniform beams, it is easy to show this equivalence (e.g., Magrab 1980).

In general, as the taper or c increases, the frequency decreases, except the fundamental frequency of the C-F case.

4.5.3 Isospectral Beams and the $m = 4$, $n = 4$ Case

Isospectral beams are two different beams that have exactly the same spectrum of frequencies. The most important work is due to Gottlieb (1987), who found seven kinds of nonuniform beams that are isospectral to a uniform beam, but only for

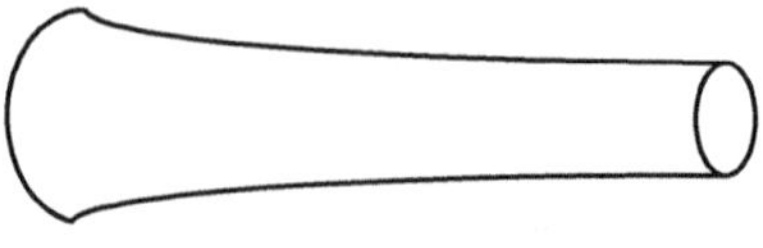

FIGURE 4.46 Variable-diameter beam ($m = 8$, $n = 4$).

TABLE 4.40
Frequencies for Variable Thickness Beam, $m = 8$, $n = 4$ Case

c	0	0.1	0.3	0.5	0.7	0.9
C-C F-F	22.373	20.206	16.298	13.030	10.618	9.4309
	61.673	55.626	44.255	33.841	24.359	15.577
	120.90	108.99	86.215	64.812	44.625	37.076
	199.86	180.12	142.11	105.98	71.472	52.293
	298.56	269.03	211.95	157.40	104.94	70.494
C-P P-F	15.418	14.505	12.678	10.985	9.7056	9.2242
	49.965	45.631	37.336	29.520	22.137	14.933
	104.25	94.544	75.850	58.043	40.933	23.649
	178.27	161.23	128.28	96.716	66.230	35.284
	272.03	245.70	194.64	145.60	98.107	49.852
C-F	3.5160	3.8303	4.5928	5.5808	6.8534	7.0875
	22.035	21.065	19.066	17.007	14.873	8.4151
	61.697	56.779	47.240	38.017	28.904	12.406
	120.90	110.12	89.212	69.044	49.296	19.110
	199.86	181.26	145.12	110.25	76.211	28.610
P-P	9.8696	8.8585	6.6952	4.3751	2.0356	0.2426
	39.478	35.599	28.255	21.497	15.473	10.770
	88.826	80.072	63.332	47.604	32.835	18.708
	157.91	142.32	112.29	83.777	56.577	29.652
	246.74	222.34	175.18	130.14	86.852	43.524

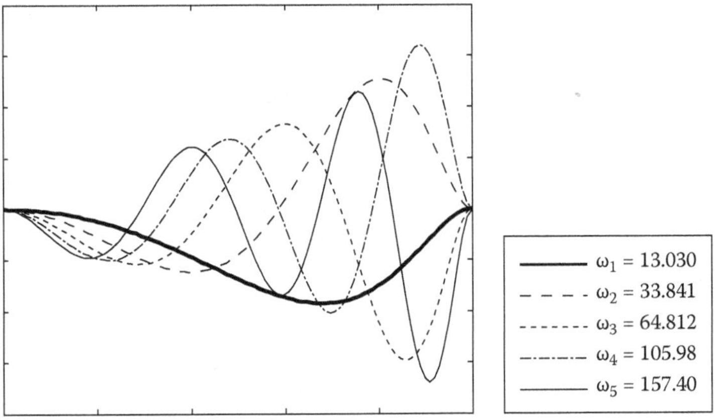

FIGURE 4.47a Vibration modes for C-C beam ($m = 8$, $n = 4$) at $c = 0.5$. (*continued*)

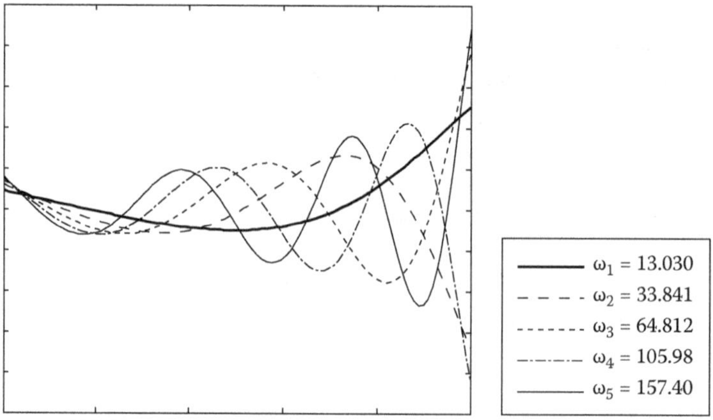

FIGURE 4.47b Vibration modes for F-F beam ($m = 8$, $n = 4$) at $c = 0.5$.

the C-C boundary conditions. His indices ν,μ correspond to our powers m,n in Table 4.41.

Notice that cases 1 and 3 in Table 4.41 belong to special cases of a family of Bessel function solutions given by Equation (4.82). Cases 2, 4, 6, 7 are special cases of exact Bessel solutions given by Cranch and Adler (1956) (see Section 4.5.1.4). These half orders of Bessel functions yield harmonic and exponential functions related to the uniform beam. The only case not already solved is case 5, where $m = n = 4$, which does have some physical meaning, and we shall analyze this in more detail.

For $m = 4$, $n = 4$, one can envision a beam with constant thickness, and the width decreases as $(1 - cx)^4$ (Figure 4.48), yielding

$$l = z^4, \quad r = z^4 \tag{4.111}$$

TABLE 4.41
Exponents for Isospectral Beams

Case	ν,μ	m,n	Subcase of
1	0, −1	3/2, −1/2	$m = n + 2$
2	−1, 2	4, −4	$m = -n$
3	−1, −1	5/2, ½	$m = n + 2$
4	−2, 3	5/2, −7/2	$m = n + 6$
5	−2, 0	4, 4	…
6	−3, 3	3/2, −9/2	$m = n + 6$
7	−3, 2	0, −8	$m = (n + 8)/3$

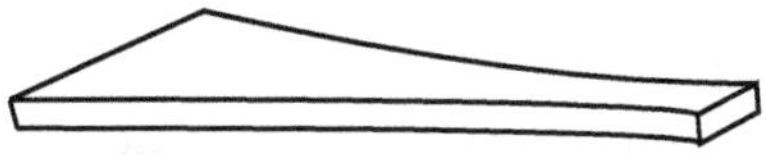

FIGURE 4.48 Beam with variable Width $m = 4$, $n = 4$.

Let

$$w = \frac{U(z)}{z^2} \tag{4.112}$$

Equation (4.78) reduces to

$$\frac{d^4U}{dx^4} - \omega^2 U = 0 \tag{4.113}$$

which governs a uniform beam with the same frequency ω, provided the ends are all clamped.

Based on the isospectral transform, we generalize the exact solutions for the beam described by Equation (4.111) with arbitrary boundary conditions. Using Equations (4.112) and (4.113), the exact solution is

$$w = \frac{1}{(1-cx)^2}\left\{C_1 \cos\left[\sqrt{\omega}x\right] + C_2 \sin\left[\sqrt{\omega}x\right] + C_3 \cosh\left[\sqrt{\omega}(1-x)\right] + C_4 \sinh\left[\sqrt{\omega}(1-x)\right]\right\} \tag{4.114}$$

This particular form is most suitable for evaluation. The boundary conditions then yield a 4 × 4 determinant for the frequencies. The results are shown in Table 4.42. Note that c does not affect the frequency of the C–C case.

4.5.4 Exponential-Type Solutions

Suppiger and Taleb (1956) and Cranch and Adler (1956) also considered exponential-type solutions, where

$$l = e^{-cx}, \quad r = e^{-cx} \tag{4.115}$$

in Equation (4.76). This represents a beam with constant height and exponentially decaying width similar to Figure 4.48. The complete solution is as follows (Wang and Wang 2012). Let

$$w = e^{\lambda x} \tag{4.116}$$

TABLE 4.42
Frequencies for a Beam with Constant Thickness and Fourth-Power Width ($m = 4$, $n = 4$)

c	0.1	0.3	0.5	0.7	0.9
	22.373	22.373	22.373	22.373	22.373
	61.673	61.673	61.673	61.673	61.673
C-C	120.90	120.90	120.90	120.90	120.90
	199.86	199.86	199.86	199.86	199.86
	298.56	298.56	298.56	298.56	298.56
	15.859	16.857	18.054	19.511	21.313
	50.394	51.476	53.005	55.294	58.970
C-P	104.68	105.82	107.55	110.43	115.98
	178.71	179.88	181.72	185.01	192.27
	272.47	273.66	275.58	279.17	287.94
	3.9944	5.3361	7.5577	11.483	18.129
	22.925	25.194	28.651	35.105	50.070
C-F	62.570	64.914	68.715	76.293	98.610
	121.78	124.19	128.19	136.42	164.22
	200.74	203.18	207.30	215.95	247.64
	9.8524	9.6759	9.1757	8.0294	5.2628
	39.495	39.660	40.116	41.090	42.969
P-P	88.852	89.115	89.868	91.620	95.606
	157.94	158.26	159.18	161.43	167.27
	246.77	247.12	248.14	250.75	258.20
	15.880	17.157	19.450	24.481	36.033
	50.427	51.884	54.702	61.165	80.841
P-F	104.72	106.26	109.34	116.54	142.28
	178.75	180.33	183.56	191.23	221.26
	272.51	274.13	277.45	285.45	318.50
	22.422	22.944	24.619	29.410	42.203
	61.739	62.434	64.565	70.549	91.190
F-F	120.98	121.74	124.07	130.68	156.96
	199.94	200.74	203.20	210.21	240.42
	298.63	299.46	302.01	309.31	342.29

The indicial equation is

$$\lambda^2(c-\lambda)^2 - \omega^2 = 0 \tag{4.117}$$

Let

$$\alpha = \frac{\sqrt{4\omega + c^2}}{2}, \qquad \beta = \frac{\sqrt{4\omega - c^2}}{2}, \qquad \gamma = \frac{\sqrt{c^2 - 4\omega}}{2} \tag{4.118}$$

If $\omega > c^2/4$, the general solution is

$$w = e^{cx/2}[C_1 \cosh(\alpha x) + C_2 \sinh(\alpha x) + C_3 \cos(\beta x) + C_4 \sin(\beta x)] \tag{4.119}$$

If $\omega = c^2/4$, the general solution is

$$w = e^{cx/2}[C_1 \cosh(cx/\sqrt{2}) + C_2 \sinh(cx/\sqrt{2}) + C_3 + C_4 x] \tag{4.120}$$

If $\omega < c^2/4$, the general solution is

$$w = e^{cx/2}[C_1 \cosh(\alpha x) + C_2 \sinh(\alpha x) + C_3 \cosh(\gamma x) + C_4 \sinh(\gamma x)] \tag{4.121}$$

Although $\omega > c^2/4$ in most cases, there are situations where $\omega < c^2/4$. Table 4.43 shows the results for some common end conditions. Mode shapes for a C-C beam with constant height and exponentially decaying width at $c = 0.5$ are shown in Figure 4.49a. The mode shapes for the F-F beam, having exactly the same frequencies, are shown in Figure 4.49b.

For the C-F beam with an end mass M on the smaller end, the boundary condition in view of Equation (4.15) is

$$w''(1) = 0, \qquad e^{-c} w'''(1) + \nu\omega^2 w(1) = 0 \tag{4.122}$$

where $\nu = M/\rho_0 L$. Table 4.44 shows the results.

TABLE 4.43
Frequencies of an Exponential Beam

c	0	0.1	0.5	1	2
C-C F-F	22.373	22.375	22.408	22.512	22.938
	61.673	61.675	61.720	61.860	62.423
	120.90	120.91	120.96	121.11	121.72
	199.86	199.86	199.91	200.07	200.72
	298.56	298.56	298.61	298.78	299.44
C-P P-F	15.418	15.524	15.955	16.512	17.720
	49.965	50.066	50.499	51.103	52.527
	104.25	104.35	104.79	105.42	106.95
	178.27	178.37	178.82	179.46	181.04
	272.03	272.13	272.58	273.23	274.85
P-P	9.8696	9.8686	9.8454	9.7729	9.4873
	39.478	39.479	39.501	39.570	39.852
	88.826	88.828	88.862	88.971	89.405
	157.91	157.92	157.96	158.08	158.60
	246.74	246.74	246.79	246.93	247.49

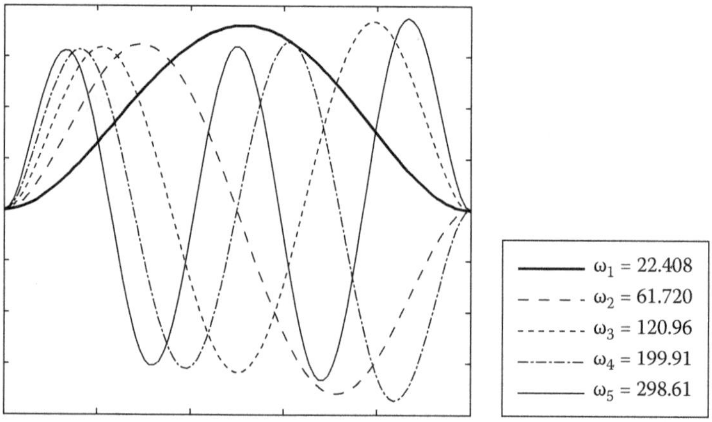

FIGURE 4.49a Mode shapes for C-C beam with constant height and exponentially decaying width at $c = 0.5$.

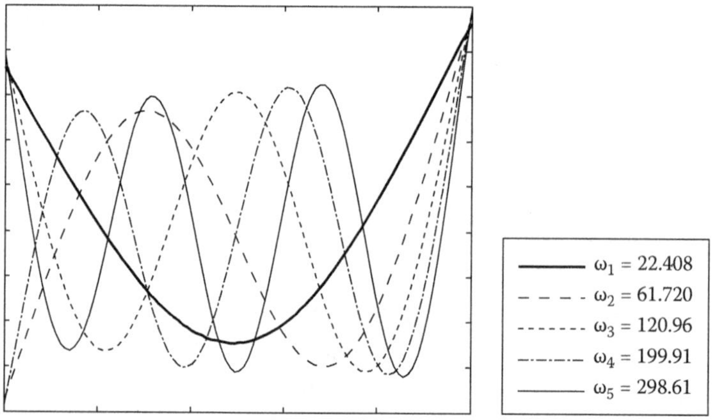

FIGURE 4.49b Mode shapes for F-F beam with constant height and exponentially decaying width at $c = 0.5$.

4.6 DISCUSSION

We did not include beam models other than the Euler-Bernoulli model. The most important is the Timoshenko beam, which admits some exact solutions.

As in other chapters, we excluded solutions that only yield a partial frequency spectrum, since such results would not be useful practically or be suitable as accuracy standards. We also excluded exact solutions limited to beams that taper to a sharp point at one end, beams with specific end loads, and beams with no physical application (see Section 4.5.1.4).

TABLE 4.44
Frequencies of the C-F Exponential Beam with Mass at Free End

$\nu \backslash c$	0.1	0.3	0.5	0.7	1	2
	3.6251	3.8516	4.0893	4.3386	4.7349	6.2626
	22.243	22.665	23.094	23.531	24.202	26.584
0	61.899	62.312	62.738	63.179	63.865	66.375
	121.10	121.52	121.95	122.40	123.10	125.69
	200.06	200.48	200.91	201.36	202.07	204.70
	3.0213	3.1221	3.2130	3.2925	3.3869	3.4540
	19.373	19.392	19.392	19.379	19.345	19.353
0.1	55.382	55.113	54.858	54.625	54.338	54.093
	110.46	109.99	109.57	109.21	108.79	108.45
	185.01	184.39	180.86	183.41	182.90	182.52
	1.5504	1.5332	1.5120	1.4873	1.4446	1.2691
	16.303	16.419	16.548	16.689	16.925	17.912
1	50.925	51.007	51.118	51.256	51.510	52.698
	105.22	105.30	105.41	105.55	105.82	107.11
	179.25	179.32	179.43	179.58	179.85	181.19
	0.5351	0.5224	0.5094	0.4963	0.4763	0.4084*
	15.611	15.812	16.018	16.229	16.555	17.740
10	50.157	50.354	50.563	50.786	51.144	52.544
	104.44	104.64	104.86	105.09	105.46	106.96
	178.46	178.66	178.88	179.12	179.50	181.06
	0.1709	0.1665	0.1622	0.1578	0.1513*	0.1294*
	15.533	15.746	15.961	16.180	16.516	17.722
100	50.075	50.284	50.505	50.737	51.107	52.529
	104.36	104.57	104.80	105.04	105.43	106.95
	178.38	178.60	178.82	179.07	179.47	181.04
	0.05408	0.05271	0.05133*	0.04994*	0.04786*	0.04092*
	15.525	15.739	15.955	16.175	16.512	17.721
1,000	50.067	50.278	50.499	50.732	51.103	52.527
	104.35	104.56	104.79	105.03	105.42	106.95
	178.37	178.59	178.82	179.07	179.46	181.04

Note: Values are from Equation (4.119) except for the values with asterisks, which are from Equation (4.121).

REFERENCES

Cranch, E. T., and A. A. Adler. 1956. Bending vibrations of variable section beams. *J Appl. Mech.* 23:103–8.

Galef, A. E. 1968. Bending frequencies of compressed beams. *J. Acoust. Soc. Am.* 44:643.

Gorman, D. J. 1975. *Free vibration analysis of beams and shafts*. New York: Wiley.

Gottlieb, H. P. W. 1987. Isospectral Euler-Bernoulli beams with continuous density and rigidity functions. *Proc. Roy. Soc. London* A413:235–50.

Karnovsky, I. A., and O. I. Lebed. 2004a. *Free vibrations of beams and frames.* New York: McGraw-Hill.

———. 2004b. *Non-classical vibrations of arches and beams.* New York: McGraw-Hill.

Kirchhoff, G. 1882. *Gesammelte abhandlungen.* Leipzig: Barth, Sec. 18.

Magrab, E. B. 1980. *Vibration of elastic structural members.* New York: Springer.

Murphy, G. M. 1960. *Ordinary differential equations and their solutions.* Princeton, NJ: Van Nostrand.

Sanger, D. J. 1968. Transverse vibration of a class of non-uniform beams. *J. Mech. Eng. Sci.* 10:111–20.

Suppiger, E. W., and N. J. Taleb. 1956. Free lateral vibration of beams with variable cross section. *Zeit. Angew. Math. Phys.* 7:501–20.

Wang, C. Y., and C. M. Wang. 2001. Vibration of a beam with an internal hinge. *Int. J. Struct. Stab. Dyn.* 1:163–67.

———. 2012. Exact vibration solution for exponentially tapered cantilever with tip mass. *J. Vibr. Acoust.* 134 (4): 041012.

———. 2013. Exact solutions for the vibration of a class of non-uniform beams. *J. Eng. Mech:* 139 (7): to appear.

5 Vibration of Isotropic Plates

5.1 INTRODUCTION

In this chapter, the governing equations for free vibration of isotropic plates of uniform thickness are presented, and the exact vibration solutions for plates of various geometries and boundary conditions are given. For a detailed derivation of the governing equations of plate motion, the reader may consult standard textbooks on plates such as Timoshenko and Woinowsky-Krieger (1959), Szilard (1974), Ugural (1981), Liew et al. (1998), Soedel (2004), and Reddy (2007).

5.2 GOVERNING EQUATIONS AND BOUNDARY CONDITIONS FOR VIBRATING THIN PLATES

The classical thin-plate theory is based on the Kirchhoff (1850) hypothesis:

- Straight lines perpendicular to the mid-surface of the plate (i.e., the transverse normals) before deformation remain straight after deformation.
- The transverse normals rotate such that they remain perpendicular to the mid-surface after deformation.
- The transverse normals do not experience any elongation (i.e., they are inextensible).

The consequence of the Kirchhoff hypothesis is that the transverse strains are zero, and hence the transverse stresses do not appear in the formulation of the classical thin-plate model.

Consider an elastic, isotropic, thin plate of uniform thickness h, Young's modulus E, Poisson ratio ν, and mass density (per unit volume) ρ. By adopting the rectangular coordinates system as shown in Figure 5.1 and assuming the natural vibration solution to be periodic, i.e.,

$$\overline{w}(x,y,t) = w(x,y)e^{i\omega t} \tag{5.1}$$

where w is the transverse displacement (i.e., displacement in the z-direction), $i=\sqrt{-1}$, ω is the frequency of natural vibration associated with the mode shape w, and t is the time. The governing equation of motion of a vibrating thin plate may be expressed as

$$\{D\nabla^4 w(x,y) - \rho h\omega^2 w(x,y)\}e^{i\omega t} = 0 \tag{5.2}$$

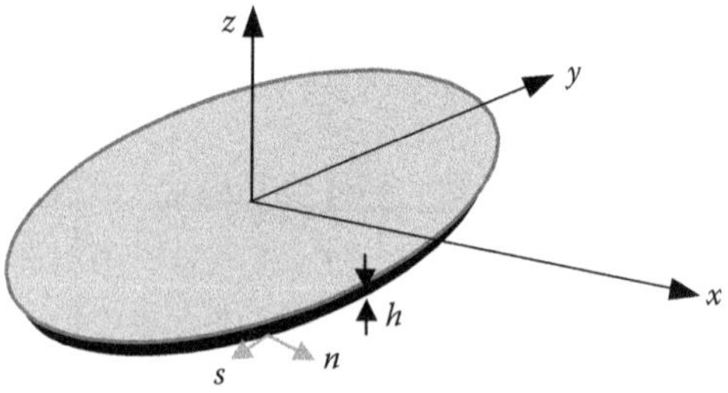

FIGURE 5.1 Plate showing the rectangular coordinates system.

In Equation (5.2), the biharmonic operator $\nabla^4(\bullet) = \partial^4(\bullet)/\partial x^4 + 2\partial^4(\bullet)/\partial x^2\partial y^2 + \partial^4(\bullet)/\partial y^4$, and $D = Eh^3/[12(1-\nu^2)]$ is the flexural rigidity. Since Equation (5.2) is valid for any time t, we have

$$\nabla^4 w(x,y) - k^2 w(x,y) = 0, \quad \text{where} \quad k^2 = \omega^2 \frac{\rho h}{D} \tag{5.3}$$

Note that the aforementioned governing equation for the classical thin-plate model does not include the effect of rotary inertia. To allow for this effect, the left-hand side of Equation (5.3) has to be augmented by $+\ (\rho h^3/12D)\nabla^2 w$, where the Laplacian operator $\nabla^2(\bullet) = \partial^2(\bullet)/\partial x^2 + \partial^2(\bullet)/\partial y^2$.

The problem at hand is to determine the natural frequencies ω of the thin plate such that Equation (5.3) has a nontrivial solution w. In order to solve Equation (5.3), it is necessary to specify the boundary conditions for the plate.

The boundary conditions may be classified as essential or natural. Essential boundary conditions involve deflection w and slope $\partial w/\partial n$, where n is the direction normal to the plate edge (see Figure 5.1). Natural boundary conditions involve the normal bending moment M_{nn} and the effective shear force V_n, where (for a straight edge)

$$M_{nn} = -D\left(\frac{\partial^2 w}{\partial n^2} + \nu \frac{\partial^2 w}{\partial s^2}\right) \tag{5.4a}$$

and

$$V_n = -D\left(\frac{\partial^3 w}{\partial n^3} + (2-\nu)\frac{\partial^3 w}{\partial n\,\partial s^2}\right) \tag{5.4b}$$

where s is the direction tangential to the plate edge (see Figure 5.1). The common types of boundary conditions for a plate are (a) a free edge, (b) a clamped edge, and (c) a simply supported edge. At a free edge, the boundary conditions are $M_{nn} = 0$ and $V_n = 0$. At a clamped edge, $w = 0$ and $\partial w/\partial n = 0$, and at a simply supported edge, $w = 0$ and $M_{nn} = 0$.

For circular and annular plates, the analysis is more expediently carried out in the polar coordinates (r,θ). The governing equation of plate motion, in polar coordinates, takes the form of

$$\nabla^4 w(r,\theta) - k^2 w(r,\theta) = 0 \tag{5.5}$$

where $k^2 = \omega^2 \rho h/D$, the biharmonic operator $\nabla^4(\bullet) = \nabla^2\nabla^2(\bullet)$, and the Laplacian operator $\nabla^2(\bullet) = (1/r)\partial(r\,\partial(\bullet)/\partial r)/\partial r + (1/r^2)\partial^2(\bullet)/\partial\theta^2$. At a free edge, the boundary conditions are $M_{rr} = 0$ and $V_r = 0$. At a clamped edge, $w = 0$ and $\partial w/\partial r = 0$, and at a simply supported edge, $w = 0$ and $M_{rr} = 0$. The bending moment M_{rr} and the effective shear force V_r are, respectively, given by

$$M_{rr} = -D\left[\frac{\partial^2 w}{\partial r^2} + \nu\left(\frac{1}{r^2}\frac{\partial^2 w}{\partial\theta^2} + \frac{1}{r}\frac{\partial w}{\partial r}\right)\right] \tag{5.6a}$$

and

$$V_r = -D\left[\frac{\partial}{\partial r}\left(\nabla^2 w\right) + \frac{1-\nu}{r}\frac{\partial}{\partial\theta}\left(\frac{1}{r}\frac{\partial^2 w}{\partial r\,\partial\theta} - \frac{1}{r^2}\frac{\partial w}{\partial\theta}\right)\right] \tag{5.6b}$$

5.3 EXACT VIBRATION SOLUTIONS FOR THIN PLATES

Except for some plate shapes and boundary conditions that will be discussed here for cases where exact vibration solutions are obtainable, the vibration problems of plates are usually solved via numerical techniques such as the finite element method, the finite difference method, the differential quadrature method, and the Ritz method. (For details of these methods, one may refer to the book edited by Shanmugam and Wang [2007].)

5.3.1 Rectangular Plates with Four Edges Simply Supported

Consider a rectangular plate of length a and width b with its four edges simply supported, as shown in Figure 5.2.

The solution (or vibration mode shape) that satisfies the boundary conditions $w = 0$ and $M_{nn} = 0$ for all four edges is given by Navier (1823)

$$w(x,y) = A_{mn}\sin\frac{m\pi x}{a}\sin\frac{n\pi y}{b} \tag{5.7}$$

where A_{mn} are the unknown coefficients (or amplitudes of vibration), m is the number of half waves in the x-direction, and n is the number of half waves in the y-direction. By substituting Equation (5.7) into Equation (5.3), we obtain the exact vibration frequency solution for simply supported rectangular plates in terms of m and n

$$\omega_{mn} = \sqrt{\frac{D}{\rho h}}\left[\left(\frac{m\pi}{a}\right)^2 + \left(\frac{n\pi}{b}\right)^2\right] \tag{5.8}$$

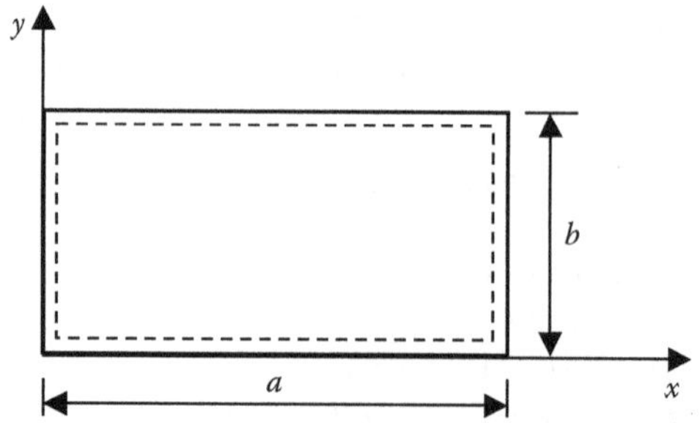

FIGURE 5.2 Rectangular plate with all edges simply supported.

The fundamental frequency of vibration is obtained by minimizing ω_{mn} with respect to m and n. For a square plate ($a = b$), it can be readily shown that the fundamental frequency is $\omega_{11} = \frac{2\pi^2}{b^2}\sqrt{\frac{D}{\rho h}}$. Table 5.1 presents the first six natural frequencies values $\bar{\omega}_{mn} = \omega_{mn} b^2 \sqrt{\rho h / D}$ and the corresponding mode shapes for rectangular plates with all edges simply supported. The number of half waves (m, n) in the x- and y-directions is given in parentheses.

5.3.2 Rectangular Plates with Two Parallel Sides Simply Supported

Exact vibration solutions are also possible for rectangular plates with two parallel sides simply supported, while the other two sides can take any combination of clamped, simply supported, and free boundary conditions. As pointed out by Levy (1899), the partial differential equation for such plates may be converted into an ordinary differential equation, since the two simply supported parallel edges (say, parallel to x-axis) allow the mode shape to be made separable in the form

$$w(x,y) = W_n(x)\sin\frac{n\pi y}{b} \tag{5.9}$$

where it satisfies the simply supported boundary conditions $w = 0$ and $M_{yy} = 0$ at $y = 0$ edge and $y = b$ edge.

By substituting Equation (5.9) into Equation (5.3), we obtain the following fourth-order ordinary differential equation:

$$\frac{d^4 W_n}{dx^4} - 2\left(\frac{n\pi}{b}\right)^2 \frac{d^2 W_n}{dx^2} + \left[\left(\frac{n\pi}{b}\right)^4 - k^2\right] W_n = 0, \quad \text{where} \quad k^2 = \omega^2 \frac{\rho h}{D} \tag{5.10}$$

The form of the solution to Equation (5.10) depends on the nature of the roots λ of the equation

$$\lambda^4 - 2\left(\frac{n\pi}{b}\right)^2 \lambda^2 + \left[\left(\frac{n\pi}{b}\right)^4 - k^2\right] = 0 \tag{5.11}$$

TABLE 5.1
Natural Frequencies $\bar{\omega}_{mm} = \omega_{mn} b^2\sqrt{\rho h/D}$ and Mode Shapes for Rectangular Plates with Simply Supported Edges

a/b	Mode 1	Mode 2	Mode 3	Mode 4	Mode 5	Mode 6
0.5	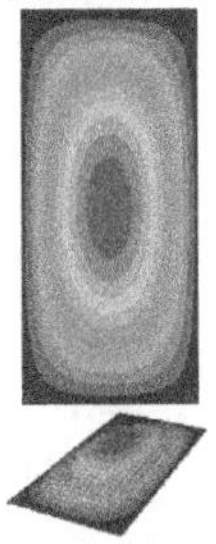 49.348 (1, 1)	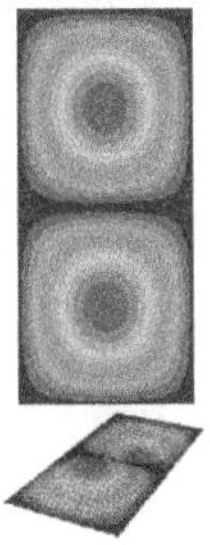 78.957 (1, 2)	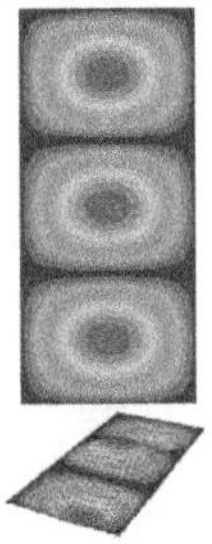 128.305 (1, 3)	 167.783 (2, 1)	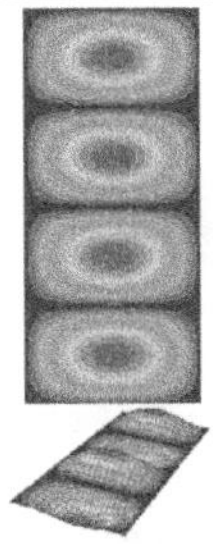 197.392 (1, 4)	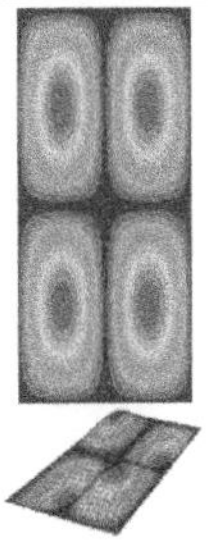 197.392 (2, 2)
1.0	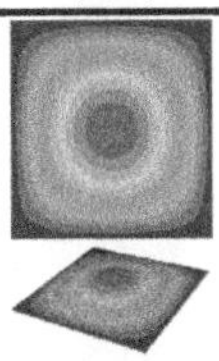 19.739 (1, 1)	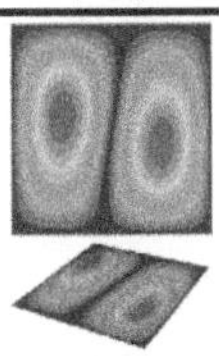 49.348 (1, 2)	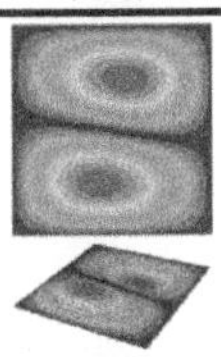 49.348 (2, 1)	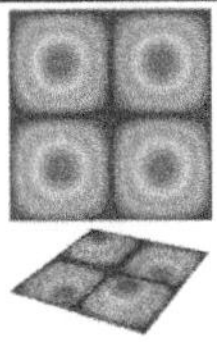 78.957 (2, 2)	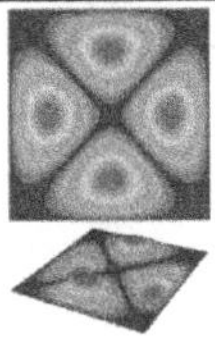 98.696 (1, 3)	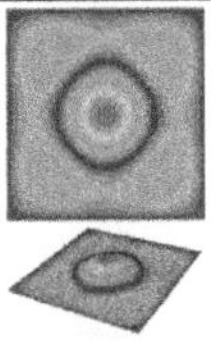 98.696 (3, 1)
2.0	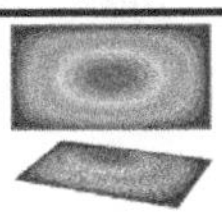 12.337 (1, 1)	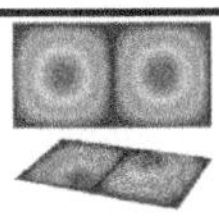 19.739 (2, 1)	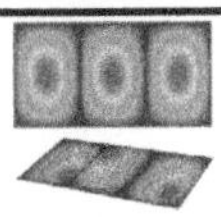 32.076 (3, 1)	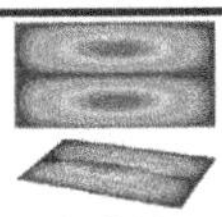 41.946 (1, 2)	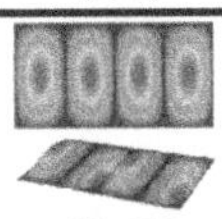 49.348 (4, 1)	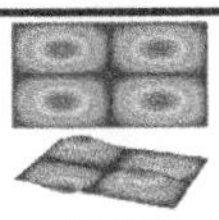 49.348 (2, 2)

There are two distinct cases. The first case is when $k^2 \geq (n\pi/b)^4$, and the second case is when $k^2 < (n\pi/b)^4$. The general solution for Equation (5.10) when $k^2 \geq (n\pi/b)^4$ is given by Voigt (1893)

$$W_n(x) = A_1 \cosh\lambda_1 x + A_2 \sinh\lambda_1 x + A_3 \cos\lambda_2 x + A_4 \sin\lambda_2 x \tag{5.12}$$

whereas the solution for Equation (5.10) when $k^2 < (n\pi/b)^4$ is given by

$$W_n(x) = A_1 \cosh\lambda_1 x + A_2 \sinh\lambda_1 x + A_3 \cosh\lambda_2 x + A_4 \sinh\lambda_2 x \tag{5.13}$$

where A_i, $i = 1,2,3,4$ are the integration constants and

$$(\lambda_1)^2 = k + \left(\frac{n\pi}{b}\right)^2 \tag{5.14}$$

$$(\lambda_2)^2 = \left| k - \left(\frac{n\pi}{b}\right)^2 \right| \tag{5.15}$$

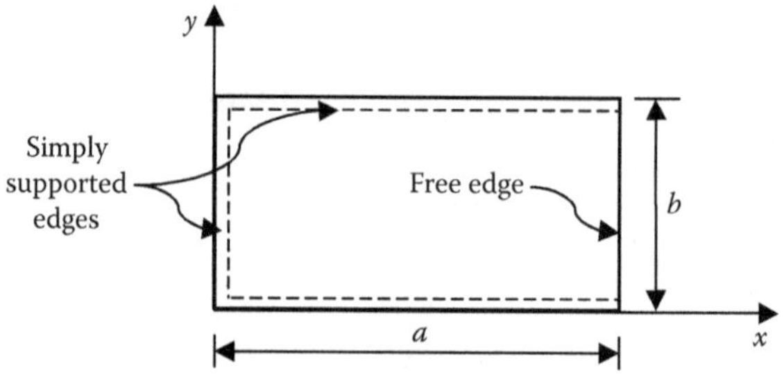

FIGURE 5.3 Rectangular plate with three simply supported edges and one free edge.

Consider a rectangular plate with three edges—$x = 0$, $y = 0$, and $y = b$—that are simply supported, whereas the edge $x = a$ is free, as shown in Figure 5.3. The boundary conditions on the edges are

$$w = 0 \text{ and } M_{yy} = -D\left(\frac{\partial^2 w}{\partial y^2} + \nu\frac{\partial^2 w}{\partial x^2}\right) = 0 \text{ for simply supported edges}$$

$$y = 0 \text{ and } y = b \tag{5.16}$$

$$w = 0 \text{ and } M_{xx} = -D\left(\frac{\partial^2 w}{\partial x^2} + \nu\frac{\partial^2 w}{\partial y^2}\right) = 0 \text{ for simply supported edge } x = 0 \tag{5.17}$$

$$M_{xx} = 0 \quad \text{and} \quad V_x = -D\left(\frac{\partial^3 w}{\partial x^3} + (2-\nu)\frac{\partial^3 w}{\partial y^2\,\partial x}\right) = 0 \quad \text{for free edge } x = \text{a} \tag{5.18}$$

By substituting Equation (5.12) into Equations (5.17) and (5.18) and solving the equations for nontrivial solutions, we obtain the following exact characteristic equation:

$$\lambda_2\left[k + (1-\nu)\left(\frac{n\pi}{b}\right)^2\right]^2 \sinh(\lambda_1 a)\cos(\lambda_2 a) - \lambda_1\left[k - (1-\nu)\left(\frac{n\pi}{b}\right)^2\right]^2$$

$$\times\cosh(\lambda_1 a)\sin(\lambda_2 a) = 0 \tag{5.19}$$

The substitution of Equation (5.13) into Equations (5.17) and (5.18) yields the following exact characteristic equation

$$\lambda_2\left[k + (1-\nu)\left(\frac{n\pi}{b}\right)^2\right]^2 \sinh(\lambda_1 a)\cosh(\lambda_2 a) - \lambda_1\left[k - (1-\nu)\left(\frac{n\pi}{b}\right)^2\right]^2$$

$$\times\cosh(\lambda_1 a)\sinh(\lambda_2 a) = 0 \tag{5.20}$$

Equations (5.19) and (5.20) are to be solved by using, for example, the false-position method for the roots ω for a given number of half waves n in the y-direction with the prescribed geometrical and material properties of the plate.

The mode shape for the case $k^2 \geq (n\pi/b)^4$ is given by

$$w(x,y) = k[\sinh(\lambda_1 a)\sin(\lambda_2 x) + \sinh(\lambda_1 x)\sin(\lambda_2 a)]\sin\frac{n\pi y}{b} \tag{5.21}$$

and for the case $k^2 < (n\pi/b)^4$, the mode shape is given by

$$w(x,y) = k[\sinh(\lambda_1 a)\sinh(\lambda_2 x) + \sinh(\lambda_1 x)\sinh(\lambda_2 a)]\sin\frac{n\pi y}{b} \tag{5.22}$$

Table 5.2 presents the first six natural frequencies $\bar{\omega} = \omega b^2\sqrt{\rho h/D}$ and modes shapes for rectangular plates with three edges simply supported and one edge free. The number of half waves (n) is given in parentheses, and the Poisson ratio is taken as $\nu = 0.3$. All the frequency values presented in Table 5.2 have been obtained from solving Equation (5.19).

Next, we consider a rectangular plate with two edges ($y = 0$, $y = b$) simply supported, the edge $x = 0$ clamped, and the remaining edge $x = a$ free, as shown in Figure 5.4.

TABLE 5.2
Natural Frequencies $\bar{\omega} = \omega b^2\sqrt{\rho h/D}$ for Rectangular Plate with Three Edges Simply Supported and One Edge Free

a/b	Mode 1	Mode 2	Mode 3	Mode 4	Mode 5	Mode 6
0.5	 16.135 (1)	 46.738 (2)	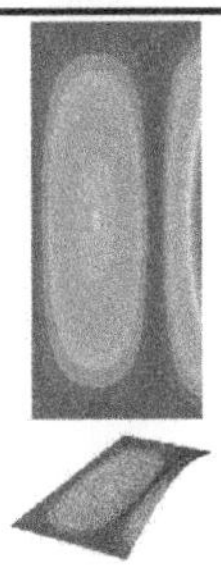 75.283 (1)	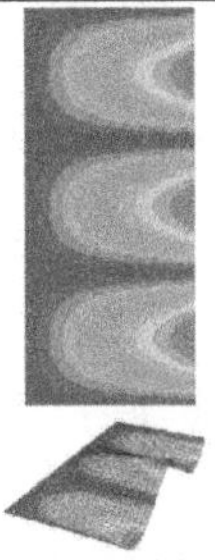 96.041 (3)	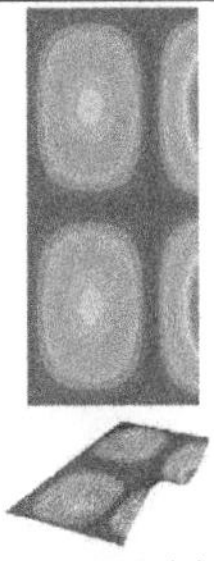 111.025 (2)	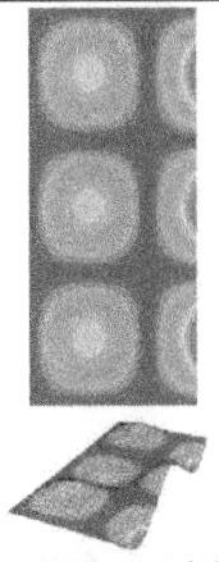 164.696 (3)
1.0	 11.685 (1)	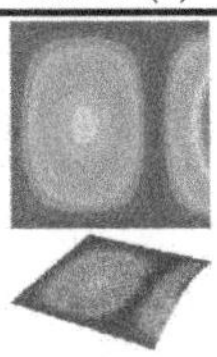 27.756 (1)	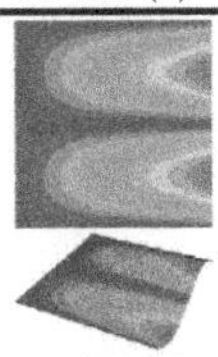 41.197 (2)	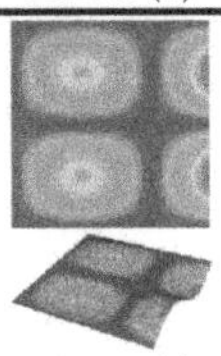 59.066 (2)	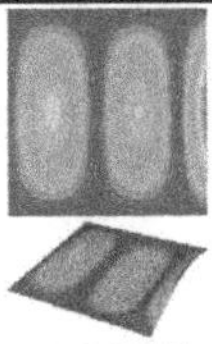 61.861 (1)	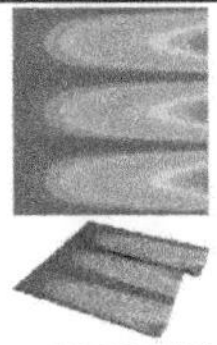 90.294 (3)
2.0	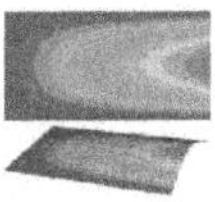 10.299 (1)	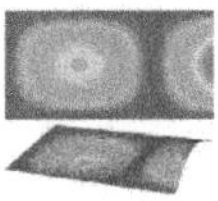 14.766 (1)	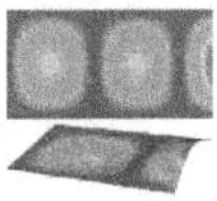 23.621 (1)	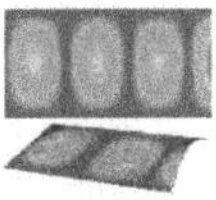 37.127 (1)	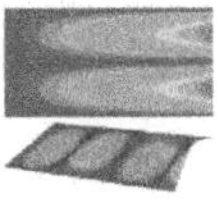 39.770 (2)	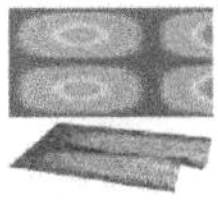 44.524 (2)

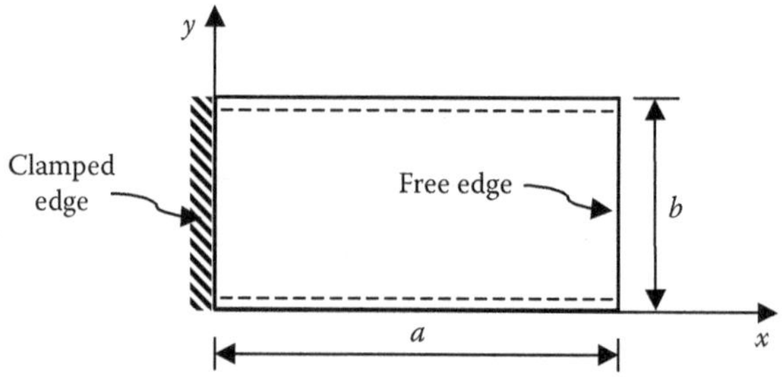

FIGURE 5.4 Rectangular plate with two simply supported edges, a clamped edge, and a free edge

The boundary conditions on the edges are

$$w = 0 \text{ and } M_{yy} = -D\left(\frac{\partial^2 w}{\partial y^2} + \nu\frac{\partial^2 w}{\partial x^2}\right) = 0 \text{ for simply supported edges } y = 0 \text{ and } y = b \tag{5.23}$$

$$w = 0 \text{ and } \frac{\partial w}{\partial x} = 0 \quad \text{for clamped edge } x = 0 \tag{5.24}$$

$$M_{xx} = -D\left(\frac{\partial^2 w}{\partial x^2} + \nu\frac{\partial^2 w}{\partial y^2}\right) = 0 \quad \text{and} \quad V_x = -D\left(\frac{\partial^3 w}{\partial x^3} + (2-\nu)\frac{\partial^3 w}{\partial y^2\,\partial x}\right) = 0$$

for free edge $x = a$ (5.25)

The substitution of Equation (5.12) into Equations (5.24) and (5.25) yields the following exact characteristic equations

$$\lambda_1\lambda_2\left[k^2 - (1-\nu)^2\left(\frac{n\pi}{b}\right)^4\right] + \left(\frac{n\pi}{b}\right)^2\left[(1-2\nu)k^2 - (1-\nu)^2\left(\frac{n\pi}{b}\right)^4\right]\sinh(\lambda_1 a)\sin(\lambda_2 a)$$
$$+\lambda_1\lambda_2\left[k^2 + (1-\nu)^2\left(\frac{n\pi}{b}\right)^4\right]\cosh(\lambda_1 a)\cos(\lambda_2 a) = 0 \quad \text{for } k^2 \geq \left(\frac{n\pi}{b}\right)^4 \tag{5.26a}$$

$$\lambda_1\lambda_2\left[k^2 - (1-\nu)^2\left(\frac{n\pi}{b}\right)^4\right] + \left(\frac{n\pi}{b}\right)^2\left[(1-2\nu)k^2 - (1-\nu)^2\left(\frac{n\pi}{b}\right)^4\right]\sinh(\lambda_1 a)\sinh(\lambda_2 a)$$
$$+\lambda_1\lambda_2\left[k^2 + (1-\nu)^2\left(\frac{n\pi}{b}\right)^4\right]\cosh(\lambda_1 a)\cosh(\lambda_2 a) = 0 \quad \text{for } k^2 < \left(\frac{n\pi}{b}\right)^4 \tag{5.26b}$$

Table 5.3 presents the first six natural frequencies values $\bar{\omega} = \omega b^2\sqrt{\rho h/D}$ and mode shapes for rectangular plates with two parallel edges simply supported, one edge clamped, and one edge free. The number of half waves (n) is given in

TABLE 5.3
Natural Frequencies $\bar{\omega} = \omega b^2 \sqrt{\rho h / D}$ for Rectangular Plate with Two Parallel Edges Simply Supported, One Edge Clamped, and One Edge Free

a/*b*	Mode 1	Mode 2	Mode 3	Mode 4	Mode 5	Mode 6
0.5	 22.815 (1)	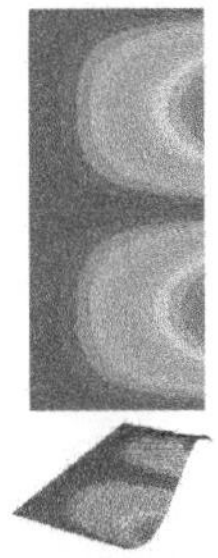 50.748 (2)	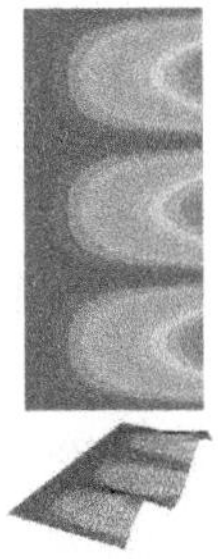 98.778 (3)	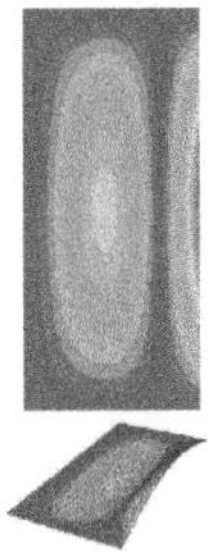 99.777 (1)	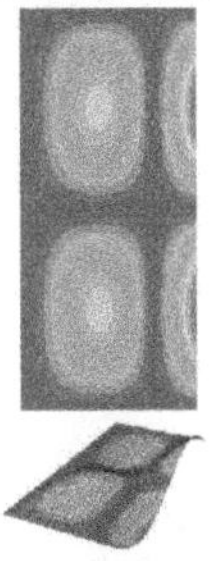 132.255 (2)	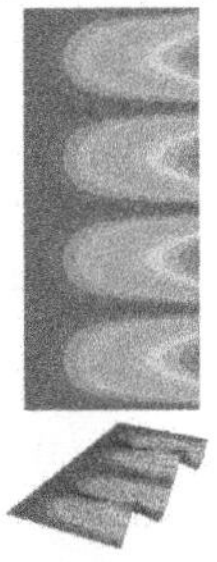 166.812 (4)
1.0	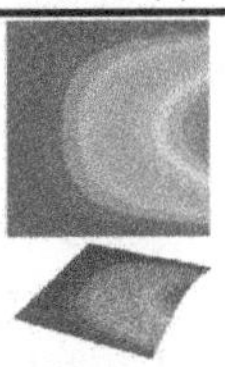 12.687 (1)	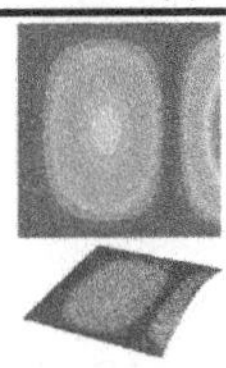 33.064 (1)	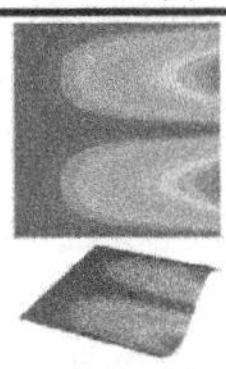 41.702 (2)	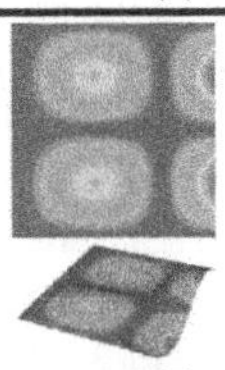 63.014 (2)	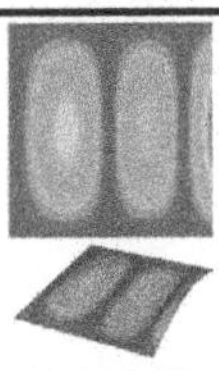 72.395 (1)	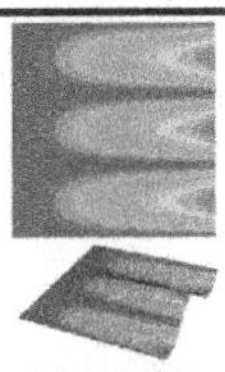 90.610 (3)
2.0	 10.426 (1)	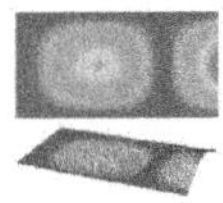 15.753 (1)	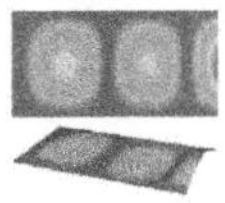 25.790 (1)	 39.826 (2)	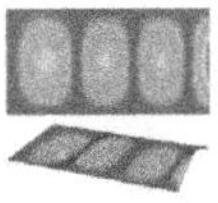 40.592 (1)	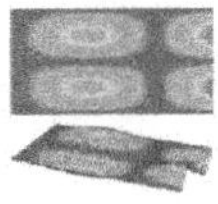 45.106 (2)

parentheses, and the Poisson ratio is taken as ν = 0.3. All the frequency values presented in Table 5.3 are obtained from solving Equation (5.26a).

Let us consider a rectangular plate with two parallel edges ($y = 0$, $y = b$) simply supported and the other two parallel edges ($x = 0$, $x = a$) clamped, as shown in Figure 5.5. The boundary conditions on the edges are

$$w = 0 \text{ and } M_{yy} = -D\left(\frac{\partial^2 w}{\partial y^2} + \nu \frac{\partial^2 w}{\partial x^2}\right) = 0 \text{ for simply supported edges } y = 0 \text{ and } y = b \tag{5.27}$$

$$w = 0 \text{ and } \frac{\partial w}{\partial x} = 0 \quad \text{for clamped edges } x = 0 \text{ and } x = a \tag{5.28}$$

The substitution of Equation (5.12) into Equation (5.28) yields the following exact characteristic equation

$$2\lambda_1\lambda_2[1 - \cosh(\lambda_1 a)\cos(\lambda_2 a)] + \left(\lambda_1^2 - \lambda_2^2\right)\sinh(\lambda_1 a)\sin(\lambda_2 a) = 0 \tag{5.29}$$

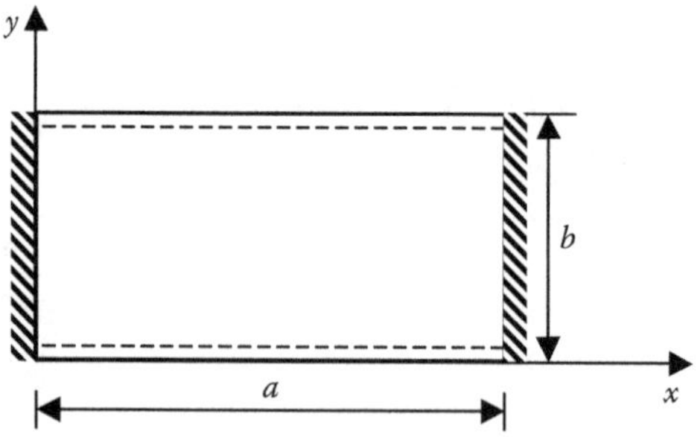

FIGURE 5.5 Rectangular plate with two simply supported edges and two clamped edges.

whereas the substitution of Equation (5.13) into Equation (5.28) yields the following characteristic equation

$$2\lambda_1\lambda_2[1-\cosh(\lambda_1 a)\cosh(\lambda_2 a)]+\left(\lambda_1^2+\lambda_2^2\right)\sinh(\lambda_1 a)\sinh(\lambda_2 a)=0 \quad (5.30)$$

Table 5.4 presents the first six natural frequencies values $\bar{\omega}=\omega b^2\sqrt{\rho h/D}$ and mode shapes for rectangular plates with two parallel edges simply supported and the other two parallel edges clamped. The number of half waves (n) is given in parentheses. All the frequencies presented in Table 5.4 are obtained from solving Equation (5.29).

For the next case, consider a rectangular plate with three edges ($y = 0$, $y = b$, and $x = a$) simply supported and the other remaining edge ($x = 0$) clamped as shown in Figure 5.6. The boundary conditions on the edges are

$$w=0 \text{ and } M_{yy}=0 \text{ for simply supported edges } y=0,\ y=b \quad (5.31a)$$

$$w=0 \text{ and } M_{xx}=0 \text{ for simply supported edge } x=a \quad (5.31b)$$

$$w=0 \text{ and } \frac{\partial w}{\partial x}=0 \text{ for clamped edge } x=0 \quad (5.32)$$

The substitution of Equation (5.12) into Equations (5.31b) and (5.32) yields the following exact characteristic equation

$$\lambda_1\cosh(\lambda_1 a)\sin(\lambda_2 a)-\lambda_2\sinh(\lambda_1 a)\cos(\lambda_2 a)=0 \quad (5.33)$$

The characteristic equation obtained from using Equation (5.13) is

$$\lambda_1\cosh(\lambda_1 a)\sinh(\lambda_2 a)-\lambda_2\sinh(\lambda_1 a)\cosh(\lambda_2 a)=0 \quad (5.34)$$

Table 5.5 presents the first six natural frequencies values and mode shapes for rectangular plates with three edges simply supported and one edge clamped. The number of half waves (n) is given in parentheses. All the frequency values presented in Table 5.5 are obtained from solving Equation (5.33).

TABLE 5.4
Natural Frequencies $\bar{\omega} = \omega b^2 \sqrt{\rho h/D}$ for Rectangular Plate with Two Parallel Edges Simply Supported and the Other Two Parallel Edges Clamped

a/b	Mode 1	Mode 2	Mode 3	Mode 4	Mode 5	Mode 6
0.5	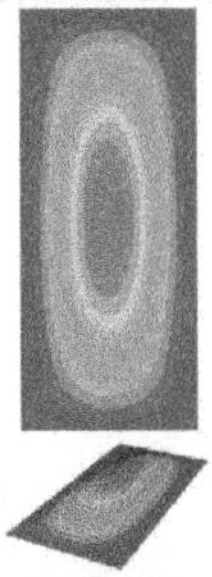 95.263 (1)	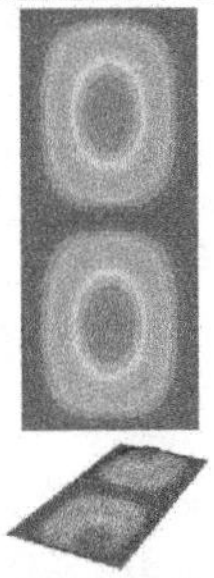 115.803 (2)	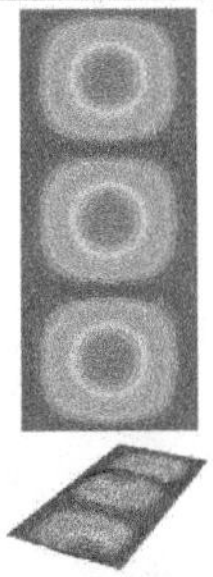 156.357 (3)	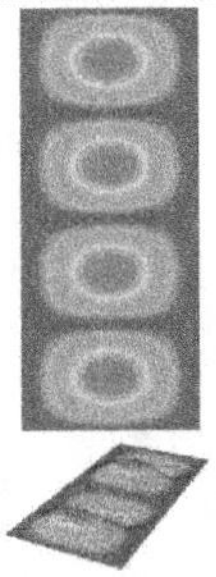 218.972 (4)	 254.138 (1)	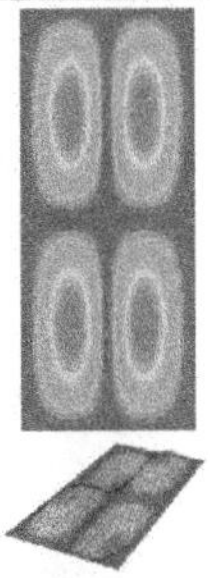 277.308 (2)
1.0	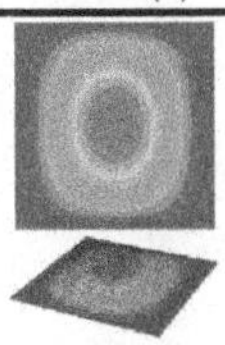 28.951 (1)	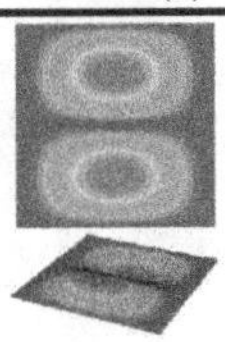 54.743 (2)	 69.327 (1)	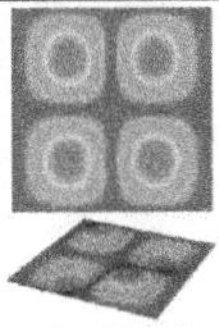 94.585 (2)	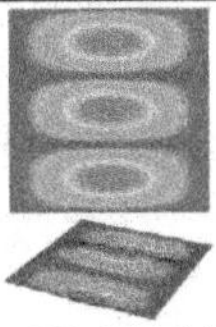 102.216 (3)	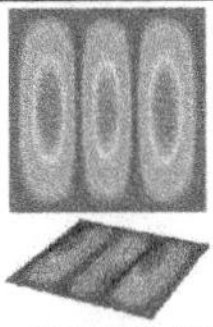 129.100 (1)
2.0	 13.686 (1)	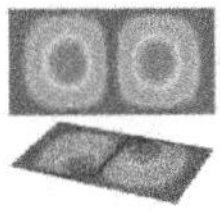 23.646 (1)	 38.694 (1)	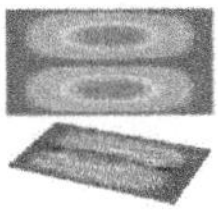 42.587 (2)	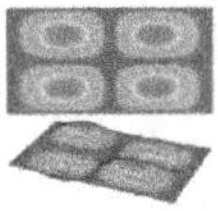 51.674 (2)	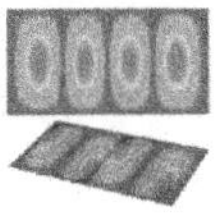 58.646 (1)

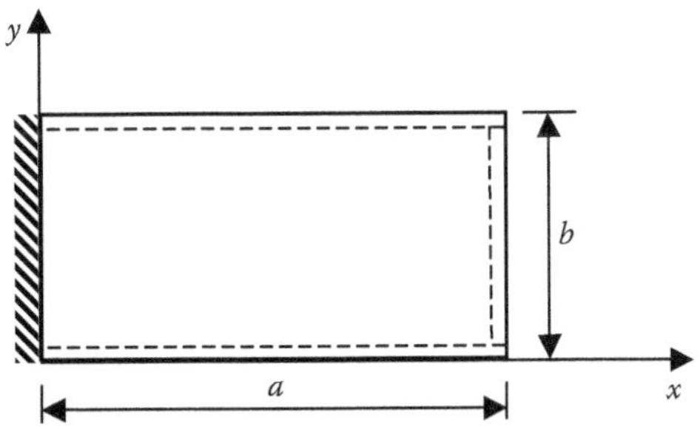

FIGURE 5.6 Rectangular plate with three simply supported edges and a clamped edge.

TABLE 5.5
Natural Frequencies $\bar{\omega} = \omega b^2 \sqrt{\rho h/D}$ for Rectangular Plate with Three Edges Simply Supported and One Edge Clamped

a/b	Mode 1	Mode 2	Mode 3	Mode 4	Mode 5	Mode 6
0.5	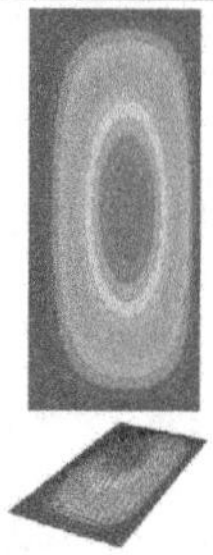 69.329 (1)	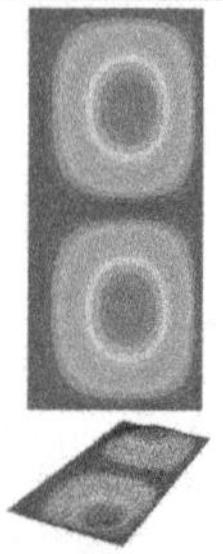 94.581 (2)	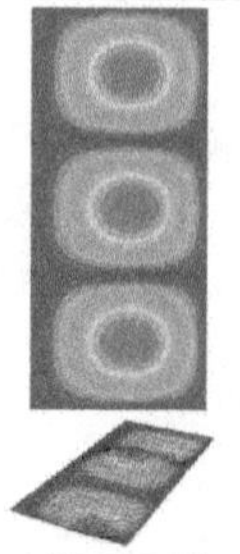 140.203 (3)	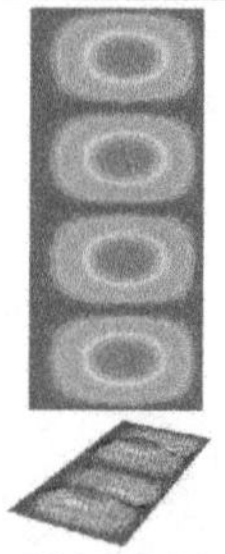 206.698 (4)	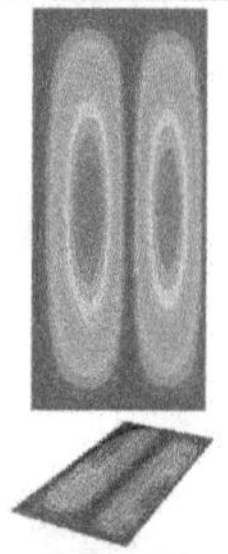 208.407 (1)	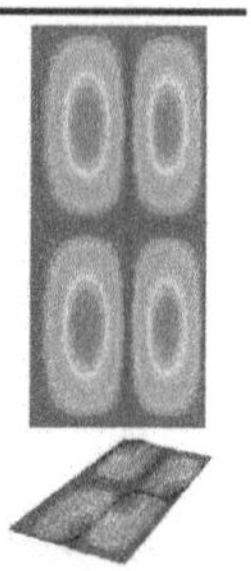 234.589 (2)
1.0	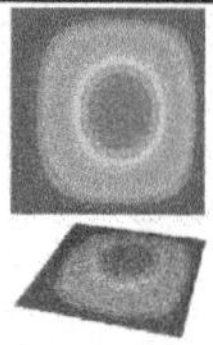 23.646 (1)	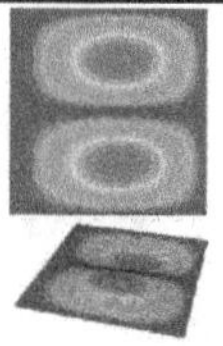 51.674 (2)	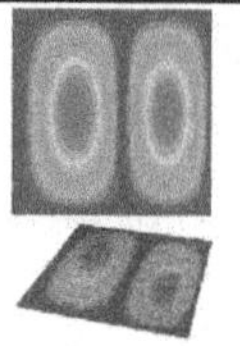 58.646 (1)	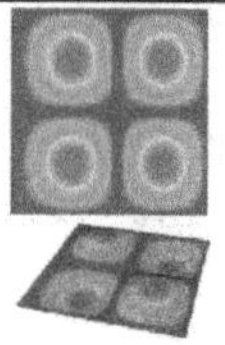 86.130 (2)	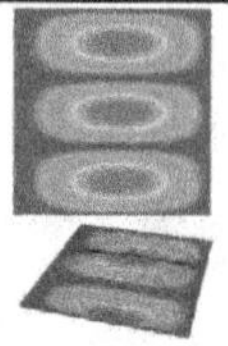 100.267 (3)	 113.229 (1)
2.0	 12.918 (1)	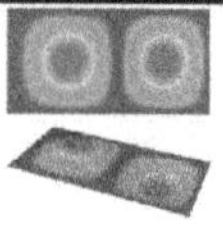 21.533 (1)	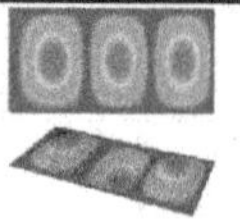 35.211 (1)	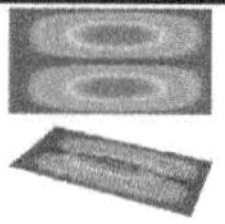 42.239 (2)	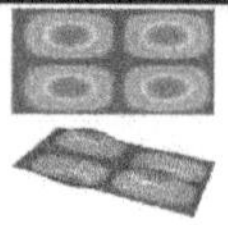 50.431 (2)	 53.823 (1)

Finally, we consider a rectangular plate with two parallel edges ($y = 0$, $y = b$) simply supported, while the other two parallel edges ($x = 0$, $x = a$) are free, as shown in Figure 5.7. The boundary conditions on the edges are

$$w = 0 \text{ and } M_{yy} = -D\left(\frac{\partial^2 w}{\partial y^2} + \nu\frac{\partial^2 w}{\partial x^2}\right) = 0 \text{ for simply supported edges } y = 0,\ y = b, \tag{5.35}$$

$$M_{xx} = 0 \text{ and } V_x = -D\left(\frac{\partial^3 w}{\partial x^3} + (2-\nu)\frac{\partial^3 w}{\partial y^2\,\partial x}\right) = 0 \text{ for free edges } x = 0,\ x = a \tag{5.36}$$

The substitution of Equation (5.12) into Equation (5.36) yields the following exact characteristic equation

$$\left\{\lambda_1^2\left[k - (1-\nu)\left(\frac{n\pi}{b}\right)^2\right]^4 - \lambda_2^2\left[k + (1-\nu)\left(\frac{n\pi}{b}\right)^2\right]^4\right\}\sinh(\lambda_1 a)\sin(\lambda_2 a)$$

$$+ 2\lambda_1\lambda_2\left[(\lambda_1\lambda_2)^2 - \nu(2-\nu)\left(\frac{n\pi}{b}\right)^4\right]^2(1 - \cosh\lambda_1 a\cos\lambda_2 a) = 0 \tag{5.37a}$$

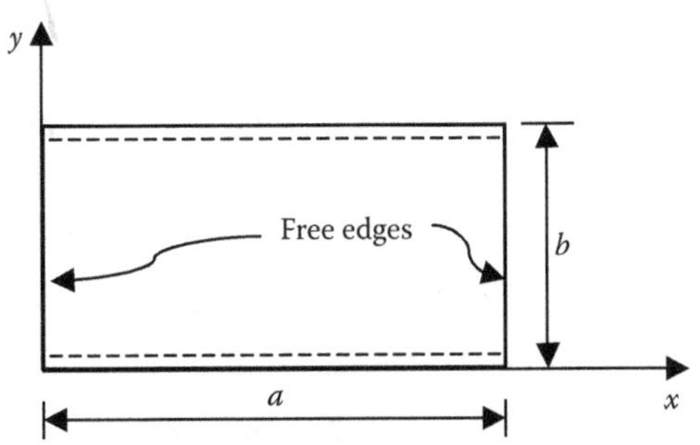

FIGURE 5.7 Rectangular plate with two simply supported edges and two free edges.

whereas the substitution of Equation (5.13) into Equation (5.36) yields the following characteristic equation:

$$\left\{\lambda_1^2\left[k-(1-\nu)\left(\frac{n\pi}{b}\right)^2\right]^4+\lambda_2^2\left[k+(1-\nu)\left(\frac{n\pi}{b}\right)^2\right]^4\right\}\sinh(\lambda_1 a)\sinh(\lambda_2 a)$$

$$+2\lambda_1\lambda_2\left[(\lambda_1\lambda_2)^2-\nu(2-\nu)\left(\frac{n\pi}{b}\right)^4\right]^2(1-\cosh\lambda_1 a\cosh\lambda_2 a)=0 \qquad (5.37\text{b})$$

Table 5.6 presents the first six natural frequencies $\bar{\omega}=\omega b^2\sqrt{\rho h/D}$ and mode shapes for rectangular plates with two parallel edges simply supported and the other two parallel edges free. The number of half waves (n) is given in parentheses, and the Poisson ratio is taken as ν = 0.3. Note that the frequency values without the asterisk are obtained from Equation (5.37a), whereas the frequency values with the asterisk are obtained from Equation (5.37b).

It is worth noting that there are 21 cases involving the possible combinations of free, simply supported, and clamped edge conditions for the free vibration of rectangular plates. Exact characteristic equations are given for these six cases having two opposite sides simply supported. By using the symplectic dual method, Xing and Liu (2009a) were able to obtain the exact vibration solutions for rectangular plates with any combination of clamped and simply supported edges (i.e., two adjacent edges simply supported and the other two edges clamped, one edge simply supported and the other three edges clamped, and all four edges clamped). Vibration results for the remaining cases have to be determined using numerical techniques such as the Ritz method and the finite element method. Sample vibration frequencies for all 21 cases are presented in a paper by Leissa (1973).

5.3.3 Rectangular Plates with Clamped but Vertical Sliding Edges

It is well known that the hollow TM (transverse magnetic) waveguide problem is analogous to the vibration problems of polygonal thin plates with simply supported,

TABLE 5.6
Natural Frequencies $\bar{\omega} = \omega b^2 \sqrt{\rho h/D}$ for Rectangular Plate with Two Parallel Edges Simply Supported and Two Parallel Edges Free

a/*b*	Mode 1	Mode 2	Mode 3	Mode 4	Mode 5	Mode 6
0.5	 9.513* (1)	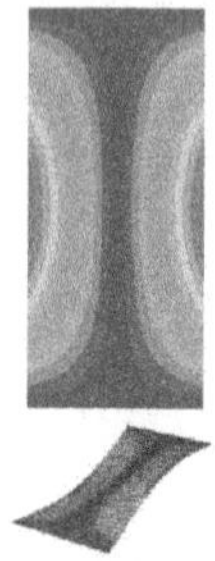 27.522 (1)	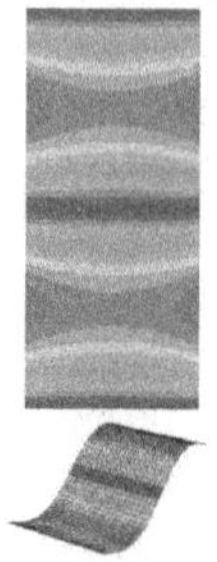 38.526* (2)	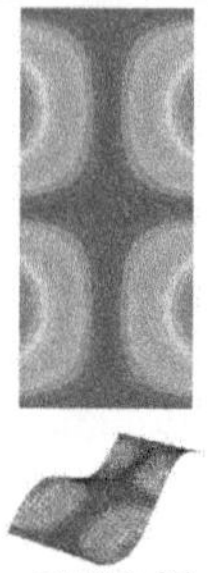 64.539 (2)	 87.285* (3)	 105.490 (1)
1.0	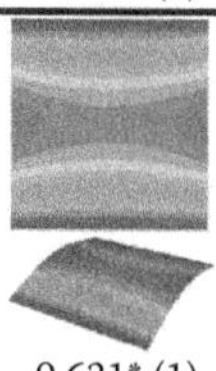 9.631* (1)	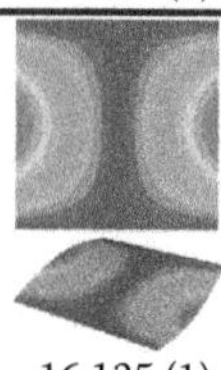 16.135 (1)	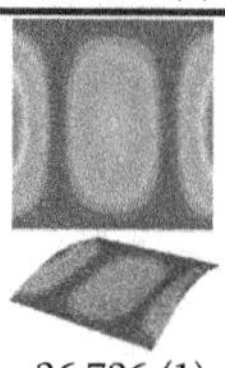 36.726 (1)	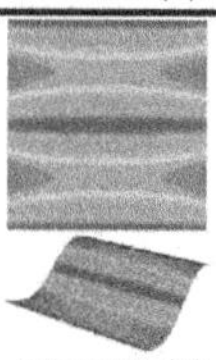 38.945* (2)	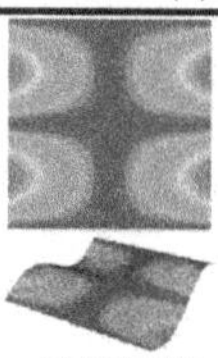 46.737 (2)	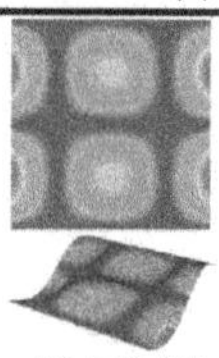 70.740 (2)
2.0	 9.736* (1)	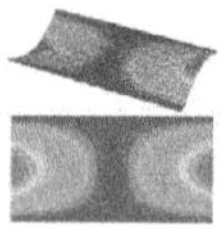 11.685 (1)	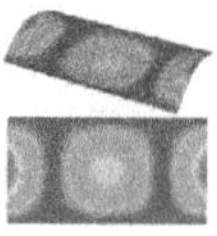 17.685 (1)	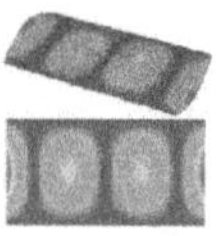 27.756 (1)	 39.188* (2)	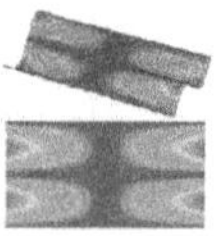 41.197 (2)

straight edges and of prestressed membranes (e.g., see Conway [1960] and Ng [1974]). Recently, Wang, Wang, and Tay (2013) pointed out that the TE (transverse electric) waveguide problem is analogous to the vibration problem of thin plates with zero slope and zero twisting moment at the edges, which are allowed to slide freely in the vertical (transverse) direction. Such a plate problem is found in vibrating piston heads. Following is the proof for this analogy.

The wave propagation of the TE waveguide is governed by the Helmholtz equation (Conway 1960; Pnueli 1975), i.e.,

$$\nabla^2\phi + k^2\phi = 0 \tag{5.38}$$

where $\nabla^2 = \frac{\partial^2}{\partial x^2} + \frac{\partial^2}{\partial y^2}$ is the Laplacian operator. Here, ϕ is the scalar wave amplitude function, and k is the wave frequency parameter (Wang 2010)

$$k = 2\pi(f)\sqrt{c} \tag{5.39}$$

where f is the frequency of the wave propagation, and c is the inductivity capacity. The boundary condition for the TE modes is (Pnueli 1975)

$$\frac{\partial \phi}{\partial n} = 0 \text{ at the edge} \tag{5.40}$$

where n is the normal to the cross section of the waveguide. Equation (5.40) is also known as the Neumann boundary condition. The eigenvalues k and the corresponding mode shapes ϕ are sought by solving the Helmholtz equation (Equation [5.38]) with the boundary condition of Equation (5.40).

The aforementioned TE waveguide problem will be shown to be analogous to a classical thin-plate vibration problem with zero slope and zero twisting moment at the edges that are allowed to slide freely in the vertical (transverse) direction. According to the classical thin-plate theory, the governing equation for a vibrating plate with thickness h, mass density ρ, Young's modulus E, and Poisson ratio ν, is given by (Szilard 1974).

$$\nabla^4 w - \chi^2 w = 0 \tag{5.41}$$

where w is the transverse displacement, and χ is the frequency parameter of the thin plate defined by

$$\chi = \omega \sqrt{\frac{\rho h}{D}} \tag{5.42}$$

in which ω is the angular frequency of the thin plate, and $D = Eh^3/[12(1-\nu^2)]$, the flexural rigidity. At the plate edges, the boundary conditions are

$$\frac{\partial w}{\partial n} = 0 \tag{5.43}$$

$$Q_n = -D\frac{\partial}{\partial n}(\nabla^2 w) = 0 \tag{5.44}$$

where Q_n is the transverse shear force. Note that the additional shearing force at the edge $\partial M_{ns}/\partial s$ is zero because the twisting moment is zero along the edge of the plate.

The fourth-order governing differential Equation (5.41) may be factorized as (Liew et al. 1998)

$$(\nabla^2 - \chi)(\nabla^2 + \chi)w = 0 \tag{5.45}$$

Alternatively, Equation (5.45) may be written as two second-order differential equations

$$(\nabla^2 - \chi)w = \bar{w} \tag{5.46}$$

$$(\nabla^2 + \chi)\bar{w} = 0 \tag{5.47}$$

If we differentiate Equation (5.46) with respect to n, we get

$$\frac{\partial}{\partial n}(\nabla^2 w) - \chi \frac{\partial w}{\partial n} = \frac{\partial \bar{w}}{\partial n} \tag{5.48}$$

By substituting the boundary conditions given in Equations (5.43) and (5.44) into Equation (5.48), we get

$$\frac{\partial \bar{w}}{\partial n} = 0 \quad \text{at the edge} \tag{5.49}$$

This implies that the governing equation and boundary condition of the aforementioned vibration thin-plate problem may be expressed by Equations (5.47) and (5.49), respectively. The governing equation and boundary condition are mathematically similar to their counterparts for the TE waveguide problem (see Equations [5.38] and [5.40]). Therefore, it has been proved that the TE waveguide problem is analogous to the vibration problem of plates with a zero slope and zero twisting moment at the freely vertically sliding edges. This analogy implies that

$$k = \sqrt{\chi} \tag{5.50}$$

i.e., the wave frequency parameters of the waveguides are equal to the square root of the corresponding plate vibration frequency parameters, and the TE mode shapes correspond to their thin-plate vibration mode shape counterparts.

Now, consider a rectangular domain enclosed by the lines

$$x = -b/2,\ x = b/2,\ y = -a/2, \text{ and } y = a/2 \tag{5.51}$$

where the origin is located at the centroid of the rectangular domain. The exact solutions for TE modes ϕ for the rectangular waveguide with an aspect ratio $b/a \geq 1$ are given as follows (Schelkunoff 1943):

- For S-S mode (i.e., symmetric about both x- and y-axes),

$$\phi = \cos(\alpha x)\cos(\beta y) \text{ where } \alpha = \frac{2m\pi}{b} \text{ and } \beta = \frac{2n\pi}{a},\ (m,n) \text{ are integers except zero.} \tag{5.52}$$

- For A-S mode (i.e., antisymmetric about x-axis and symmetric about y-axis),

$$\phi = \cos(\alpha x)\sin(\beta y) \text{ where } \alpha = \frac{2m\pi}{b} \text{ and } \beta = \frac{2(n+1/2)\pi}{a},\ (m,n) \text{ are integers} \tag{5.53}$$

- For S-A mode (i.e., symmetric about x-axis and antisymmetric about y-axis),

$$\phi = \sin(\alpha x)\cos(\beta y) \text{ where } \alpha = \frac{2(m+1/2)\pi}{b},\ \beta = \frac{2n\pi}{a},\ (m,n) \text{ are integers} \tag{5.54}$$

- For A-A mode (i.e., antisymmetric about both x- and y-axes),

$$\phi = \sin(\alpha x)\sin(\beta y) \quad \text{where } \alpha = \frac{2(m+1/2)\pi}{b} \text{ and } \beta = \frac{2(n+1/2)\pi}{a}, (m,n) \text{ are integers} \tag{5.55}$$

and the exact frequency parameter is given by

$$k = \sqrt{\alpha^2 + \beta^2} \tag{5.56}$$

The foregoing exact mode shapes and frequency parameter are also valid for the corresponding vibrating thin plates with sliding boundary conditions. Tables 5.7 and 5.8 present the exact mode shapes and the frequency values $ka = \sqrt{\chi}a$ of TE square and rectangular waveguides or clamped square and rectangular plates with sliding edges. The bracketed values represent (m, n). The symbol S denotes symmetrical mode and the symbol A the antisymmetrical mode.

5.3.4 Triangular Plates with Simply Supported Edges

Conway (1960) pointed out the analogies between the vibration and buckling problems of simply supported polygonal plates and the vibration problem of uniformly prestressed membranes. Owing to these analogies, one may obtain the exact solutions for any two of the problems upon having derived the exact solution of one of these three problems using the following relationships:

$$\omega_n \sqrt{\frac{\rho h}{D}} = \frac{N_n}{D} = \hat{\omega}_n \frac{\mu}{T} \tag{5.57}$$

TABLE 5.7
Exact Mode Shapes and Frequency Parameters for Square Thin Plate with Sliding Boundary Condition

Mode 1 (S-A)	Mode 2 (A-S)	Mode 3 (A-A)	Mode 4 (S-S)	Mode 5 (S-S)	Mode 6 (S-A)	Mode 7 (A-S)
		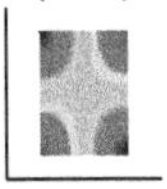	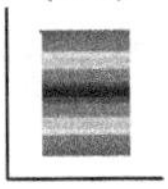	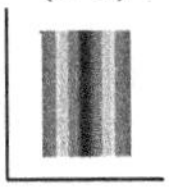	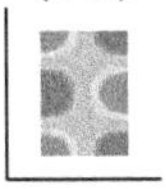	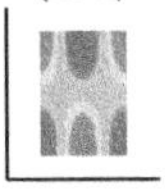
(−1, 0) (0, 0)	(0, −1) (0, 0)	(−1, −1) (−1, 0) (0, −1) (0, 0)	(0, −1) (0, 1)	(−1, 0) (1, 0)	(−1, −1) (−1, 1) (0, −1) (0, 1)	(−1, −1) (−1, 0) (1, −1) (1, 0)
3.1416	3.1416	4.4429	6.2832	6.2832	7.0248	7.0248

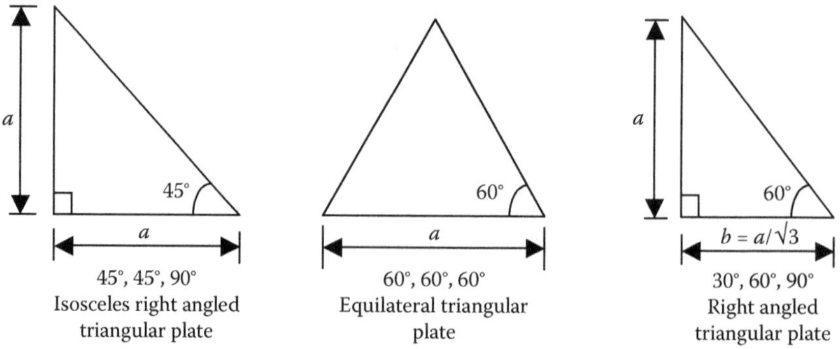

FIGURE 5.8 Triangular plates.

where ω_n is the nth angular frequency of vibration of the simply supported polygonal plate, N_n is the buckling load for the nth mode of the simply supported plate under hydrostatic in-plane load, $\hat{\omega}_n$ is the nth angular frequency of vibration of the prestressed membrane, μ is the mass density per unit area of membrane, and T is the uniform tension per unit length of membrane.

In Section 3.3.2, exact solutions for prestressed membranes are presented for three triangular shapes (45°, 45°, 90°), (60°, 60°, 60°), and (30°, 60°, 90°), as shown in Figure 5.8. In view of Section 3.3.2 and the membrane frequencies, one may obtain the exact frequencies for the three triangular plates with simply supported edges. Table 5.9 furnishes the first six natural frequencies and their corresponding mode shapes for the three aforementioned simply supported triangular plates.

Some authors patched right triangles to form a larger shape. All the little triangular edges are nodal lines, and thus the frequencies are the same. It would give some degenerate higher exact frequencies for the composite shape, but never the important lower ones, such as the fundamental frequency.

TABLE 5.8
Exact Mode Shapes and Frequency Parameters for Rectangular Thin Plate with Sliding Boundary Condition (aspect ratio b/a = 2)

Mode 1 (S-A)	Mode 2 (S-S)	Mode 3 (A-S)	Mode 4 (A-A)	Mode 5 (A-S)	Mode 6 (S-A)
	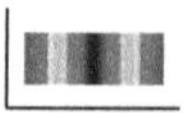		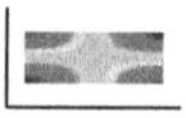		
(−1, 0) (0, 0)	(−1, 0) (1, 0)	(0, −1) (0, 0)	(−1, −1) (−1, 0) (0, −1) (0, 0)	(−1, −1) (−1, 0) (1, −1) (1, 0)	(−2, 0) (1, 0)
1.5708	3.1416	3.1416	3.5124	4.4429	4.7124

TABLE 5.9
Natural Frequencies for (45°, 45°, 90°), (60°, 60°, 60°), and (30°, 60°, 90°) Triangular Plates with Simply Supported Edges

Triangular Plate	Mode 1	Mode 2	Mode 3	Mode 4	Mode 5	Mode 6
45°, 45°, 90°	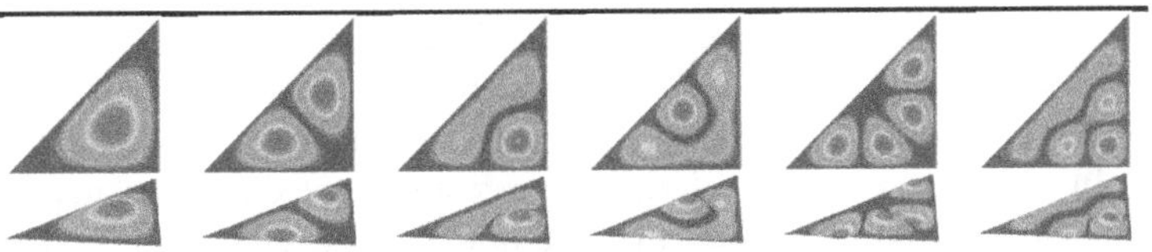					
	$5\pi^2$	98.696	128.303	167.792	197.405	246.747
60°, 60°, 60°	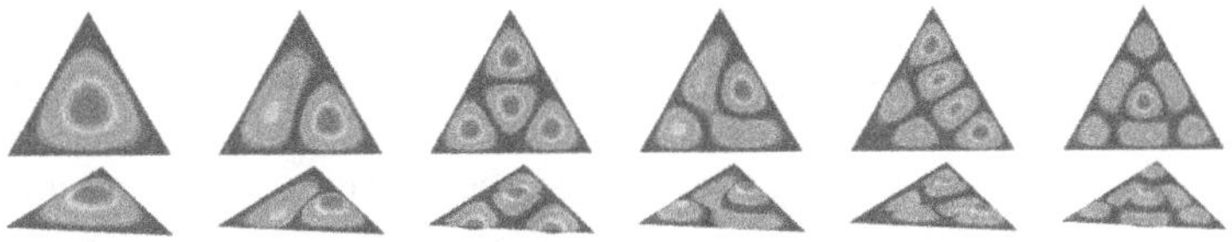					
	$16\pi^2/3$	122.824	210.568	228.111	333.436	368.490
30°, 60°, 90°						
	92.181	171.167	250.209	276.492	368.666	408.149

5.3.5 Circular Plates

For a circular thin plate with radius R, the vibration mode shape may be expressed as

$$w(r,\theta) = W_n(r)\cos n\theta \tag{5.58}$$

By substituting Equation (5.58) into Equation (5.5), the fourth-order partial differential equation may be converted into the following two second-order ordinary differential equations:

$$\frac{d^2W_{n1}}{dr^2} + \frac{1}{r}\frac{dW_{n1}}{dr} - \left(\frac{n^2}{r^2} - k\right)W_{n1} = 0 \tag{5.59}$$

$$\frac{d^2W_{n2}}{dr^2} + \frac{1}{r}\frac{dW_{n2}}{dr} - \left(\frac{n^2}{r^2} + k\right)W_{n2} = 0 \tag{5.60}$$

The exact solutions for Equations (5.59) and (5.60) are, respectively,

$$W_{n1}(r) = A_1 J_n(r\sqrt{k}) + A_2 Y_n(r\sqrt{k}) \tag{5.61}$$

$$W_{n2}(r) = A_3 I_n(r\sqrt{k}) + A_4 K_n(r\sqrt{k}) \tag{5.62}$$

where J_n and Y_n are the Bessel function of the first kind and second kind, respectively; I_n and K_n are the modified Bessel function of the first kind and the second kind, respectively; A_i, $i = 1,2,3,4$ are the unknown coefficients that are to be solved from the boundary conditions; and n is the circumferential wave number.

In order to avoid a singularity at the center of the circular plate, the Bessel function and the modified Bessel function of the second kind must be dropped. In view of this, the typical vibration mode shape for a circular vibrating plate is given by

$$w_n(r,\theta) = W_n(r)\cos n\theta = [A_1 J_n(r\sqrt{k}) + A_3 I_n(r\sqrt{k})]\cos n\theta \tag{5.63}$$

For a circular plate with a *clamped edge*, the boundary conditions are

$$W_n = 0 \quad \text{and} \quad \frac{dW_n}{dr} = 0 \quad \text{at } r = R \tag{5.64}$$

By substituting Equation (5.63) into the boundary conditions of Equation (5.64), and for nontrivial solution, we obtain the following exact characteristic equation

$$J_n(\alpha)I_{n+1}(\alpha) + I_n(\alpha)J_{n+1}(\alpha) = 0, \quad \text{where} \quad \alpha = R\sqrt{k} \tag{5.65}$$

The roots of the characteristic equation—Equation (5.65)—furnish the vibration frequencies. It is interesting to note that the nondimensional frequency parameter α of the clamped circular plate is independent of the Poisson ratio, but this is not the case for a simply supported circular plate, as shown in the following discussion. Table 5.10 presents the first six natural frequencies $\bar{\omega}_{sn} = \omega_{sn}R^2\sqrt{\rho h/D}$ and mode shapes for a clamped circular plate. Note that s represents the number of nodal circles and n the number of nodal diameters.

TABLE 5.10
Natural Frequencies $\bar{\omega}_{sn} = \omega_{sn}R^2\sqrt{\rho h/D}$ for a Clamped Circular Plate

Mode 1	Mode 2	Mode 3	Mode 4	Mode 5	Mode 6
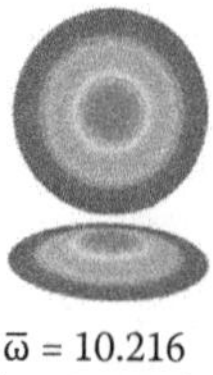	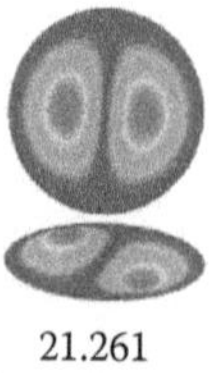	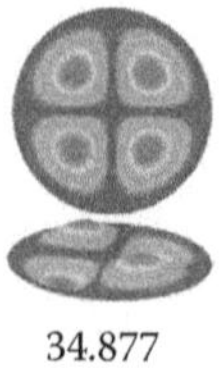	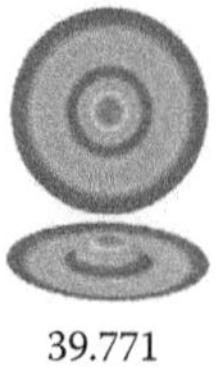	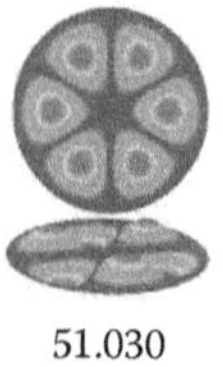	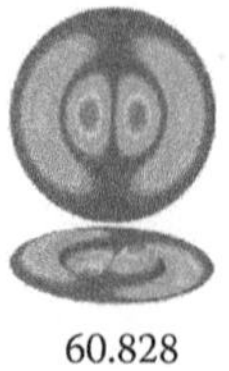
$\bar{\omega} = 10.216$	21.261	34.877	39.771	51.030	60.828
($s = 0, n = 0$)	($s = 0, n = 1$)	($s = 0, n = 2$)	($s = 1, n = 0$)	($s = 0, n = 3$)	($s = 1, n = 1$)

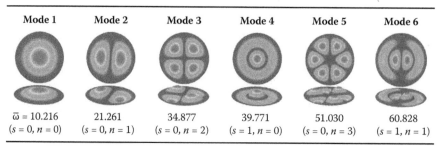

For a circular plate with a *simply supported edge*, the boundary conditions are

$$W_n = 0 \quad \text{and} \quad M_{rr} = -D\left[\frac{\partial^2 w_n}{\partial r^2} + \nu\left(\frac{1}{r}\frac{\partial w_n}{\partial r} + \frac{1}{r^2}\frac{\partial^2 w_n}{\partial \theta^2}\right)\right] = 0 \quad \text{at } r = R \tag{5.66}$$

By substituting Equation (5.63) into the boundary conditions of Equation (5.66), and for nontrivial solution, we obtain the following exact characteristic equation:

$$\frac{J_{n+1}(\alpha)}{J_n(\alpha)} + \frac{I_{n+1}(\alpha)}{I_n(\alpha)} - \frac{2\alpha}{1-\nu} = 0 \quad \text{where} \quad \alpha = R\sqrt{k} \tag{5.67}$$

The roots of the characteristic equation—Equation (5.67)—furnish the vibration frequencies of the simply supported circular plates. Table 5.11 presents the first six natural frequencies $\bar{\omega}_{sn} = \omega_{sn} R^2 \sqrt{\rho h/D}$ and mode shapes for a simply supported circular plate with Poisson ratio $\nu = 0.3$.

For a circular plate with a completely *free edge*, the boundary conditions are

$$M_{rr} = -D\left[\frac{\partial^2 w_n}{\partial r^2} + \nu\left(\frac{1}{r}\frac{\partial w_n}{\partial r} + \frac{1}{r^2}\frac{\partial^2 w_n}{\partial \theta^2}\right)\right] = 0$$

and

$$V_r = -D\left[\frac{\partial}{\partial r}(\nabla^2 w) + \frac{1-\nu}{r}\frac{\partial}{\partial \theta}\left(\frac{1}{r}\frac{\partial^2 w}{\partial r \partial \theta} - \frac{1}{r^2}\frac{\partial w}{\partial \theta}\right)\right] = 0 \text{ at } r = R \tag{5.68}$$

Examples of a completely free circular plate are a flying circular plate and a pontoon-type circular floating structure. By substituting Equation (5.63) into Equation (5.68), we obtain the following characteristic equation:

$$\frac{\alpha^2 J_n(\alpha) + (1-\nu)[\alpha J_n'(\alpha) - n^2 J_n(\alpha)]}{\alpha^2 I_n(\alpha) - (1-\nu)[\alpha I_n'(\alpha) - n^2 I_n(\alpha)]} - \frac{\alpha^3 J_n'(\alpha) + (1-\nu)n^2[\alpha J_n'(\alpha) - J_n(\alpha)]}{\alpha^3 I_n'(\alpha) - (1-\nu)n^2[\alpha I_n'(\alpha) - I_n(\alpha)]} = 0 \tag{5.69}$$

TABLE 5.11
Natural Frequencies $\bar{\omega}_{sn} = \omega_{sn} R^2 \sqrt{\rho h/D}$ for a Simply Supported Circular Plate

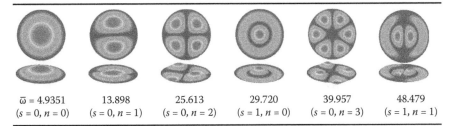

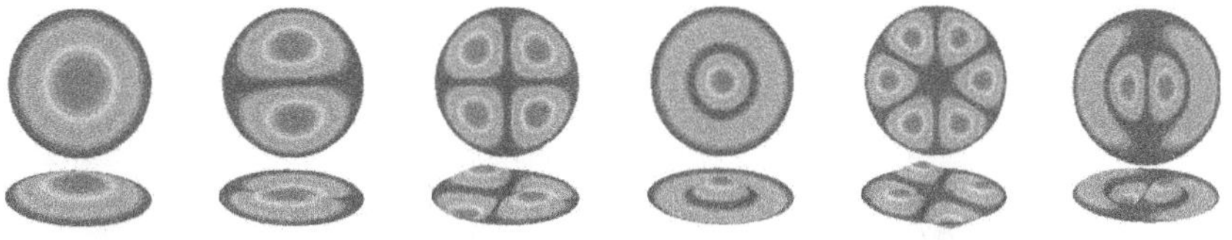

$\bar{\omega}$ = 4.9351	13.898	25.613	29.720	39.957	48.479
(s = 0, n = 0)	(s = 0, n = 1)	(s = 0, n = 2)	(s = 1, n = 0)	(s = 0, n = 3)	(s = 1, n = 1)

TABLE 5.12
Natural Frequencies $\bar{\omega}_{sn} = \omega_{sn} R^2 \sqrt{\rho h/D}$ for a Free Circular Plate

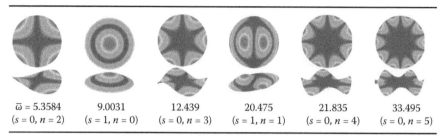

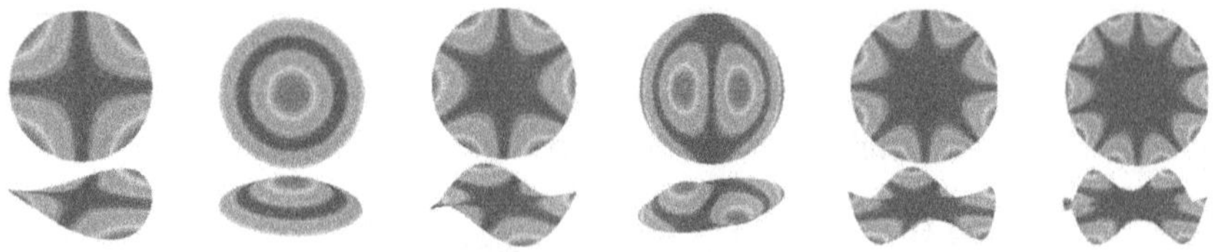

$\bar{\omega}$ = 5.3584	9.0031	12.439	20.475	21.835	33.495
(s = 0, n = 2)	(s = 1, n = 0)	(s = 0, n = 3)	(s = 1, n = 1)	(s = 0, n = 4)	(s = 0, n = 5)

Itao and Crandall (1979) gave some values for the exact frequencies of a completely free circular plate with Poisson ratio $\nu = 0.33$. Table 5.12 presents the first six natural frequencies $\bar{\omega}_{sn} = \omega_{sn} R^2 \sqrt{\rho h/D}$ and mode shapes for a free circular plate with $\nu = 0.3$.

5.3.6 Annular Plates

As for the circular plates, exact vibration solutions are possible for annular plates. The solution for Equation (5.5) is given by

$$\begin{aligned} w_n(r,\theta) &= W_n(r)\cos n\theta \\ &= \left[A_1 J_n\left(r\sqrt{k}\right) + A_2 Y_n\left(r\sqrt{k}\right) + A_3 I_n\left(r\sqrt{k}\right) + A_4 K_n\left(r\sqrt{k}\right)\right]\cos n\theta \end{aligned} \tag{5.70}$$

The substitution of Equation (5.70) into the boundary conditions for the inner edge ($r = b$) and outer edge ($r = a$) of the annular plates yields four equations. The frequency determinant is constructed from these four equations. The characteristic equation becomes rather lengthy and complex to display. We shall adopt Vogel and Skinner's (1965) method of presentation of these equations for nine combinations of inner edge and outer edge boundary conditions. Since the boundary conditions involve $M_{rr} = 0$, $V_r = 0$, $w = 0$, and $\partial w/\partial r = 0$, we first note that

$$W_n(r) = A_1 J_n(r\sqrt{k}) + A_2 Y_n(r\sqrt{k}) + A_3 I_n(r\sqrt{k}) + A_4 K_n(r\sqrt{k}) \tag{5.71}$$

$$\begin{aligned} \frac{1}{\sqrt{k}}\frac{dW_n(r)}{dr} &= A_1\left[\frac{n}{r\sqrt{k}} J_n(r\sqrt{k}) - J_{n+1}(r\sqrt{k})\right] + A_2\left[\frac{n}{r\sqrt{k}} Y_n(r\sqrt{k}) - Y_{n+1}(r\sqrt{k})\right] \\ &+ A_3\left[\frac{n}{r\sqrt{k}} I_n(r\sqrt{k}) + I_{n+1}(r\sqrt{k})\right] + A_4\left[\frac{n}{r\sqrt{k}} K_n(r\sqrt{k}) - K_{n+1}(r\sqrt{k})\right] \end{aligned} \tag{5.72}$$

$$\frac{1}{Dk\cos n\theta}M_{rr}(r)=A_1\left\{J_n(r\sqrt{k})-(1-\nu)\left[\frac{n(n-1)}{kr^2}J_n(r\sqrt{k})+\frac{1}{r\sqrt{k}}J_{n+1}(r\sqrt{k})\right]\right\}$$
$$+A_2\left\{Y_n(r\sqrt{k})-(1-\nu)\left[\frac{n(n-1)}{kr^2}Y_n(r\sqrt{k})+\frac{1}{r\sqrt{k}}Y_{n+1}(r\sqrt{k})\right]\right\}$$
$$-A_3\left\{I_n(r\sqrt{k})+(1-\nu)\left[\frac{n(n-1)}{kr^2}I_n(r\sqrt{k})-\frac{1}{r\sqrt{k}}I_{n+1}(r\sqrt{k})\right]\right\}$$
$$-A_4\left\{K_n(r\sqrt{k})+(1-\nu)\left[\frac{n(n-1)}{kr^2}K_n(r\sqrt{k})+\frac{1}{r\sqrt{k}}K_{n+1}(r\sqrt{k})\right]\right\} \quad (5.73)$$

$$\frac{r}{D\sqrt{k}\cos n\theta}V_r(r)$$
$$=A_1\left\{nJ_n(r\sqrt{k})-r\sqrt{k}J_{n+1}(r\sqrt{k})+\frac{n^2(1-\nu)}{kr^2}[(n-1)J_n(r\sqrt{k})-r\sqrt{k}J_{n+1}(r\sqrt{k})]\right\}$$
$$+A_2\left\{nY_n(r\sqrt{k})-r\sqrt{k}Y_{n+1}(r\sqrt{k})+\frac{n^2(1-\nu)}{kr^2}[(n-1)Y_n(r\sqrt{k})-r\sqrt{k}Y_{n+1}(r\sqrt{k})]\right\}$$
$$-A_3\left\{nI_n(r\sqrt{k})+r\sqrt{k}I_{n+1}(r\sqrt{k})-\frac{n^2(1-\nu)}{kr^2}[(n-1)I_n(r\sqrt{k})+r\sqrt{k}I_{n+1}(r\sqrt{k})]\right\}$$
$$-A_4\left\{nK_n(r\sqrt{k})-r\sqrt{k}K_{n+1}(r\sqrt{k})-\frac{n^2(1-\nu)}{kr^2}[(n-1)K_n(r\sqrt{k})-r\sqrt{k}K_{n+1}(r\sqrt{k})]\right\} \quad (5.74)$$

In view of Equations (5.71) to (5.74), we develop the eigenvalue equation for the annular plates with nine combinations of inner and outer boundary conditions, as shown in Table 5.13. The first description indicates the boundary condition for the outer edge, and the second description is the boundary condition for the inner edge.

Tables 5.14 to 5.22 present the first six natural frequencies $\bar{\omega}=\omega a^2\sqrt{\rho h/D}$ and mode shapes of annular plates with inner radius to outer radius ratios b/a = 0.3, 0.5, 0.7, Poisson ratio ν = 0.3, and various combinations of inner to outer edge boundary conditions.

5.3.7 Annular Sector Plates

For circular or annular sector plates, exact solutions exist only for simply supported radial edges. Leissa (1969) noted that if the opening angle of the sector is π/N, where N is an integer, then all the vibration modes of the sector plate can be found from the higher modes of the full plate. For circular sector plates with opening angle larger than 180°, or with a reentrant vertex, Huang, Leissa, and McGee (1993) showed that

TABLE 5.13
Nine Combinations of Boundary Conditions for Annular Plates and Equations Forming the Eigenvalue Problem

Annular Plates		Four Equations That Make Up the Eigenvalue Problem			
1	Free-Clamped	$M_{rr}(a)=0$	$V_r(a)=0$	$w(b)=0$	$\frac{dw(b)}{dr}=0$
2	Clamped-Clamped	$w(a)=0$	$\frac{dw(a)}{dr}=0$	$w(b)=0$	$\frac{dw(b)}{dr}=0$
3	Free-Free	$M_{rr}(a)=0$	$V_r(a)=0$	$M_{rr}(b)=0$	$V_r(b)=0$
4	Simply Supported	$w(a)=0$	$M_{rr}(a)=0$	$w(b)=0$	$M_{rr}(b)=0$
5	Free-Simply Supported	$M_{rr}(a)=0$	$V_r(a)=0$	$w(b)=0$	$M_{rr}(b)=0$
6	Simply Supported-Free	$w(a)=0$	$M_{rr}(a)=0$	$M_{rr}(b)=0$	$V_r(b)=0$
7	Simply Supported-Clamped	$w(a)=0$	$M_{rr}(a)=0$	$w(b)=0$	$\frac{dw(b)}{dr}=0$
8	Clamped-Free	$w(a)=0$	$\frac{dw(a)}{dr}=0$	$M_{rr}(b)=0$	$V_r(b)=0$
9	Clamped-Simply Supported	$w(a)=0$	$\frac{dw(a)}{dr}=0$	$w(b)=0$	$M_{rr}(b)=0$

it is necessary to appropriately retain the two singular Bessel functions (ignored for full plates) for the correct frequency.

In this section, we present the exact solutions for the vibration of the annular sectorial plates with simply supported radial edges (Figure 5.9a) for opening angles of 50°, 90°, 180°, 270°, 360°, 720°, and 1440°. Note that except for the 90° and 180° sectors, the frequencies of the other shapes are not a subset of those of the full annulus. For annular sector plates with angles more than 360°, one can envision a helical coil of low pitch (Figure 5.9b), important in enhancing heat transfer.

All length variables are normalized by the outer radius of curvature R. By using cylindrical coordinates, the plate is bounded by $r=b$, $r=1$, $\theta=\pm\alpha/2$, where α is the opening angle. Let the plate have a uniform density ρ, thickness h, and flexural rigidity D and vibrating with a circular frequency ω. The governing equation for vibration is given by

$$\nabla^4 w - k^4 w = 0 \tag{5.75}$$

where $\nabla^4=\nabla^2\nabla^2$, $\nabla^2=\frac{\partial^2}{\partial r^2}+\frac{1}{r}\frac{\partial}{\partial r}+\frac{1}{r^2}\frac{\partial^2}{\partial\theta^2}$, $w(r,\theta)$ is the vibration amplitude, and $k^2=\omega R^2\sqrt{\rho h/D}$ is the normalized frequency. The general solution to Equation (5.75) is given by

$$w=[C_1\cos(\mu\theta)+C_2\sin(\mu\theta)]u(r) \tag{5.76}$$

$$u=A_1J_\mu(kr)+A_2Y_\mu(kr)+A_3I_\mu(kr)+A_4K_\mu(kr) \tag{5.77}$$

Here J, Y are Bessel functions, and K, I are modified Bessel functions. For simply supported radial edges and if the vibration is symmetric with respect to θ, $\cos(\mu\theta)$ and $\mu=(2n-1)\pi/\alpha$ are chosen, where n is a nonzero positive integer. For vibrations antisymmetric with respect to θ, $\sin(\mu\theta)$ and $\mu=2n\pi/\alpha$ are selected instead.

TABLE 5.14
Natural Frequencies $\bar{\omega} = \omega a^2\sqrt{\rho h/D}$ for Annular Plate with Free Outer Edge and Clamped Inner Edge

$\frac{b}{a}$	Mode 1	Mode 2	Mode 3	Mode 4	Mode 5	Mode 6
0.7	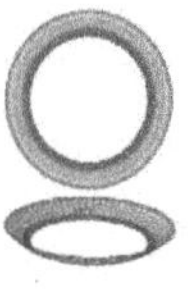				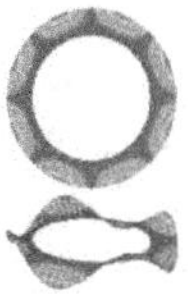	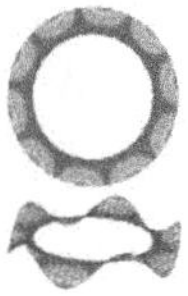
	36.953	37.498	39.277	42.654	48.071	55.880
0.5	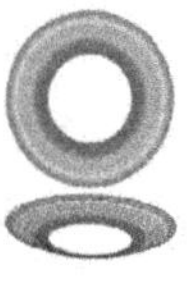	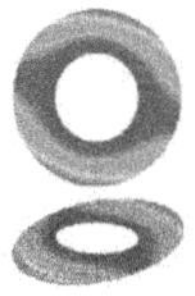	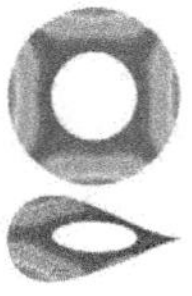		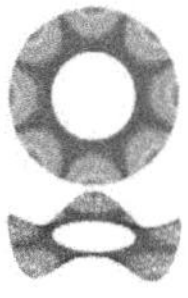	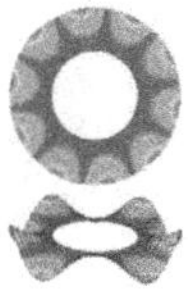
	13.024	13.290	14.704	18.562	25.596	35.730
0.3	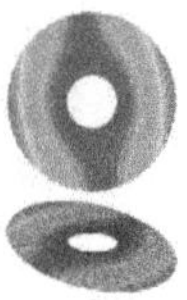	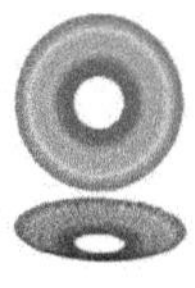	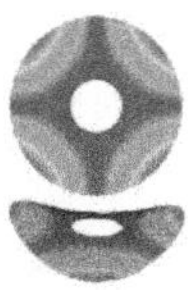		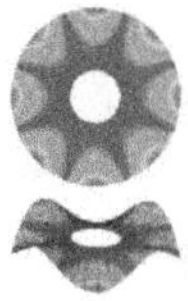	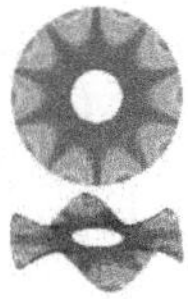
	6.552	6.660	7.956	13.276	22.076	33.567

The boundary conditions on the curved boundaries could be clamped (C), simply supported (S), or free (F). The first letter will be used for the outer boundary at $r = 1$, and the second letter for the inner boundary at $r = b$. There are nine different cases: C-C, S-S, F-F, C-S, S-C, C-F, F-C, S-F, and F-S.

The normalized radial bending moment is given by

$$M(u) = \frac{d^2u}{dr^2} + \nu\left[\frac{1}{r}\frac{du}{dr} - \frac{\mu^2}{r^2}u\right] \tag{5.78}$$

The normalized effective shear force is given by

$$V(u) = \frac{d^3u}{dr^3} + \frac{1}{r}\frac{d^2u}{dr^2} - [1+\mu^2(2-\nu)]\frac{1}{r^2}\frac{du}{dr} + \mu^2(3-\nu)\frac{\mu}{r^3} \tag{5.79}$$

TABLE 5.15
Natural Frequencies $\bar{\omega} = \omega a^2 \sqrt{\rho h / D}$ for Annular Plate with Clamped Outer Edge and Clamped Inner Edge

$\frac{b}{a}$	Mode 1	Mode 2	Mode 3	Mode 4	Mode 5	Mode 6
0.7	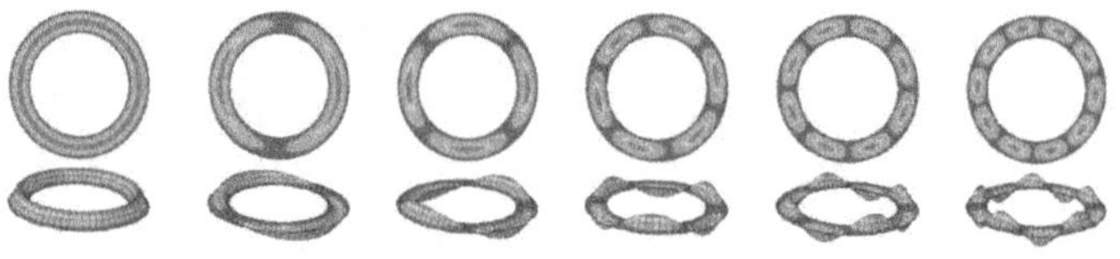					
	248.402	249.164	251.481	255.444	261.197	268.921
0.5	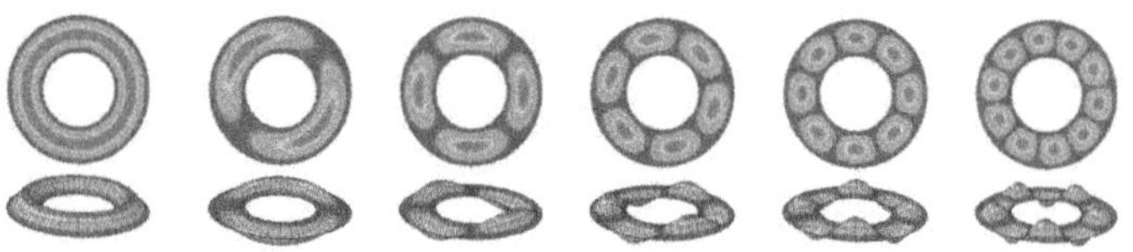					
	89.251	90.230	93.321	98.928	107.567	119.697
0.3	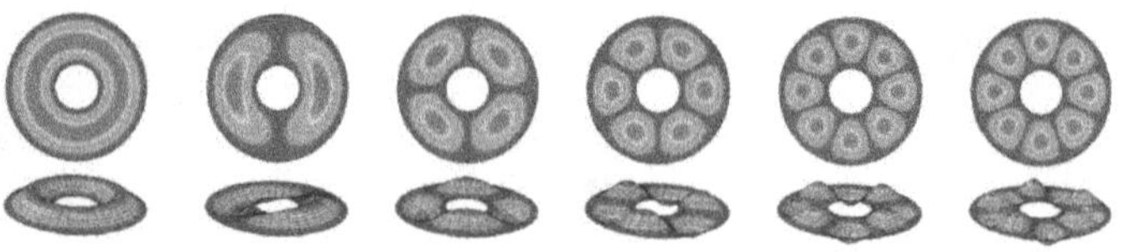					
	45.346	46.644	51.139	60.033	73.945	92.495

Here, ν is the Poisson ratio, taken to be 0.3 in our computations. For a clamped edge, let the submatrix be

$$Q_C(r) = \begin{pmatrix} J_\mu(kr) & Y_\mu(kr) & I_\mu(kr) & K_\mu(kr) \\ J'_\mu(kr) & Y'_\mu(kr) & I'_\mu(kr) & K'_\mu(kr) \end{pmatrix} \tag{5.80}$$

For a simply supported edge, the submatrix is

$$Q_S(r) = \begin{pmatrix} J_\mu(kr) & Y_\mu(kr) & I_\mu(kr) & K_\mu(kr) \\ M(J_\mu(kr)) & M(Y_\mu(kr)) & M(I_\mu(kr)) & M(K_\mu(kr)) \end{pmatrix} \tag{5.81}$$

TABLE 5.16
Natural Frequencies $\bar{\omega} = \omega a^2 \sqrt{\rho h/D}$ for Annular Plate with Free Outer Edge and Free Inner Edge

$\frac{b}{a}$	Mode 1	Mode 2	Mode 3	Mode 4	Mode 5	Mode 6
0.7						
	3.573	9.859	13.163	18.697	21.914	30.025
0.5		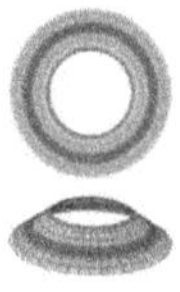			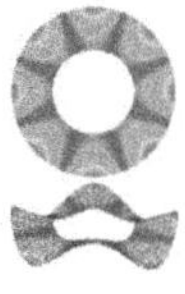	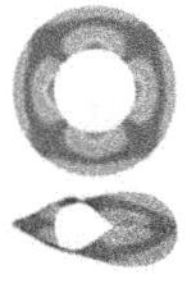
	4.271	9.313	11.425	17.198	21.067	31.115
0.3	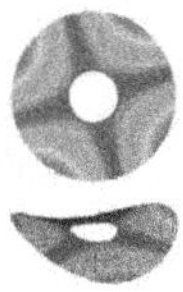	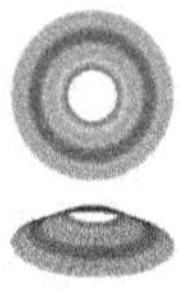	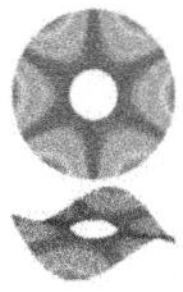	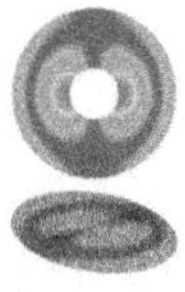	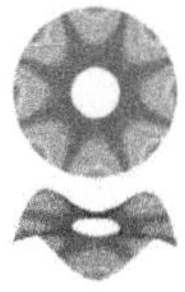	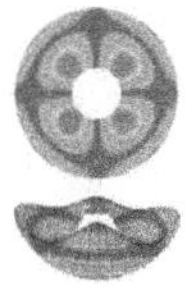
	4.906	8.353	12.266	18.292	21.783	32.973

For a free edge, the submatrix is

$$Q_F(r) = \begin{pmatrix} M(J_\mu(kr)) & M(Y_\mu(kr)) & M(I_\mu(kr)) & M(K_\mu(kr)) \\ V(J_\mu(kr)) & V(Y_\mu(kr)) & V(I_\mu(kr)) & V(K_\mu(kr)) \end{pmatrix} \tag{5.82}$$

Thus the exact characteristic equations for the aforementioned nine cases are

$$\begin{vmatrix} Q_C(b) \\ Q_C(1) \end{vmatrix} = 0, \quad \begin{vmatrix} Q_S(b) \\ Q_S(1) \end{vmatrix} = 0, \quad \begin{vmatrix} Q_F(b) \\ Q_F(1) \end{vmatrix} = 0, \quad \begin{vmatrix} Q_C(b) \\ Q_S(1) \end{vmatrix} = 0,$$

$$\begin{vmatrix} Q_S(b) \\ Q_C(1) \end{vmatrix} = 0, \quad \begin{vmatrix} Q_C(b) \\ Q_F(1) \end{vmatrix} = 0, \quad \begin{vmatrix} Q_F(b) \\ Q_C(1) \end{vmatrix} = 0, \quad \begin{vmatrix} Q_S(b) \\ Q_F(1) \end{vmatrix} = 0, \quad \begin{vmatrix} Q_F(b) \\ Q_5(1) \end{vmatrix} = 0. \tag{5.83}$$

TABLE 5.17
Natural Frequencies $\bar{\omega} = \omega a^2 \sqrt{\rho h/D}$ for Annular Plate with Simply Supported Outer Edge and Simply Supported Inner Edge

$\frac{b}{a}$	Mode 1	Mode 2	Mode 3	Mode 4	Mode 5	Mode 6
0.7				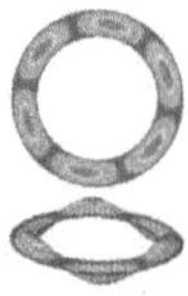	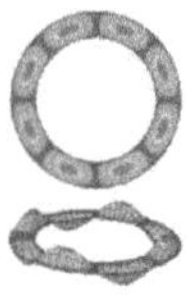	
	110.063	111.443	115.585	122.493	132.173	144.626
0.5	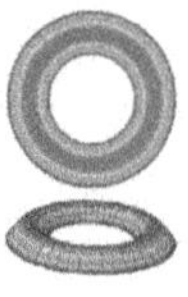	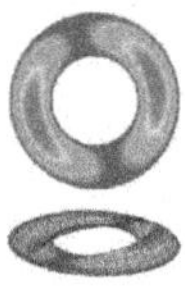	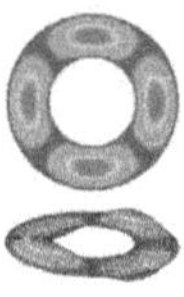	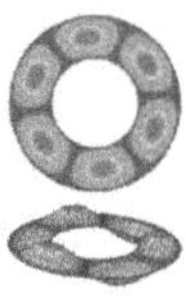	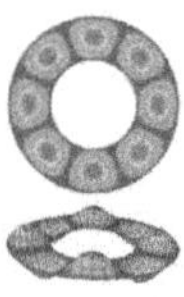	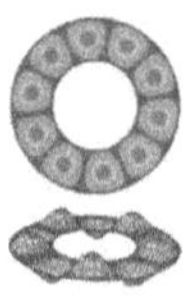
	40.043	41.797	47.089	55.957	68.379	84.257
0.3		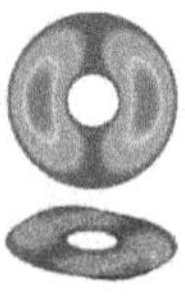			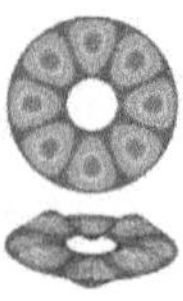	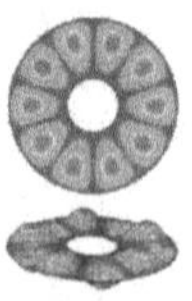
	21.079	23.317	30.273	41.910	57.546	76.427

We choose b, n and thus μ. Then from Equation (5.83), the normalized frequencies k^2 are obtained by a root search algorithm to any desired accuracy.

The frequency equations do not include the circular sector plate, for which b = 0. However, by decreasing the inner radius $b = 10^{-2}, 10^{-4}, 10^{-6}, 10^{-8}$, one finds that the results become independent of the inner edge conditions. Table 5.23 shows all three cases F-C, F-S, F-F converged to the same values, which are comparable to the exact results of the free outer edge circular sector plate of Huang, Leissa, and McGee (1993). The results for the clamped or simply supported outer edges are similar. Thus, the adopted root search algorithm is accurate to at least five significant figures.

The first five natural frequencies for three representative inner radii of b = 0.1, 0.5, 0.9 will be presented. Let n be the nth harmonic in the θ direction and m be the mth root of k in the radial direction. The mode shapes are given as (n,m) or $[n,m]$, where the parentheses denote symmetric mode, and the brackets denote antisymmetric mode. Figure 5.9c shows schematically the mode shapes. Tables 5.24 to 5.32 show the frequencies of the nine cases.

TABLE 5.18
Natural Frequencies $\bar{\omega} = \omega a^2 \sqrt{\rho h/D}$ for Annular Plate with Free Outer Edge and Simply Supported Inner Edge

$\frac{b}{a}$	Mode 1	Mode 2	Mode 3	Mode 4	Mode 5	Mode 6
0.7	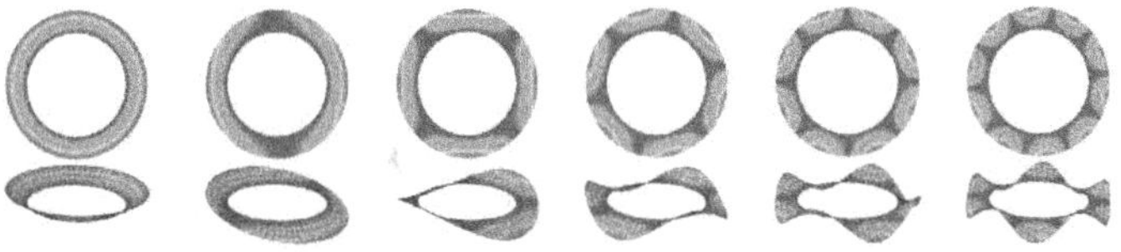					
	6.187	8.351	13.427	20.454	29.403	40.387
0.5	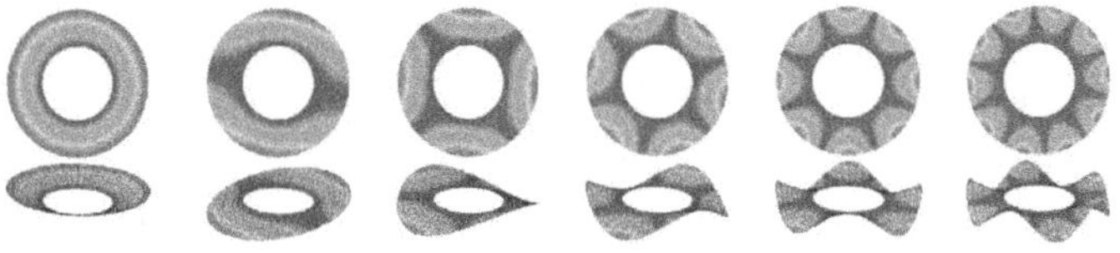					
	4.121	4.862	7.986	14.035	22.788	34.047
0.3	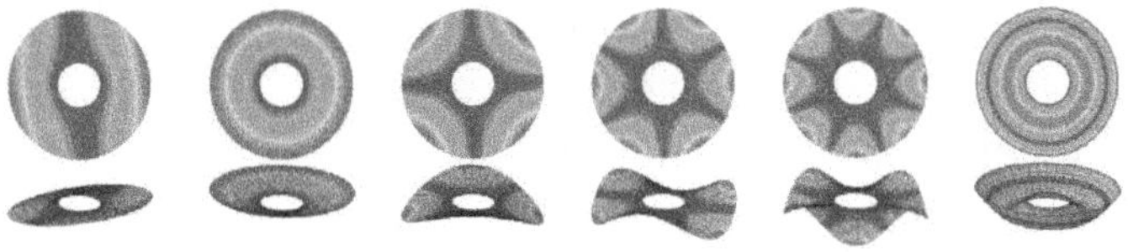					
	3.374	3.422	6.080	12.611	21.876	31.603

Note that since frequencies are based on the outer radius, in general they increase with a smaller area, i.e., larger b and smaller α. Frequencies also increase with stiffer boundary constraints. From high to low, the frequencies of the following curved edge conditions are ordered as follows: C-C, C-S, S-C, S-S, C-F, S-F, F-C, F-S, F-F (there are a few exceptions). The frequencies of C-S are larger than those of S-C, since the former has a longer clamped boundary. Similarly, the frequencies of C-F and S-F are larger than F-C and F-S, respectively. These differences are peculiar to annular sector plates and do not apply to rectangular plates. Exact solutions for other opening angles are given by Ramakrishnan and Kunukkasseril (1973).

Does a very slender annular sector mimic a long straight strip? The characteristic equation of a straight strip with C-C edges is

$$\cos \lambda \cosh \lambda - 1 = 0 \tag{5.84}$$

TABLE 5.19
Natural Frequencies $\bar{\omega} = \omega a^2\sqrt{\rho h/D}$ for Annular Plate with Simply Supported Outer Edge and Free Inner Edge

$\frac{b}{a}$	Mode 1	Mode 2	Mode 3	Mode 4	Mode 5	Mode 6
0.7	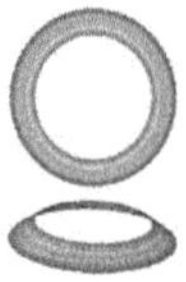		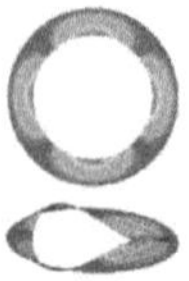	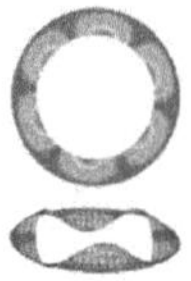	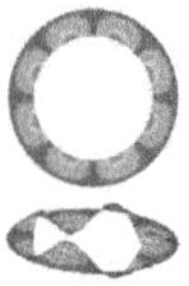	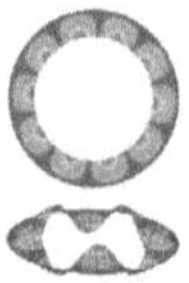
	6.931	13.311	24.329	37.097	51.713	68.502
0.5	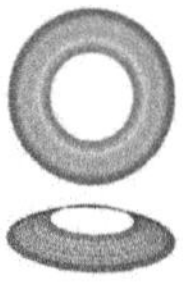	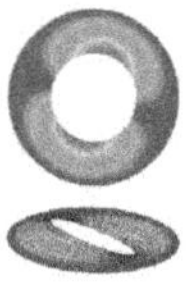	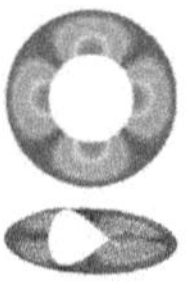	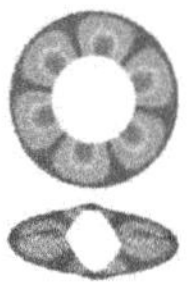	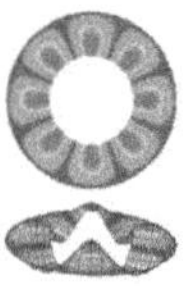	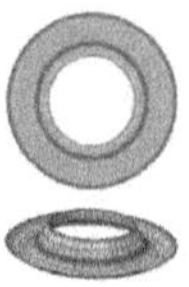
	5.077	11.607	22.357	35.636	52.032	65.842
0.3	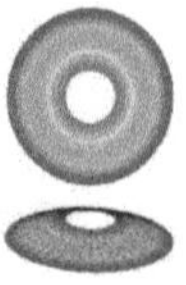		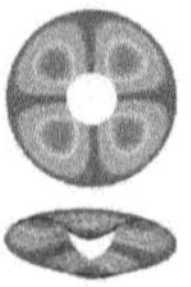	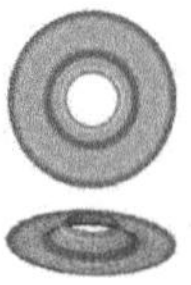	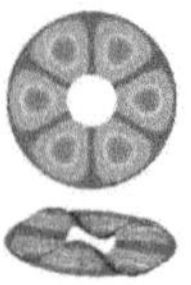	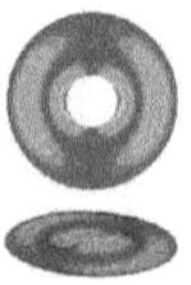
	4.664	12.816	24.116	37.042	38.775	45.837

By using our normalization, the frequency is given by

$$k^2 = \frac{\lambda^2}{(1-b)^2} \tag{5.85}$$

The first nonzero root of Equation (5.84) is $\lambda = 4.7300$. For $b = 0.9$, Equation (5.85) yields 2237.3, which compares favorably with the large α values in Table 5.24, where the first five frequencies cluster together. For the C-F or F-C cases, the straight strip equation is

$$\cos\lambda \cosh\lambda + 1 = 0 \tag{5.86}$$

TABLE 5.20
Natural Frequencies $\bar{\omega} = \omega a^2 \sqrt{\rho h/D}$ for Annular Plate with Simply Supported Outer Edge and Clamped Inner Edge

$\frac{b}{a}$	Mode 1	Mode 2	Mode 3	Mode 4	Mode 5	Mode 6
0.7	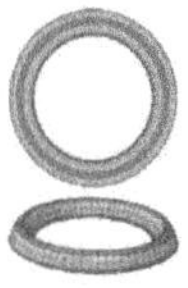	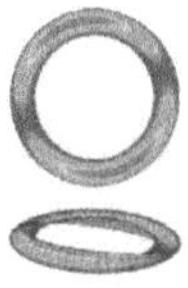	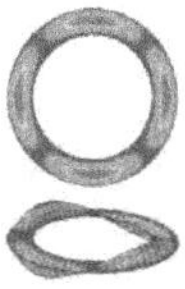	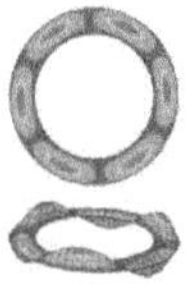	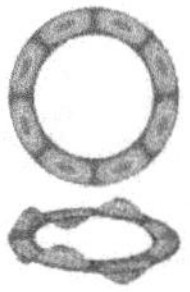	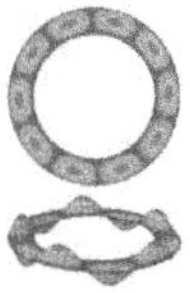
	168.524	169.489	172.414	177.380	184.511	193.952
0.5	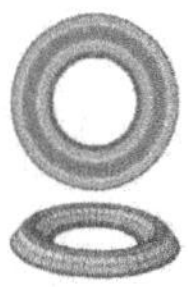		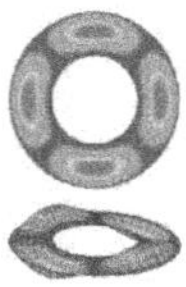	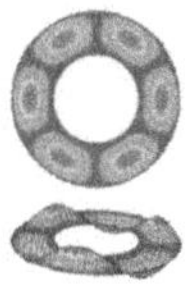	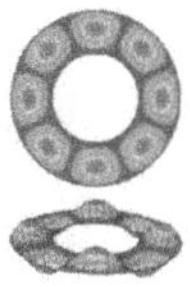	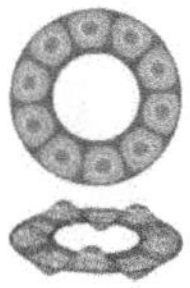
	59.820	60.987	64.631	71.107	80.802	93.988
0.3	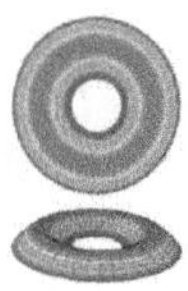	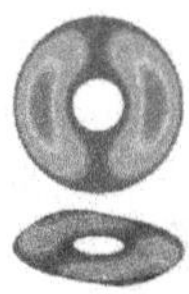		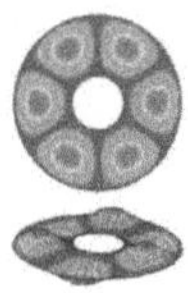	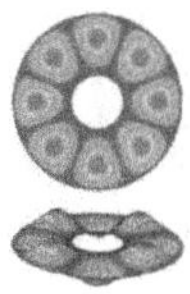	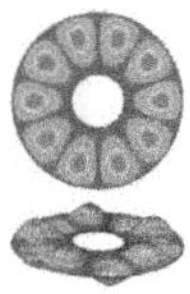
	29.978	31.403	36.243	45.459	59.273	77.133

The first root is $\lambda = 1.8751$, giving a frequency of 351.60 for $b = 0.9$. This approximates the values in Tables 5.29 and 5.30, with the difference attributed to the curvature of the sector. For C-S and S-C cases, the equation is

$$\sin\lambda \cosh\lambda - \cos\lambda \sinh\lambda = 0 \tag{5.87}$$

with the first root $\lambda = 3.9266$. The frequency for large α and $b = 0.9$ is estimated to be 1541.8. For the S-S case, $\lambda = n\pi$, which furnishes a frequency of 986.96 for slender sectors. Most interesting are the F-F, F-S, S-F cases for which the strip would give a zero frequency, which is often discarded due to zero deformation. But this zero frequency corresponds to small, nontrivial vibrations for the slender sector (see Tables 5.26, 5.31, 5.32).

TABLE 5.21
Natural Frequencies $\bar{\omega} = \omega a^2 \sqrt{\rho h / D}$ for Annular Plate with Clamped Outer Edge and Free Inner Edge

$\frac{b}{a}$	Mode 1	Mode 2	Mode 3	Mode 4	Mode 5	Mode 6
0.7	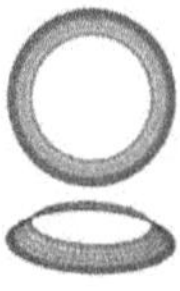 43.142	 45.332	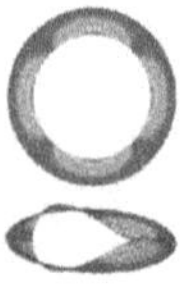 51.585	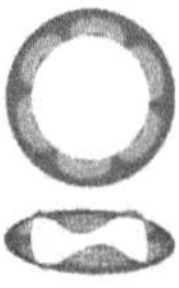 61.290	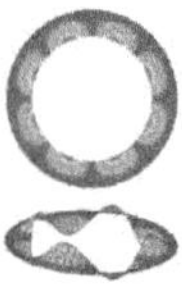 74.009	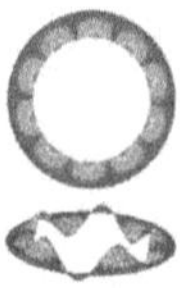 89.579
0.5	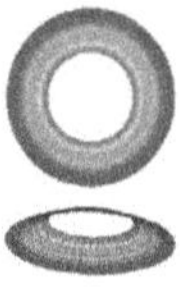 17.715	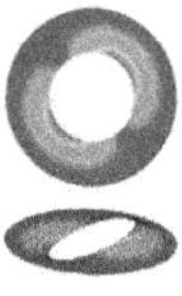 22.015	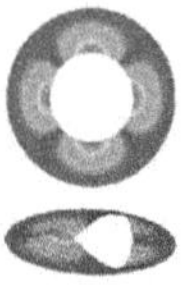 32.116	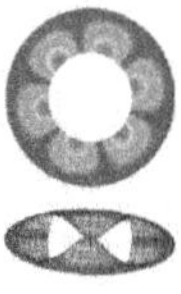 45.812	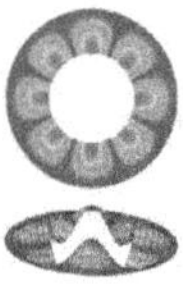 63.018	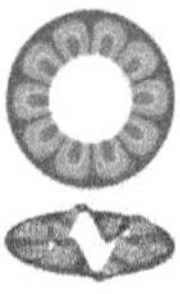 83.814
0.3	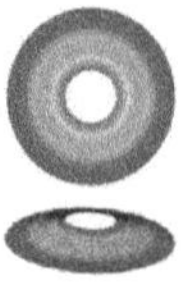 11.424	 19.540	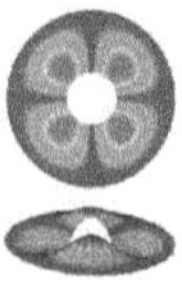 32.594	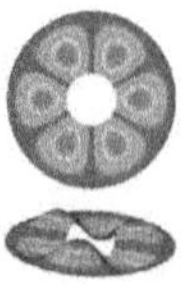 49.069	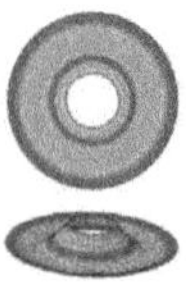 51.745	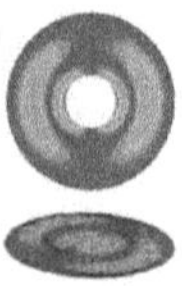 59.759

Does the simply supported condition on the radial edges of a slender sector influence the frequencies? The answer is probably very little. One can either invoke Saint Venant's principle or actually compute the frequencies of a long rectangle, with various boundary conditions on the short edges. Note that the Ritz method can be used on long rectangular plates but not on sector plates with $\alpha > 360°$.

Consider the mode shapes. It is generally believed that the fundamental mode (corresponding to the lowest frequency, including zero) does not have internal nodal lines. The work of Wang and Wang (2005) showed that, for a full annular plate, the fundamental mode may have two internal radial nodes. Although the fundamental mode with no internal nodes (1,1) is prevalent, there are exceptions. For example, the $\alpha > 180°$ of the F-F case has [1,1] or one internal radial node as fundamental mode. The fundamental [1,1] mode also occurs in F-S and F-C cases. They also find fundamental modes of (2,1), [2,1], (3,1), and (5,1), especially at higher α and lower b situations. Since the fundamental mode may have internal nodal lines, it is concluded

TABLE 5.22
Natural Frequencies $\bar{\omega} = \omega a^2 \sqrt{\rho h/D}$ for Annular Plate with Clamped Outer Edge and Simply Supported Inner Edge

$\frac{b}{a}$	Mode 1	Mode 2	Mode 3	Mode 4	Mode 5	Mode 6
0.7	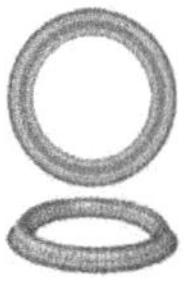		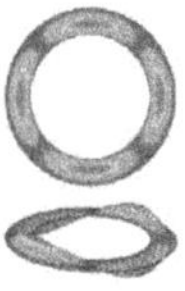		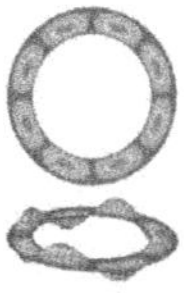	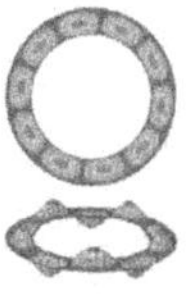
	174.408	175.516	178.866	184.530	192.612	203.233
0.5	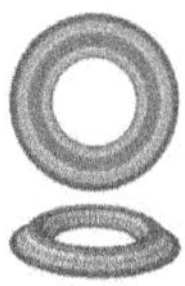	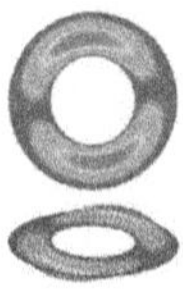	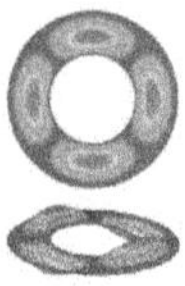	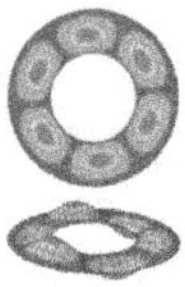	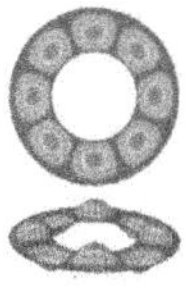	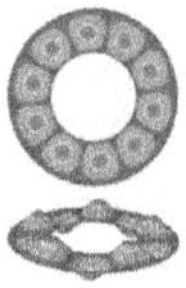
	63.973	65.486	70.136	78.182	89.857	105.253
0.3	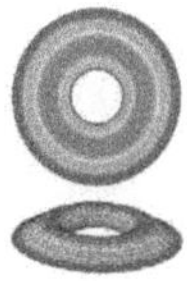	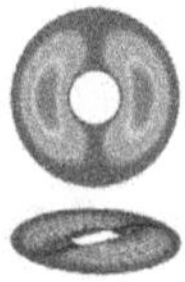	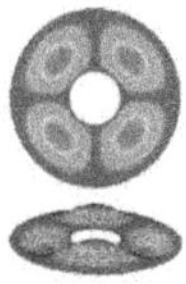	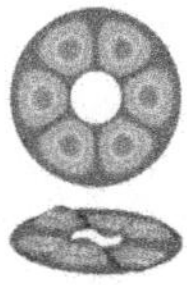	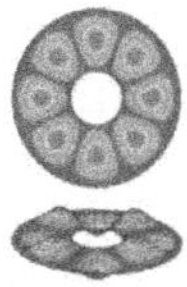	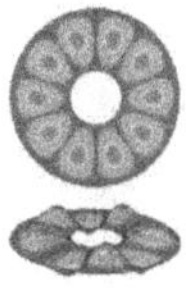
	33.765	35.906	42.731	54.609	71.063	91.216

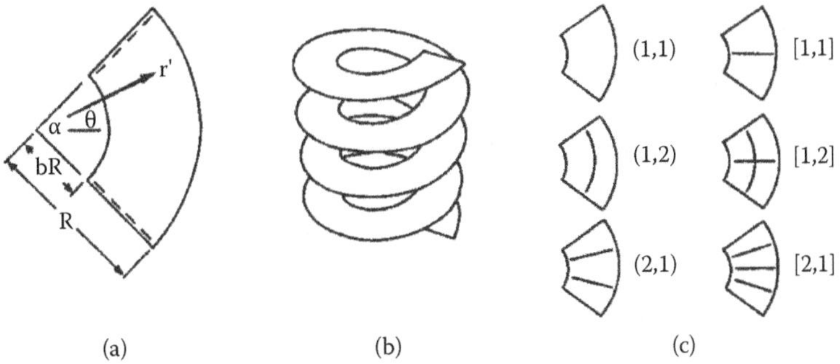

FIGURE 5.9 (a) The annular sector plate. (b) A coil with $\alpha = 1440°$, $b = 0.5$. (c) Mode shapes with internal nodal lines and their designations.

TABLE 5.23
First Two Roots of k^2 Using $b = 10^{-8}$ as Inner Radius

α	F-C, F-S, F-F	Free Outer Edge[a]
270°	2.75856	2.75864
	20.7230	20.7233
360°	3.22484	3.22484
	20.8057	20.8057

[a] Data for free outer edge is from Huang, Leissa, and McGee (1993).

that, for the same plate, an increase in nodal lines does not necessarily imply an increase in the frequency.

If the opening angle of the annular sector is π/N, where N is an integer, can all the frequencies of the annular sector plate be found from those of the full annular plate? This seems to be true. Consider an F-F plate with $\alpha = 90°$ and $b = 0.5$. The results are given in Table 5.33. In the table, the full annular plate frequency values with asterisks agree with Gabrielson (1999), which are more accurate than any of the results collated in Leissa's book (1969). The results for the annular sector plate are from Table 5.26, which are slightly more accurate than those of Ramakrishnan and Kunukkasseril (1973), whose results are in parentheses.

Table 5.34 lists the frequencies of the C-C full annular plate and those of the $\alpha = 360°$ sector plate, both with $b = 0.5$. The shapes are the same, except the sector has a simply supported radial slit. We find the fundamental frequencies are different. Also, the sector plate has a lot more frequencies than the full annulus for the same range, opposite to Table 5.33.

The F-F sector plate has an interesting application. Transverse vibrations of the F-F long sector are equivalent to the longitudinal vibrations of a coil. Consider the case where $\alpha = 1440°$ and $b = 0.5$ coil, as shown in Figure 5.9b. The fundamental frequency (Table 5.26) is 0.2861 (1,1). This means that the lowest frequency for longitudinal oscillations of the coil is 0.2861, with no fixed points. The second frequency is 0.5184 [1,1], for which the longitudinal wave has a fixed point at the middle, separating waves propagating in opposite directions.

Annular sector plates are important as curved girders and heat-transfer fins (inside or outside a cylinder). Their frequencies and mode shapes are decidedly different from those of the full annular plate.

5.4 GOVERNING EQUATIONS AND BOUNDARY CONDITIONS FOR VIBRATING THICK PLATES

For thick plates, it is necessary to allow for the effects of transverse deformation and rotary inertia due to their significant influences on the vibration frequencies. In order to accommodate the allowance of the former effect, the Kirchhoff hypothesis

TABLE 5.24
Frequencies for the C-C Sector Plate with Mode Shapes

α\b	0.1	0.5	0.9
50°	61.955 (1,1)	103.72 (1,1)	2245.1 (1,1)
	128.46 (1,2)	159.74 [1,1]	2269.2 [1,1]
	145.51 [1,1]	262.67 (2,1)	2310.7 (2,1)
	214.75 (1,3)	264.97 (1,2)	2371.0 [2,1]
	250.54 [1,2]	326.15 [1,2]	2452.2 (3,1)
90°	36.617 (1,1)	93.321 (1,1)	2239.6 (1,1)
	69.678 [1,1]	107.57 [1,1]	2247.0 [1,1]
	90.448 (1,2)	135.60 (2,1)	2259.3 (2,1)
	114.21 (2,1)	178.82 [2,1]	2276.9 [2,1]
	140.23 [1,2]	236.18 (3,1)	2299.9 (3,1)
180°	28.916 (1,1)	90.230 (1,1)	2237.8 (1,1)
	36.617 [1,1]	93.321 [1,1]	2239.6 [1,1]
	51.219 (2,1)	98.928 (2,1)	2242.3 (2,1)
	69.678 [2,1]	107.57 [2,1]	2247.0 [2,1]
	90.448 [1,2]	119.70 (3,1)	2252.5 (3,1)
270°	27.919 (1,1)	89.683 (1,1)	2237.5 (1,1)
	30.614 [1,1]	91.010 [1,1]	2238.3 [1,1]
	36.617 (2,1)	93.321 (2,1)	2239.6 (2,1)
	45.792 [2,1]	96.748 [2,1]	2241.5 [2,1]
	57.048 (3,1)	101.45 (3,1)	2244.0 (3,1)
360°	27.621 (1,1)	84.493 (1,1)	2237.3 (1,1)
	28.916 [1,1]	90.230 [1,1]	2237.8 [1,1]
	31.778 (2,1)	91.491 (2,1)	2238.6 (2,1)
	36.617 [2,1]	93.321 [2,1]	2239.6 [2,1]
	43.261 (3,1)	95.778 (3,1)	2241.0 (3,1)
720°	27.361 (1,1)	89.311 (1,1)	2237.2 (1,1)
	27.621 [1,1]	89.493 [1,1]	2237.3 [1,1]
	28.114 (2,1)	89.799 (2,1)	2237.5 (2,1)
	28.916 [2,1]	90.230 [2,1]	2237.8 [2,1]
	30.113 (3,1)	90.793 (3,1)	2238.1 (3,1)
1440°	27.300 (1,1)	89.266 (1,1)	2237.2 (1,1)
	27.361 [1,1]	89.311 [1,1]	2237.2 [1,1]
	27.466 (2,1)	89.387 (2,1)	2237.3 (2,1)
	27.621 [2,1]	89.493 [2,1]	2237.3 [2,1]
	27.834 (3,1)	89.630 (3,1)	2237.4 (3,1)

Note: The parentheses denote symmetric mode; the brackets denote antisymmetric mode.

TABLE 5.25

Frequencies for the S-S Sector Plate

α\b	0.1	0.5	0.9
50°	49.790 (1,1)	62.988 (1,1)	1001.6 (1,1)
	110.55 (1,2)	130.56 [1,1]	1044.7 [1,1]
	127.30 [1,1]	183.69 (1,2)	1116.5 (2,1)
	190.84 (1,3)	235.98 (2,1)	1217.0 [2,1]
	226.26 [1,2]	259.72 [1,2]	1346.3 (3,1)
90°	25.936 (1,1)	47.089 (1,1)	991.70 (1,1)
	56.842 [1,1]	68.379 [1,1]	1005.0 [1,1]
	76.687 (1,2)	103.44 (2,1)	1027.2 (2,1)
	97.995 (2,1)	166.35 (1,2)	1058.2 [2,1]
	121.71 [1,2]	189.60 [1,2]	1098.1 (3,1)
180°	16.776 (1,1)	41.797 (1,1)	988.38 (1,1)
	25.936 [1,1]	47.089 [1,1]	991.70 [1,1]
	39.976 (2,1)	55.957 (2,1)	997.24 (2,1)
	56.507 (1,2)	68.379 [2,1]	1005.0 [2,1]
	56.842 [2,1]	84.257 (3,1)	1015.0 (3,1)
270°	15.374 (1,1)	40.822 (1,1)	987.76 (1,1)
	19.043 [1,1]	43.166 [1,1]	989.24 [1,1]
	25.936 (2,1)	47.089 (2,1)	991.70 (2,1)
	34.940 [2,1]	52.603 [2,1]	995.15 [2,1]
	45.313 (3,1)	59.706 (3,1)	999.58 (3,1)
360°	14.953 (1,1)	40.481 (1,1)	987.55 (1,1)
	16.776 [1,1]	41.797 [1,1]	988.38 [1,1]
	20.499 (2,1)	43.998 (2,1)	989.76 (2,1)
	25.936 [2,1]	47.089 [2,1]	991.70 [2,1]
	32.545 (3,1)	51.075 (3,1)	994.20 (3,1)
720°	14.594 (1,1)	40.153 (1,1)	987.34 (1,1)
	14.953 [1,1]	40.481 [1,1]	987.55 [1,1]
	15.649 (2,1)	41.029 (2,1)	987.90 (2,1)
	16.776 [2,1]	41.797 [2,1]	988.38 [2,1]
	18.314 (3,1)	42.787 (3,1)	989.00 (3,1)
1440°	14.511 (1,1)	40.071 (1,1)	987.29 (1,1)
	14.594 [1,1]	40.153 [1,1]	987.34 [1,1]
	14.738 (2,1)	40.289 (2,1)	987.43 (2,1)
	14.953 [2,1]	40.481 [2,1]	987.55 [2,1]
	15.253 (3,1)	40.728 (3,1)	987.70 (3,1)

Note: The parentheses denote symmetric mode; the brackets denote antisymmetric mode.

TABLE 5.26
Frequencies for the F-F Sector Plate

α\b	0.1	0.5	0.9
50°	17.802 (1,1)	16.929 (1,1)	12.303 (1,1)
	64.993 (1,2)	58.629 (1,2)	53.353 [1,1]
	66.932 [1,1]	66.769 [1,1]	122.10 (2,1)
	129.83 (1,3)	137.20 (1,3)	161.03 (1,2)
	144.22 (2,1)	144.21 (2,1)	218.81 [2,1]
90°	5.3034 (1,1)	4.2711 (1,1)	2.9338 (1,1)
	21.835 [1,1]	21.067 [1,1]	15.503 [1,1]
	34.931 (1,2)	31.115 (1,2)	36.604 (2,1)
	47.378 (2,1)	47.063 (2,1)	66.228 [2,1]
	73.540 [1,2]	66.722 [1,2]	93.796 (1,3)
180°	5.3034 [1,1]	4.2711 [1,1]	2.9338 [1,1]
	12.437 (2,1)	11.425 (2,1)	8.1440 (2,1)
	20.406 (1,1)	17.198 (1,1)	15.503 [2,1]
	21.835 [2,1]	21.067 [2,1]	24.990 (3,1)
	33.495 (3,1)	32.982 (3,1)	55.720 (1,1)
270°	5.3034 (2,1)	0.6115 (1,1)	0.3744 (1,1)
	9.8117 [2,1]	1.0613 [1,1]	0.7023 [1,1]
	15.316 (3,1)	4.2711 (2,1)	2.9338 (2,1)
	16.942 (1,1)	8.7499 [2,1]	6.1663 [2,1]
	21.835 [3,1]	13.346 (1,2)	10.360 (3,1)
360°	1.8005 (1,1)	0.7044 (1,1)	0.4183 (1,1)
	2.4814 (2,1)	1.7353 (2,1)	1.1615 (2,1)
	5.3034 [2,1]	4.2711 [2,1]	2.9338 [2,1]
	8.5939 (3,1)	7.5187 (3,1)	5.2673 (3,1)
	12.437 [3,1]	11.425 [3,1]	8.1440 [3,1]
720°	1.0504 (2,1)	0.5075 (2,1)	0.2942 (1,1)
	1.1989 (3,1)	0.5184 (1,1)	0.3150 (2,1)
	1.8005 [1,1]	0.7044 [1,1]	0.4183 [1,1]
	1.8813 (1,1)	0.7588 (3,1)	0.4989 (3,1)
	2.4814 [3,1]	1.7353 [3,1]	1.1615 [3,1]
1440°	1.2861 (1,1)	0.2861 (1,1)	0.1598 (1,1)
	1.4769 (3,1)	0.5184 [1,1]	0.2942 [1,1]
	1.8005 [2,1]	0.6498 (3,1)	0.3838 (2,1)
	1.8813 [1,1]	0.6619 (2,1)	0.3949 (3,1)
	1.9714 (2,1)	0.7044 [2,1]	0.4183 [2,1]

Note: The parentheses denote symmetric mode; the brackets denote antisymmetric mode.

TABLE 5.27
Frequencies for the C-S Sector Plate

α\b	0.1	0.5	0.9
50°	213.78 (1,1)	84.742 (1,1)	1560.7 (1,1)
	250.54 [1,1]	151.76 [1,1]	1593.9 [1,1]
	259.48 (2,1)	225.44 (1,2)	1650.3 (2,1)
	319.35 (1,2)	260.62 (2,1)	1731.0 [2,1]
	372.97 [1,2]	250.54 [1,2]	1837.5 (3,1)
90°	35.406 (1,1)	70.136 (1,1)	1553.1 (1,1)
	69.667 [1,1]	89.858 [1,1]	1563.3 [1,1]
	86.716 (1,2)	124.31 (2,1)	1580.3 (2,1)
	114.21 (2,1)	172.68 [2,1]	1604.4 [2,1]
	140.13 [1,2]	209.21 (1,2)	1635.7 (3,1)
180°	25.283 (1,1)	65.486 (1,1)	1550.6 (1,1)
	35.406 [1,1]	70.136 [1,1]	1553.1 [1,1]
	51.065 (2,1)	78.182 (2,1)	1557.4 (2,1)
	69.667 [2,1]	89.858 [2,1]	1563.3 [2,1]
	90.448 [1,2]	105.25 (3,1)	1571.0 (3,1)
270°	23.724 (1,1)	64.643 (1,1)	1550.1 (1,1)
	27.775 [1,1]	66.676 [1,1]	1551.2 [1,1]
	35.406 (2,1)	70.136 (2,1)	1553.1 (2,1)
	45.459 [2,1]	75.106 [2,1]	1555.8 [2,1]
	56.982 (3,1)	81.663 (3,1)	1559.1 (3,1)
360°	23.247 (1,1)	64.349 (1,1)	1550.0 (1,1)
	25.283 [1,1]	65.486 [1,1]	1550.6 [1,1]
	29.378 (2,1)	67.404 (2,1)	1551.6 (2,1)
	35.406 [2,1]	70.136 [2,1]	1553.1 [2,1]
	42.785 (3,1)	73.718 (3,1)	1555.0 (3,1)
720°	22.830 (1,1)	64.067 (1,1)	1549.8 (1,1)
	23.247 [1,1]	64.349 [1,1]	1550.0 [1,1]
	24.033 (2,1)	64.821 (2,1)	1550.2 (2,1)
	25.283 [2,1]	65.486 [2,1]	1550.6 [2,1]
	27.061 (3,1)	66.345 (3,1)	1551.1 (3,1)
1440°	22.733 (1,1)	63.997 (1,1)	1549.8 (1,1)
	22.830 [1,1]	64.067 [1,1]	1549.8 [1,1]
	22.998 (2,1)	64.185 (2,1)	1549.9 (2,1)
	23.247 [2,1]	64.349 [2,1]	1550.0 [2,1]
	23.588 (3,1)	64.562 (3,1)	1550.1 (3,1)

Note: The parentheses denote symmetric mode; the brackets denote antisymmetric mode.

TABLE 5.28
Frequencies for the S-C Sector Plate

α\b	0.1	0.5	0.9
50°	49.807 (1,1)	76.517 (1,1)	1554.7 (1,1)
	110.72 (1,2)	135.50 [1,1]	1576.6 [1,1]
	127.30 [1,1]	217.89 (1,2)	1630.8 (2,1)
	191.56 (1,3)	237.10 (2,1)	1708.5 [2,1]
	226.26 [1,2]	282.25 [1,2]	1811.4 (3,1)
90°	26.717 (1,1)	64.631 (1,1)	1537.4 (1,1)
	56.848 [1,1]	80.803 [1,1]	1547.1 [1,1]
	74.631 (1,2)	110.75 (2,1)	1563.5 (2,1)
	97.995 (2,1)	154.68 [2,1]	1586.7 [2,1]
	121.78 [1,2]	204.05 (1,2)	1616.7 (3,1)
180°	26.717 [1,1]	60.987 (1,1)	1535.0 (1,1)
	40.062 (2,1)	64.631 [1,1]	1537.4 [1,1]
	56.848 [2,1]	71.107 (2,1)	1541.4 (2,1)
	63.243 (1,1)	80.803 [2,1]	1547.1 [2,1]
	74.631 [1,2]	93.988 (3,1)	1554.5 (3,1)
270°	18.418 (1,1)	60.336 (1,1)	1534.5 (1,1)
	21.041 [1,1]	61.912 [1,1]	1535.6 [1,1]
	26.717 (2,1)	64.631 (2,1)	1537.4 (2,1)
	35.135 [2,1]	68.608 [2,1]	1539.9 [2,1]
	45.349 (3,1)	73.964 (3,1)	1543.2 (3,1)
360°	18.125 (1,1)	60.109 (1,1)	1534.3 (1,1)
	19.394 [1,1]	60.987 [1,1]	1535.0 [1,1]
	22.159 (2,1)	62.481 (2,1)	1536.0 (2,1)
	26.717 [2,1]	64.631 [2,1]	1537.4 [2,1]
	32.830 (3,1)	67.489 (3,1)	1539.2 (3,1)
720°	17.869 (1,1)	59.892 (1,1)	1534.2 (1,1)
	18.125 [1,1]	60.109 [1,1]	1534.3 [1,1]
	18.609 (2,1)	60.473 (2,1)	1534.6 (2,1)
	19.394 [2,1]	60.987 [2,1]	1535.0 [2,1]
	20.558 (3,1)	61.655 (3,1)	1535.4 (3,1)
1440°	17.809 (1,1)	59.838 (1,1)	1534.2 (1,1)
	17.869 [1,1]	59.892 [1,1]	1534.2 [1,1]
	17.972 (2,1)	59.983 (2,1)	1534.3 (2,1)
	18.125 [2,1]	60.109 [2,1]	1534.3 [2,1]
	18.335 (3,1)	60.273 (3,1)	1534.5 (3,1)

Note: The parentheses denote symmetric mode; the brackets denote antisymmetric mode.

TABLE 5.29
Frequencies for the C-F Sector Plate

$\alpha \backslash b$	0.1	0.5	0.9
50°	61.909 (1,1)	55.702 (1,1)	375.25 (1,1)
	128.09 (1,2)	134.56 (1,2)	419.55 [1,1]
	145.51 [1,1]	140.73 [1,1]	492.59 (2,1)
	213.22 (1,3)	234.61 [1,2]	594.13 [2,1]
	259.48 (2,1)	258.36 (2,1)	724.43 (3,1)
90°	34.535 (1,1)	32.116 (1,1)	364.96 (1,1)
	69.660 [1,1]	63.018 [1,1]	378.73 [1,1]
	83.478 (1,2)	107.49 (1,2)	401.56 (2,1)
	114.21 (2,1)	107.96 (2,1)	433.31 [2,1]
	140.08 [1,2]	143.04 [1,2]	473.90 (3,1)
180°	21.195 (1,1)	22.015 (1,1)	361.50 (1,1)
	34.535 [1,1]	32.116 [1,1]	364.96 [1,1]
	50.990 (2,1)	45.812 (2,1)	370.70 (2,1)
	60.061 (1,2)	63.018 [2,1]	378.73 [2,1]
	69.663 [2,1]	83.814 (3,1)	389.02 (3,1)
270°	18.077 (1,1)	19.728 (1,1)	360.86 (1,1)
	24.997 [1,1]	24.918 [1,1]	362.40 [1,1]
	34.535 (2,1)	32.116 (2,1)	364.96 (2,1)
	45.277 [2,1]	40.868 [2,1]	368.53 [2,1]
	52.681 (1,2)	51.148 (3,1)	373.12 (3,1)
360°	16.409 (1,1)	18.871 (1,1)	360.64 (1,1)
	21.195 [1,1]	22.015 [1,1]	361.50 [1,1]
	27.195 (2,1)	26.559 (2,1)	362.94 (2,1)
	34.535 [2,1]	32.116 [2,1]	364.96 [2,1]
	42.508 (3,1)	38.539 (3,1)	367.54 (3,1)
720°	12.959 (1,1)	18.010 (1,1)	360.42 (1,1)
	16.409 [1,1]	18.871 [1,1]	360.64 [1,1]
	18.837 (2,1)	20.233 (2,1)	361.00 (2,1)
	21.195 [2,1]	22.015 [2,1]	361.50 [2,1]
	23.967 (3,1)	24.142 (3,1)	362.15 (3,1)
1440°	11.051 (1,1)	17.789 (1,1)	360.37 (1,1)
	12.959 [1,1]	18.010 [1,1]	360.42 [1,1]
	14.856 (2,1)	18.373 (2,1)	360.51 (2,1)
	16.409 [2,1]	18.871 [2,1]	360.64 [2,1]
	17.687 (3,1)	19.494 (3,1)	360.80 (3,1)

Note: The parentheses denote symmetric mode; the brackets denote antisymmetric mode.

TABLE 5.30
Frequencies for the F-C Sector Plate

$\alpha \backslash b$	0.1	0.5	0.9
50°	17.804 (1,1)	22.395 (1,1)	354.39 (1,1)
	65.044 (1,2)	67.581 [1,1]	385.36 [1,1]
	66.932 [1,1]	106.88 (1,2)	439.68 (2,1)
	130.20 (1,3)	144.29 (2,1)	519.78 [2,1]
	144.22 (2,1)	173.88 [1,2]	627.36 (3,1)
90°	5.6627 (1,1)	14.704 (1,1)	347.47 (1,1)
	21.836 [1,1]	25.596 [1,1]	356.76 [1,1]
	47.378 (2,1)	48.673 (2,1)	372.57 (2,1)
	73.557 [1,2]	82.106 [2,1]	395.32 [2,1]
	81.704 [2,1]	91.738 (1,2)	425.45 (3,1)
180°	3.4781 (1,1)	13.290 (1,1)	345.17 (1,1)
	5.6227 [1,1]	14.704 [1,1]	347.47 [1,1]
	12.451 (2,1)	18.562 (2,1)	351.32 (2,1)
	27.673 (1,2)	25.596 [2,1]	356.76 [2,1]
	36.941 [1,2]	35.730 (3,1)	363.83 (3,1)
270°	3.4847 [1,1]	13.128 (1,1)	344.74 (1,1)
	3.8100 (1,1)	13.572 [1,1]	345.76 [1,1]
	5.6227 (2,1)	14.704 (2,1)	347.47 (2,1)
	9.8528 [2,1]	16.938 [2,1]	349.86 [2,1]
	15.320 (3,1)	20.547 (3,1)	352.96 (3,1)
360°	3.4781 [1,1]	13.080 (1,1)	344.59 (1,1)
	3.7434 (2,1)	13.290 [1,1]	345.17 [1,1]
	3.9821 (1,1)	13.774 (2,1)	346.12 (2,1)
	5.6227 [2,1]	14.704 [2,1]	347.47 [2,1]
	8.6634 (3,1)	16.257 (3,1)	349.20 (3,1)
720°	3.4267 (3,1)	13.038 (1,1)	344.45 (1,1)
	3.4781 [2,1]	13.080 [1,1]	344.59 [1,1]
	3.7185 (2,1)	13.160 (2,1)	344.83 (2,1)
	3.7434 [3,1]	13.290 [2,1]	345.17 [2,1]
	3.9821 [1,1]	13.487 (3,1)	345.60 (3,1)
1440°	3.4165 (5,1)	13.028 (1,1)	344.41 (1,1)
	3.4267 [5,1]	13.038 [1,1]	344.45 [1,1]
	3.4781 [4,1]	13.055 (2,1)	344.51 (2,1)
	3.5310 (6,1)	13.080 [2,1]	344.59 [2,1]
	3.5865 (4,1)	13.115 (3,1)	344.70 (3,1)

Note: The parentheses denote symmetric mode; the brackets denote antisymmetric mode.

on the normality assumption has to be relaxed so that the normals to the undeformed mid-surface remain straight and unstretched in length but not necessarily normal to the deformed mid-surface. This assumption implies a nonzero transverse shear strain, but it also leads to the statical violation of the zero shear-stress condition at the free surfaces, since the shear stress becomes constant through the plate

TABLE 5.31
Frequencies for the S-F Sector Plate

$\alpha\backslash b$	0.1	0.5	0.9
50°	49.783 (1,1)	45.082 (1,1)	89.072 (1,1)
	110.48 (1,2)	109.59 (1,2)	182.75 [1,1]
	127.30 [1,1]	124.62 [1,1]	290.78 (2,1)
	190.44 (1,3)	211.36 [1,2]	417.84 [2,1]
	226.26 [1,2]	234.90 (2,1)	567.53 (3,1)
90°	25.394 (1,1)	22.357 (1,1)	51.180 (1,1)
	56.840 [1,1]	52.032 [1,1]	98.914 [1,1]
	69.260 (1,2)	81.100 (1,2)	150.17 (2,1)
	97.995 (2,1)	94.183 (2,1)	205.34 [2,1]
	121.69 [1,2]	118.33 [1,2]	265.28 (3,1)
180°	13.872 (1,1)	11.607 (1,1)	29.769 (1,1)
	25.394 [1,1]	22.357 [1,1]	51.180 [1,1]
	39.935 (2,1)	35.636 (2,1)	74.562 (2,1)
	48.012 (2,1)	52.032 [2,1]	98.914 [2,1]
	56.840 [2,1]	71.638 (3,1)	124.10 (3,1)
270°	11.414 (1,1)	8.5788 (1,1)	23.828 (1,1)
	17.124 [1,1]	14.944 [1,1]	36.514 [1,1]
	25.394 (2,1)	22.357 (2,1)	51.180 (2,1)
	34.838 [2,1]	30.884 [2,1]	66.642 [2,1]
	41.706 (1,3)	40.739 (3,1)	82.585 (3,1)
360°	10.224 (1,1)	7.2502 (1,1)	21.366 (1,1)
	13.872 [1,1]	11.607 [1,1]	29.769 [1,1]
	19.027 (2,1)	16.705 (2,1)	40.069 (2,1)
	25.394 [2,1]	22.357 [2,1]	51.180 [2,1]
	32.385 (3,1)	28.635 (3,1)	62.725 (3,1)
720°	7.6085 (1,1)	5.6963 (1,1)	18.690 (1,1)
	10.224 [1,1]	7.2502 [1,1]	21.366 [1,1]
	11.980 (2,1)	9.2979 (2,1)	25.206 (2,1)
	13.872 [2,1]	11.607 [2,1]	29.769 [2,1]
	16.235 (3,1)	14.086 (3,1)	34.776 (3,1)
1440°	5.8376 (1,1)	5.2385 (1,1)	17.959 (1,1)
	7.6085 [1,1]	5.6963 [1,1]	18.690 [1,1]
	9.1052 (2,1)	6.3884 (2,1)	19.849 (2,1)
	10.224 [2,1]	7.2502 [2,1]	21.366 [2,1]
	11.131 (3,1)	8.2315 (3,1)	23.172 (3,1)

Note: The parentheses denote symmetric mode; the brackets denote antisymmetric mode.

TABLE 5.32
Frequencies for the F-S Sector Plate

α\b	0.1	0.5	0.9
50°	17.803 (1,1)	18.977 (1,1)	73.178 (1,1)
	65.008 (1,2)	67.081 [1,1]	150.75 [1,1]
	66.932 [1,1]	88.589 (1,2)	242.84 (2,1)
	129.95 (1,3)	144.24 (2,1)	353.60 [2,1]
	144.22 (2,1)	164.88 [1,2]	486.08 (3,1)
90°	5.4277 (1,1)	7.9861 (1,1)	42.619 (1,1)
	21.835 [1,1]	22.788 [1,1]	81.210 [1,1]
	35.774 (1,2)	47.688 (2,1)	123.51 (2,1)
	47.378 (2,1)	69.766 (1,2)	169.80 [2,1]
	73.544 [1,2]	81.794 [2,1]	220.89 (3,1)
180°	2.4377 (1,1)	4.8616 (1,1)	25.894 (1,1)
	5.4277 [1,1]	7.9861 [1,1]	42.619 [1,1]
	12.441 (2,1)	14.035 (2,1)	61.397 (2,1)
	21.835 [2,1]	22.788 [2,1]	81.210 [2,1]
	24.244 (1,2)	34.047 (3,1)	101.90 (3,1)
270°	2.6524 [1,1]	4.4183 (1,1)	21.458 (1,1)
	2.8451 (1,1)	5.5806 [1,1]	31.076 [1,1]
	5.4277 (2,1)	7.9861 (2,1)	42.619 (2,1)
	9.8250 [2,1]	11.705 [2,1]	55.003 [2,1]
	15.317 (1,3)	16.664 (3,1)	67.901 (3,1)
360°	2.4377 [1,1]	4.2812 (1,1)	19.674 (1,1)
	3.0852 (1,1)	4.8616 [1,1]	25.894 [1,1]
	3.1054 (2,1)	6.0576 (2,1)	33.847 (2,1)
	5.4277 [2,1]	7.9861 [2,1]	42.619 [2,1]
	8.6173 (3,1)	10.656 (3,1)	51.852 (3,1)
720°	2.4377 [2,1]	4.1592 (1,1)	17.788 (1,1)
	2.5113 (3,1)	4.2812 [1,1]	19.674 [1,1]
	2.7214 (2,1)	4.5065 (2,1)	22.473 (2,1)
	3.0852 [1,1]	4.8616 [2,1]	25.894 [2,1]
	3.1054 [3,1]	5.3723 (3,1)	29.730 (3,1)
1440°	2.4129 (5,1)	4.1304 (1,1)	17.285 (1,1)
	2.4377 [4,1]	4.1592 [1,1]	17.788 [1,1]
	2.5113 [5,1]	4.2087 (2,1)	18.597 (2,1)
	2.5535 (4,1)	4.2812 [2,1]	19.674 [2,1]
	2.7214 [3,1]	4.3793 (3,1)	20.979 (3,1)

Note: The parentheses denote symmetric mode; the brackets denote antisymmetric mode.

TABLE 5.33
Comparison of Frequencies for the Full Annulus and $\alpha = 90°$ Sector Plate with F-F Curved Boundaries and $b = 0.5$

Full Annulus	Annular Sector
4.2711*	4.2711 (4.2708)
9.3135*	21.067 (21.068)
11.425*	31.115 (31.119)
17.198*	47.063 (47.062)
21.067*	66.722
31.115*	
32.982*	
47.063	
63.273	
66.722	

Note: The full-annulus plate frequency values with asterisks agree with Gabrielson (1999). The results for the annular sector plate are from Table 5.24, which are slightly more accurate than those of Ramakrishnan and Kunukkasseril (1973), whose results are in parentheses.

thickness. In order to compensate for this error, Mindlin (1951) proposed a shear correction factor κ^2 to be applied to the transverse shear force. Therefore, in the open literature, vibrating plates based on the first-order shear deformation plate theory assumptions, including the effect of rotary inertia, are widely referred to as Mindlin plates.

TABLE 5.34
Frequencies of the Full Annulus and Those of the 360° Sector Plate with C-C edges and $b = 0.5$

Full Annulus	Annular Sector
89.251	84.493
90.230	90.230
93.321	91.491
98.928	93.321
	95.778
	98.928

The governing equations of motion of Mindlin plates are given by

$$\kappa^2 Gh\left[\nabla^2 w + \left(\frac{\partial \psi_x}{\partial x} + \frac{\partial \psi_y}{\partial y}\right)\right] + \rho h \omega^2 w = 0 \tag{5.88}$$

$$\frac{D(1-\nu)}{2}\nabla^2\psi_y + \frac{D(1+\nu)}{2}\left(\frac{\partial^2 \psi_x}{\partial x \partial y} + \frac{\partial^2 \psi_y}{\partial y^2}\right) - \kappa^2 Gh\left(\psi_y + \frac{\partial w}{\partial y}\right) + \frac{\rho h^3}{12}\omega^2 \psi_y = 0 \tag{5.89}$$

$$\frac{D(1-\nu)}{2}\nabla^2\psi_x + \frac{D(1+\nu)}{2}\left(\frac{\partial^2 \psi_y}{\partial x \partial y} + \frac{\partial^2 \psi_x}{\partial x^2}\right) - \kappa^2 Gh\left(\psi_x + \frac{\partial w}{\partial x}\right) + \frac{\rho h^3}{12}\omega^2 \psi_x = 0 \tag{5.90}$$

where G is the shear modulus, and ψ_x, ψ_y are the bending rotations of a transverse normal about the y- and x-axes, respectively.

The common edge conditions for Mindlin plates are given as follows:

- For a free edge (F)

$$M_{nn} = D\left(\frac{\partial \psi_n}{\partial n} + \nu\frac{\partial \psi_s}{\partial s}\right) = 0, \quad M_{ns} = D\left(\frac{1-\nu}{2}\right)\left(\frac{\partial \psi_n}{\partial s} + \frac{\partial \psi_s}{\partial n}\right) = 0,$$
$$Q_n = \kappa^2 Gh\left(\psi_n + \frac{\partial w}{\partial n}\right) = 0 \tag{5.91}$$

- For a simply supported edge, there are two kinds.
  The first kind (S), which is referred to as the hard type simple support, requires that

$$M_{nn} = 0, \quad \psi_s = 0, \quad w = 0 \tag{5.92}$$

  The second type (S*), commonly referred to as the soft-type simple support, requires that

$$M_{nn} = 0, \quad M_{ns} = 0, \quad w = 0 \tag{5.93}$$

- For a clamped edge (C)

$$\psi_n = 0, \quad \psi_s = 0, \quad w = 0 \tag{5.94}$$

5.5 EXACT VIBRATION SOLUTIONS FOR THICK PLATES

Exact vibration solutions for thick plates are obtainable for the same shapes and boundary conditions as those considered in the thin-plate section. In the following discussion, the exact vibration solutions are presented based on the Mindlin plate theory.

5.5.1 Polygonal Plates with Simply Supported Edges

Wang (1994) presented an exact relationship between the frequencies of Mindlin plates and the Kirchhoff (classical thin) plates for simply supported polygonal plates. This relationship is given by

$$\omega_N^2 = \frac{6\kappa^2 G}{\rho h^2}\left\{\left[1+\frac{1}{12}\breve{\omega}_N h^2\sqrt{\frac{\rho h}{D}}\left(1+\frac{2}{\kappa^2(1-\nu)}\right)\right] - \sqrt{\left[1+\frac{1}{12}\breve{\omega}_N h^2\sqrt{\frac{\rho h}{D}}\left(1+\frac{2}{\kappa^2(1-\nu)}\right)\right]^2 - \frac{\rho h^2}{3\kappa^2 G}\breve{\omega}_N^2}\right\} \quad (5.95)$$

where ω is the angular frequency of the Mindlin plate, $\breve{\omega}$ is the angular frequency of the corresponding Kirchhoff plate, and $N = 1,2,\ldots,n$ corresponds to the mode sequence number. So by supplying the exact frequency value of the Kirchhoff plate, the corresponding exact frequency value for the Mindlin plate is readily obtained from the aforementioned frequency relationship. The frequency relationship between Mindlin and Kirchhoff plates is shown in a graphical form in Figure 5.10. It is clear that Mindlin plate frequency is lower than its corresponding

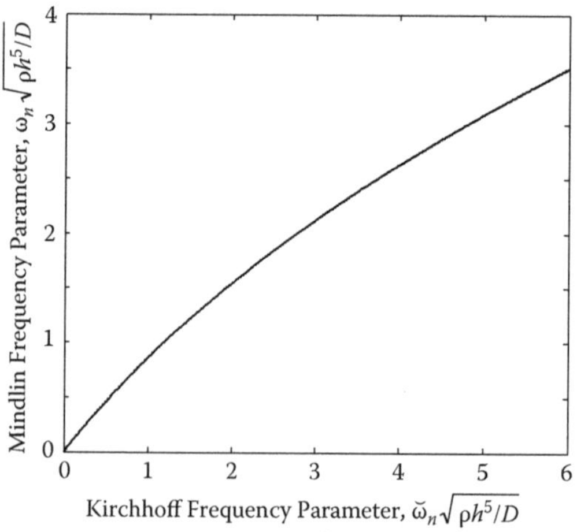

FIGURE 5.10 Frequency relationship between Mindlin and Kirchhoff plates.

TABLE 5.35
Frequencies of Triangular and Rectangular Kirchhoff Plates with Simply Supported Edges

Plate Shapes	Frequency Parameters $\tilde{\omega}_n a^2\sqrt{\rho h/D}$								
	d/a	b/a	Mode Sequence Number						Source
			1	2	3	4	5	6	
Triangle 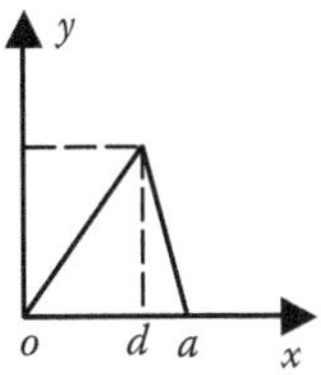	1/4	2/5	23.75	40.80	60.54	70.33	83.39	101.8	Liew (1993a)
		1/2	27.12	49.47	75.18	78.27	107.4	117.4	
		2/3	33.11	65.26	88.68	106.1	141.2	156.4	
		1.0	46.70	100.2	117.2	171.8	197.0	220.4	
		$2/\sqrt{3}$	53.78	115.9	134.7	198.1	229.2	252.0	
		2.0	101.5	195.8	275.0	315.6	427.7	464.2	
	1/2	2/5	23.61	40.70	60.55	69.78	83.42	101.5	
		1/2	26.91	49.33	76.29	76.30	108.0	116.4	
		2/3	32.72	65.22	87.83	106.0	142.5	154.8	
		1.0	45.83	102.8	111.0	177.3	199.5	203.4	
		$2/\sqrt{3}$	52.64	122.8	122.8	210.5	228.1	228.1	
		2.0	98.57	197.4	256.2	335.4	394.8	492.5	
Rectangle 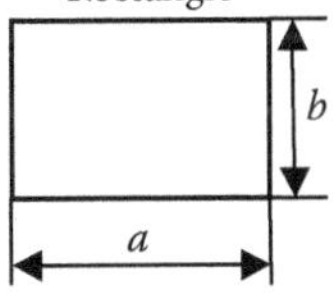	$\tilde{\omega}_n\sqrt{\frac{\rho h}{D}} = \left(\frac{m\pi}{a}\right)^2 + \left(\frac{n\pi}{b}\right)^2$ where m and n are the number of half waves								Leissa (1969)

Kirchhoff plate counterpart because of the allowance for the effect of transverse shear deformation in the former plate theory that makes the plate more flexible. Wang, Kitipornchai, and Reddy (2000) derived a similar frequency relationship for the third-order shear deformable (Reddy) plate theory and the Kirchhoff plate theory.

Some vibration frequencies (not exact solution but highly accurate) are given for simply supported Kirchhoff plates for triangular and rectangular plates in Table 5.35, parallelogram plates in Table 5.36, trapezoidal plates in Table 5.37, and regular polygonal plates in Table 5.38. When substituted in Equation (5.95), the Kirchhoff plate frequencies in these tables furnish the corresponding frequencies for Mindlin plates.

5.5.2 Rectangular Plates

Exact vibrations solutions for simply supported rectangular thick plates were obtained by Mindlin, Schacknow, and Deresiewicz (1956). Hashemi and Arsanjani (2005) derived exact vibrations solutions for rectangular thick plates with two parallel sides simply supported. Xing and Liu (2009b, 2009c) simplified the exact expressions obtained by Hashemi and Arsanjani (2005).

TABLE 5.36
Frequencies of Parallelogram Kirchhoff Plates with Simply Supported Edges

Plate Shape	Frequency Parameters $\breve{\omega}_n b^2\sqrt{\rho h/D}$								
	a/b	β	Mode Shape Number						Source
			1	2	3	4	5	6	
Parallelogram 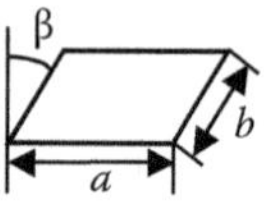	1.0	15°	20.87	48.20	56.12	79.05	104.0	108.9	Liew, Xiang, Kitipornchai, and Wang (1993)
		30°	24.98	52.63	71.87	83.86	122.8	122.8	
		45°	35.33	66.27	100.5	108.4	140.8	168.3	
		60°	66.30	105.0	148.7	196.4	213.8	250.7	
	1.5	15°	15.10	28.51	46.96	49.76	61.70	75.80	
		30°	18.17	32.94	53.48	58.02	76.05	78.61	
		45°	25.96	42.39	64.80	84.18	93.31	107.5	
		60°	48.98	70.51	96.99	127.3	162.3	171.1	
	2.0	15°	13.11	20.66	33.08	44.75	50.24	52.49	
		30°	15.90	23.95	36.82	52.64	56.63	63.26	
		45°	23.01	32.20	46.21	63.50	82.08	83.00	
		60°	44.00	56.03	72.79	92.80	117.4	151.7	

According to Liu (from Beihang University, China), the general exact solutions for vibration of rectangular Mindlin plates with two simply supported edges $x = 0$ and $x = a$ may be expressed as

$$\psi_x = g_1\frac{\partial w_1}{\partial x} + g_2\frac{\partial w_2}{\partial x} - \frac{\partial w_3}{\partial y} \tag{5.96}$$

$$\psi_y = g_1\frac{\partial w_1}{\partial y} + g_2\frac{\partial w_2}{\partial y} + \frac{\partial w_3}{\partial x} \tag{5.97}$$

$$w = w_1 + w_2 \tag{5.98}$$

TABLE 5.37
Frequencies of Symmetrical Trapezoidal Kirchhoff Plates with Simply Supported Edges

Plate Shape	Frequency Parameters $\breve{\omega}_n b^2\sqrt{\rho h/D}$								
	a/b	c/b	Mode Shape Number						Source
			1	2	3	4	5	6	
Symmetric Trapezoid 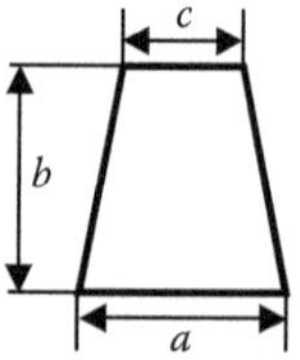	1.0	1/5	3.336	4.595	6.860	10.19	10.23	11.53	Liew and Lim (1993)
		2/5	2.198	3.479	5.499	5.789	7.737	9.027	
		3/5	1.654	3.066	3.728	5.394	6.037	6.156	
		4/5	1.356	2.833	2.879	4.560	5.086	5.192	
	1.5	1/5	6.158	7.269	9.507	12.85	17.30	20.42	
		2/5	3.703	5.175	7.390	9.272	10.63	14.22	
		3/5	2.636	4.313	5.971	6.575	9.787	10.05	
		4/5	2.089	3.802	4.494	6.121	7.257	8.000	
	2.0	1/5	9.919	10.76	13.17	16.51	20.99	26.74	
		2/5	5.351	7.575	9.633	.12.83	13.69	17.07	
		3/5	3.680	5.793	8.255	8.398	11.41	14.48	
		4/5	2.856	5.053	6.187	7.452	10.32	10.64	

TABLE 5.38
Frequencies of Regular Polygonal Plates with Simply Supported Edges

Plate Shape	Frequency Parameters $\breve{\omega}_n b^2\sqrt{\rho h/D}$							
Regular Polygon	Number of	Mode Shape Number						Source
		1	2	3	4	5	6	
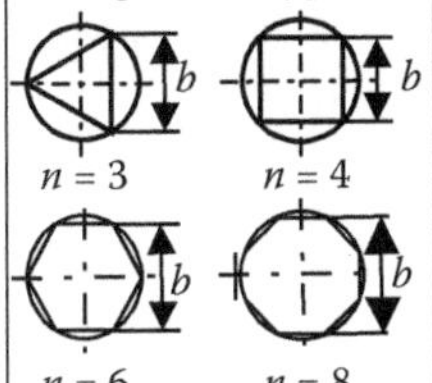	3	52.6	122	122	210	228	228	Liew (1993a)
	4	19.7	49.3	49.3	79.0	98.7	98.7	
	6	28.9	73.0	73.2	130	130	151	Liew and Lam (1991)
	8	22.2	57.8	57.8	102	102	117	

where

$$w_1 = (A_1 \sin\beta_1 y + A_2 \cos\beta_1 y)\sin\left(\frac{m\pi x}{a}\right) \tag{5.99}$$

$$w_2 = \left(A_3 \sinh\beta_2 y + A_4 \cosh\beta_2 y\right)\sin\left(\frac{m\pi x}{a}\right) \tag{5.100}$$

$$w_3 = (A_5 \sinh\beta_3 y + A_6 \cosh\beta_3 y)\cos\left(\frac{m\pi x}{a}\right) \tag{5.101}$$

$$g_1 = \left[\left(1 - \frac{\omega^2\rho h^2}{12\kappa^2 G}\right) + \frac{D}{\kappa^2 Gh}\left(\beta_1^2 + \left(\frac{m\pi}{a}\right)^2\right)\right]^{-1} \tag{5.102}$$

$$g_2 = \left[\left(1 - \frac{\omega^2\rho h^2}{12\kappa^2 G}\right) - \frac{D}{\kappa^2 Gh}\left(\beta_2^2 - \left(\frac{m\pi}{a}\right)^2\right)\right]^{-1} \tag{5.103}$$

$$\beta_1 = \sqrt{\left[\sigma + \sqrt{1 - \frac{\omega^2\rho h^2}{12\kappa^2 G} + \sigma^2}\right]\sqrt{\frac{\omega^2\rho h}{D}} - \left(\frac{m\pi}{a}\right)^2} \tag{5.104}$$

$$\beta_2 = \sqrt{\left[-\sigma + \sqrt{1 - \frac{\omega^2\rho h^2}{12\kappa^2 G} + \sigma^2}\right]\sqrt{\frac{\omega^2\rho h}{D}} + \left(\frac{m\pi}{a}\right)^2} \tag{5.105}$$

$$\beta_3 = \sqrt{\left(1 - \frac{\omega^2\rho h^2}{12\kappa^2 G}\right)\frac{2\kappa^2 Gh}{(1-\nu)D} + \left(\frac{m\pi}{a}\right)^2} \tag{5.106}$$

$$\sigma = \frac{1}{2}\left(\frac{D}{\kappa^2 Gh} + \frac{h^2}{12}\right)\sqrt{\frac{\omega^2\rho h}{D}} \tag{5.107}$$

Consider a rectangular thick plate with three edges ($x = 0$, $x = a$, $y = 0$) simply supported, whereas the edge $y = b$ is clamped. The substitution of Equations (5.96) to (5.98) into the boundary conditions (5.92) and (5.94) yields the following characteristic equation:

$$[g_2\beta_2 \sin(\beta_1 b) - g_1\beta_1 \cos(\beta_1 b)\tanh(\beta_2 b)]\beta_3 \tanh(\beta_3 b) - [g_2 - g_1]\left(\frac{m\pi}{a}\right)^2 \sin(\beta_1 b)\tanh(\beta_2 b) = 0 \tag{5.108}$$

and the following coefficients of the vibration mode shape

$$A_1 = g_2\left[\left(\frac{m\pi}{a}\right)^2 \cosh(\beta_3 b)\sinh(\beta_2 b) - \beta_2\beta_3 \cosh(\beta_2 b)\sinh(\beta_3 b)\right] \tag{5.109a}$$

$$A_2 = 0 \tag{5.109b}$$

$$A_3 = g_1\left[\beta_1\beta_3 \cos(\beta_1 b)\sinh(\beta_3 b) - \left(\frac{m\pi}{a}\right)^2 \cosh(\beta_3 b)\sin(\beta_1 b)\right] \tag{5.109c}$$

$$A_4 = 0 \tag{5.109d}$$

$$A_5 = 0 \tag{5.109e}$$

$$A_6 = g_1 g_2\left(\frac{m\pi}{a}\right)[\beta_1 \cos(\beta_1 b)\sinh(\beta_2 b) - \beta_2 \cosh(\beta_2 b)\sin(\beta_1 b)] \tag{5.109f}$$

Consider a rectangular thick plate with three edges ($x = 0$, $x = a$, $y = 0$) simply supported, whereas the edge $y = b$ is free. The characteristic equation can be obtained as

$$\begin{aligned}
&g_1\left[\beta_1^2 + \nu\left(\frac{m\pi}{a}\right)^2\right]\sin(\beta_1 b)\left[\left\{-\left(\frac{m\pi}{a}\right)^2 - \beta_3^2\right\}\{1 - g_2\}\beta_2 - 2g_2\beta_2\left(\frac{m\pi}{a}\right)^2\right] \\
&+g_2\left[\beta_2^2 - \nu\left(\frac{m\pi}{a}\right)^2\right]\cos(\beta_1 b)\tanh(\beta_2 b)\left[\left\{-\left(\frac{m\pi}{a}\right)^2 - \beta_3^2\right\}\{1 - g_1\}\beta_1 - 2g_1\beta_1\left(\frac{m\pi}{a}\right)^2\right] \\
&-2(g_1 - g_2)(1 - \nu)\beta_1\beta_2\beta_3\left(\frac{m\pi}{a}\right)^2 \cos(\beta_1 b)\tanh(\beta_3 b) = 0
\end{aligned} \tag{5.110}$$

and the coefficients of mode shape as

$$A_1 = \beta_2 \cosh(\beta_2 b)\cosh(\beta_3 b)\left[\left(\frac{m\pi}{a}\right)^2 (1 + g_2) + \beta_3^2(1 - g_2)\right] \tag{5.111a}$$

$$A_2 = 0 \tag{5.111b}$$

$$A_3 = -\beta_1 \cos(\beta_1 b)\cosh(\beta_3 b)\left[\left(\frac{m\pi}{a}\right)^2 (1+g_1)+\beta_3^2(1-g_1)\right] \tag{5.111c}$$

$$A_4 = 0 \tag{5.111d}$$

$$A_5 = 0 \tag{5.111e}$$

$$A_6 = 2(g_1 - g_2)\beta_1\beta_2\left(\frac{m\pi}{a}\right)\cos(\beta_1 b)\cosh(\beta_2 b) \tag{5.111f}$$

For rectangular plates with two edges (x = 0 and x = a) simply supported, whereas the other two edges (y = 0 and y = b) are both clamped or both free, the characteristic equation and coefficients of mode shape can be obtained in similar form as the two aforementioned cases after some simplifications. Consider shifting the origin of the y-coordinate to the center of the side x = 0, as shown in Figure 5.11.

The mode shapes of w that are antisymmetric with respect to the x-axis can be expressed as

$$w = (A_1 \sin\beta_1 y + A_3 \sinh\beta_2 y)\sin\left(\frac{m\pi x}{a}\right) \tag{5.112a}$$

$$\psi_x = \left[\frac{m\pi}{a}(g_1 A_1 \sin\beta_1 y + g_2 A_3 \sinh\beta_2 y) - \beta_3 A_6 \sinh\beta_3 y\right]\cos\left(\frac{m\pi x}{a}\right) \tag{5.112b}$$

$$\psi_y = \left[g_1\beta_1 A_1 \cos\beta_1 y + g_2\beta_2 A_3 \cosh\beta_2 y - \frac{m\pi}{a} A_6 \cosh\beta_3 y\right]\sin\left(\frac{m\pi x}{a}\right) \tag{5.112c}$$

The mode shapes of w that are symmetric with respect to the x-axis can be written as

$$w = (A_2 \cos\beta_1 y + A_4 \cosh\beta_2 y)\sin\left(\frac{m\pi x}{a}\right) \tag{5.113a}$$

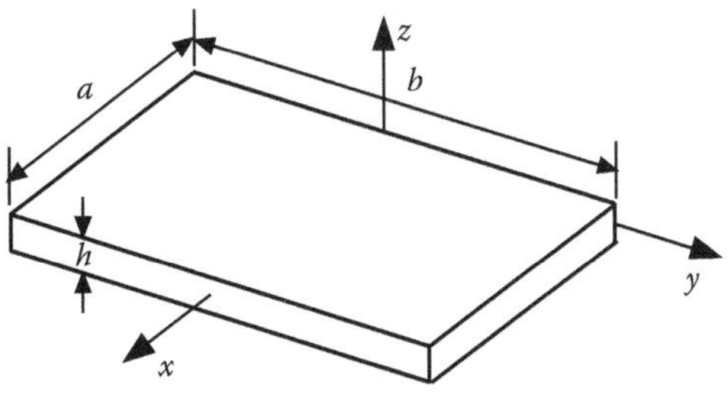

FIGURE 5.11 Mindlin plate and coordinates.

$$\psi_x = \left[\frac{m\pi}{a}(g_1 A_2 \cos\beta_1 y + g_2 A_4 \cosh\beta_2 y) - \beta_3 A_5 \cosh\beta_3 y\right]\cos\left(\frac{m\pi x}{a}\right) \tag{5.113b}$$

$$\psi_y = \left[g_1\beta_1 A_2 \sin\beta_1 y + g_2\beta_2 A_4 \sinh\beta_2 y - \frac{m\pi}{a} A_5 \sinh\beta_3 y\right]\sin\left(\frac{m\pi x}{a}\right) \tag{5.113c}$$

Consider a rectangular plate with two edges ($x = 0$ and $x = a$) simply supported, whereas the other two edges are clamped. The characteristic equation and coefficients of mode shape for the antisymmetric case are given by

$$\begin{aligned}\beta_3\left[g_1\beta_1 \cos\left(\frac{\beta_1 b}{2}\right) - g_2\beta_2 \coth\left(\frac{\beta_2 b}{2}\right)\sin\left(\frac{\beta_1 b}{2}\right)\right] \\ -\left(\frac{m\pi x}{a}\right)^2 (g_1 - g_2)\coth\left(\frac{\beta_3 b}{2}\right)\sin\left(\frac{\beta_1 b}{2}\right) = 0\end{aligned} \tag{5.114}$$

$$A_1 = g_2\left[\left(\frac{m\pi}{a}\right)^2 \cosh\left(\frac{\beta_3 b}{2}\right)\sinh\left(\frac{\beta_2 b}{2}\right) - \beta_2\beta_3 \cosh\left(\frac{\beta_2 b}{2}\right)\sinh\left(\frac{\beta_3 b}{2}\right)\right] \tag{5.115a}$$

$$A_2 = 0 \tag{5.115b}$$

$$A_3 = g_1\left[-\left(\frac{m\pi}{a}\right)^2 \cosh\left(\frac{\beta_3 b}{2}\right)\sin\left(\frac{\beta_1 b}{2}\right) + \beta_1\beta_3 \cos\left(\frac{\beta_1 b}{2}\right)\sinh\left(\frac{\beta_3 b}{2}\right)\right] \tag{5.115c}$$

$$A_4 = 0 \tag{5.115d}$$

$$A_5 = 0 \tag{5.115e}$$

$$A_6 = g_1 g_2\left(\frac{m\pi}{a}\right)\left[\beta_1 \cos\left(\frac{\beta_1 b}{2}\right)\sinh\left(\frac{\beta_2 b}{2}\right) - \beta_2 \cosh\left(\frac{\beta_2 b}{2}\right)\sin\left(\frac{\beta_1 b}{2}\right)\right] \tag{5.115f}$$

The characteristic equation and coefficients of mode shape for the symmetric case can be obtained as

$$\begin{aligned}\beta_3\left[g_1\beta_1 \sin\left(\frac{\beta_1 b}{2}\right) + g_2\beta_2 \cos\left(\frac{\beta_1 b}{2}\right)\tanh\left(\frac{\beta_2 b}{2}\right)\right] \\ +\left(\frac{m\pi x}{a}\right)^2 (g_1 - g_2)\tanh\left(\frac{\beta_3 b}{2}\right)\cos\left(\frac{\beta_1 b}{2}\right) = 0\end{aligned} \tag{5.116}$$

$$A_1 = 0 \tag{5.117a}$$

$$A_2 = g_2\left[-\left(\frac{m\pi}{a}\right)^2 \cosh\left(\frac{\beta_2 b}{2}\right)\sinh\left(\frac{\beta_3 b}{2}\right) + \beta_2\beta_3 \cosh\left(\frac{\beta_3 b}{2}\right)\sinh\left(\frac{\beta_2 b}{2}\right)\right] \tag{5.117b}$$

$$A_3 = 0 \tag{5.117c}$$

$$A_4 = g_1\left[\left(\frac{m\pi}{a}\right)^2 \sinh\left(\frac{\beta_3 b}{2}\right)\cos\left(\frac{\beta_1 b}{2}\right) + \beta_1\beta_3 \sin\left(\frac{\beta_1 b}{2}\right)\cosh\left(\frac{\beta_3 b}{2}\right)\right] \tag{5.117d}$$

$$A_5 = g_1 g_2\left(\frac{m\pi}{a}\right)\left[\beta_1 \sin\left(\frac{\beta_1 b}{2}\right)\cosh\left(\frac{\beta_2 b}{2}\right) + \beta_2 \sinh\left(\frac{\beta_2 b}{2}\right)\cos\left(\frac{\beta_1 b}{2}\right)\right] \tag{5.117e}$$

$$A_6 = 0 \tag{5.117f}$$

Consider a rectangular plate with two edges ($x = 0$ and $x = a$) simply supported, whereas the other two edges are free. The characteristic equation and coefficients of mode shape for the antisymmetric case can be obtained as

$$\begin{aligned}
&g_1\left[\beta_1^2 + \nu\left(\frac{m\pi}{a}\right)^2\right]\sin\left(\frac{\beta_1 b}{2}\right)\left[\left\{-\left(\frac{m\pi}{a}\right)^2 - \beta_3^2\right\}\{1-g_2\}\beta_2 - 2g_2\beta_2\left(\frac{m\pi}{a}\right)^2\right] \\
&+g_2\left[\beta_2^2 - \nu\left(\frac{m\pi}{a}\right)^2\right]\cos\left(\frac{\beta_1 b}{2}\right)\tanh\left(\frac{\beta_2 b}{2}\right)\left[\left\{-\left(\frac{m\pi}{a}\right)^2 - \beta_3^2\right\}\{1-g_1\}\beta_1 - 2g_1\beta_1\left(\frac{m\pi}{a}\right)^2\right] \\
&-2(g_1-g_2)(1-\nu)\beta_1\beta_2\beta_3\left(\frac{m\pi}{a}\right)^2 \cos\left(\frac{\beta_1 b}{2}\right)\tanh\left(\frac{\beta_3 b}{2}\right) = 0
\end{aligned} \tag{5.118}$$

$$A_1 = \beta_2 \cosh\left(\frac{\beta_2 b}{2}\right)\cosh\left(\frac{\beta_3 b}{2}\right)\left[\left(\frac{m\pi}{a}\right)^2 (1+g_2) + \beta_3^2(1-g_2)\right] \tag{5.119a}$$

$$A_2 = 0 \tag{5.119b}$$

$$A_3 = -\beta_1 \cos\left(\frac{\beta_1 b}{2}\right)\cosh\left(\frac{\beta_3 b}{2}\right)\left[\left(\frac{m\pi}{a}\right)^2 (1+g_1) + \beta_3^2(1-g_1)\right] \tag{5.119c}$$

$$A_4 = 0 \tag{5.119d}$$

$$A_5 = 0 \tag{5.119e}$$

$$A_6 = 2(g_1 - g_2)\beta_1\beta_2\left(\frac{m\pi}{a}\right)\cos\left(\frac{\beta_1 b}{2}\right)\cosh\left(\frac{\beta_2 b}{2}\right) \tag{5.119f}$$

The characteristic equation and coefficients of mode shape for the symmetric case can be obtained as

$$
\begin{aligned}
&-g_1\left[\beta_1^2+\nu\left(\frac{m\pi}{a}\right)^2\right]\cos\left(\frac{\beta_1 b}{2}\right)\left[\left\{-\left(\frac{m\pi}{a}\right)^2-\beta_3^2\right\}\{1-g_2\}\beta_2-2g_2\beta_2\left(\frac{m\pi}{a}\right)^2\right] \\
&+g_2\left[\beta_2^2-\nu\left(\frac{m\pi}{a}\right)^2\right]\sin\left(\frac{\beta_1 b}{2}\right)\coth\left(\frac{\beta_2 b}{2}\right)\left[\left\{-\left(\frac{m\pi}{a}\right)^2-\beta_3^2\right\}\{1-g_1\}\beta_1-2g_1\beta_1\left(\frac{m\pi}{a}\right)^2\right] \\
&-2(g_1-g_2)(1-\nu)\beta_1\beta_2\beta_3\left(\frac{m\pi}{a}\right)^2\sin\left(\frac{\beta_1 b}{2}\right)\coth\left(\frac{\beta_3 b}{2}\right)=0 \qquad (5.120)
\end{aligned}
$$

$$A_1=0 \qquad (5.121a)$$

$$A_2=-\beta_2\sinh\left(\frac{\beta_2 b}{2}\right)\sinh\left(\frac{\beta_3 b}{2}\right)\left[\left(\frac{m\pi}{a}\right)^2(1+g_2)+\beta_3^2(1-g_2)\right] \qquad (5.121b)$$

$$A_3=0 \qquad (5.121c)$$

$$A_4=-\beta_1\sin\left(\frac{\beta_1 b}{2}\right)\sinh\left(\frac{\beta_3 b}{2}\right)\left[\left(\frac{m\pi}{a}\right)^2(1+g_1)+\beta_3^2(1-g_1)\right] \qquad (5.121d)$$

$$A_5=2(g_1-g_2)\beta_1\beta_2\left(\frac{m\pi}{a}\right)\sin\left(\frac{\beta_1 b}{2}\right)\sinh\left(\frac{\beta_2 b}{2}\right) \qquad (5.121e)$$

$$A_6=0 \qquad (5.121f)$$

Consider a rectangular plate with two edges ($x = 0$ and $x = a$) simply supported, whereas the edge $y = -b/2$ is clamped and the edge $y = b/2$ is free. In order to facilitate the computation of the results, let

$$\phi_1=\frac{\sinh(\beta_3 y)}{\cosh(\frac{\beta_3 b}{2})} \quad \text{and} \quad \phi_2=\frac{\cosh(\beta_3 y)}{\cosh\left(\frac{\beta_3 b}{2}\right)} \qquad (5.122)$$

In view of Equation (5.122), Equation (5.101) can be written as

$$w_3=(A_5\phi_1+A_6\phi_2)\cosh\left(\frac{\beta_3 b}{2}\right)\cos\left(\frac{m\pi x}{a}\right) \qquad (5.123)$$

The substitution of Equations (5.96) to (5.100) and (5.123) into the boundary conditions yields the following matrix:

$$\begin{bmatrix}
-\sin\left(\frac{\beta_1 b}{2}\right) & \cos\left(\frac{\beta_1 b}{2}\right) & -\tanh\left(\frac{\beta_2 b}{2}\right) & 1 & 0 & 0 \\
-g_1\left(\frac{m\pi}{a}\right)\sin\left(\frac{\beta_1 b}{2}\right) & g_1\left(\frac{m\pi}{a}\right)\cos\left(\frac{\beta_1 b}{2}\right) & -g_2\left(\frac{m\pi}{a}\right)\tanh\left(\frac{\beta_2 b}{2}\right) & g_2\left(\frac{m\pi}{a}\right) & -\beta_3 & \beta_3\tanh\left(\frac{\beta_3 b}{2}\right) \\
g_1\beta_1\cos\left(\frac{\beta_1 b}{2}\right) & g_1\beta_1\sin\left(\frac{\beta_1 b}{2}\right) & g_2\beta_2 & -g_2\beta_2\tanh\left(\frac{\beta_2 b}{2}\right) & \left(\frac{m\pi}{a}\right)\tanh\left(\frac{\beta_3 b}{2}\right) & -\left(\frac{m\pi}{a}\right) \\
-g_1\left[\beta_1^2+\nu\left(\frac{m\pi}{a}\right)^2\right]\sin\left(\frac{\beta_1 b}{2}\right) & -g_1\left[\beta_1^2+\nu\left(\frac{m\pi}{a}\right)^2\right]\cos\left(\frac{\beta_1 b}{2}\right) & g_2\left[\beta_2^2-\nu\left(\frac{m\pi}{a}\right)^2\right]\tanh\left(\frac{\beta_2 b}{2}\right) & g_2\left[\beta_2^2-\nu\left(\frac{m\pi}{a}\right)^2\right] & (\nu-1)\left(\frac{m\pi}{a}\right)\beta_3 & (\nu-1)\left(\frac{m\pi}{a}\right)\beta_3\tanh\left(\frac{\beta_3 b}{2}\right) \\
2g_1\beta_1\left(\frac{m\pi}{a}\right)\cos\left(\frac{\beta_1 b}{2}\right) & -2g_1\beta_1\left(\frac{m\pi}{a}\right)\sin\left(\frac{\beta_1 b}{2}\right) & 2g_2\beta_2\left(\frac{m\pi}{a}\right) & 2g_2\beta_2\left(\frac{m\pi}{a}\right)\tanh\left(\frac{\beta_2 b}{2}\right) & -\left[\left(\frac{m\pi}{a}\right)^2+\beta_3^2\right]\tanh\left(\frac{\beta_3 b}{2}\right) & -\left[\left(\frac{m\pi}{a}\right)^2+\beta_3^2\right] \\
(1-g_1)\beta_1\cos\left(\frac{\beta_1 b}{2}\right) & -(1-g_1)\beta_1\sin\left(\frac{\beta_1 b}{2}\right) & (1-g_2)\beta_2 & (1-g_2)\beta_2\tanh\left(\frac{\beta_2 b}{2}\right) & \left(\frac{m\pi}{a}\right)\tanh\left(\frac{\beta_3 b}{2}\right) & \left(\frac{m\pi}{a}\right)
\end{bmatrix} \tag{5.124}$$

The natural frequencies are obtained by setting the determinant of the matrix of Equation (5.124) to zero and solving the characteristic equation for the frequencies.

Nondimensional frequency parameters $\bar{\omega}$ are presented in Tables 5.39 to 5.43 for rectangular thick plates with an aspect ratio $b/a = 0.5$; thickness-to-length ratios $h/a = 0.01$, 0.1, and 0.2; and various combinations of boundary conditions. The frequency parameters match with those obtained by Hashemi and Arsanjani (2005). It may be seen from the tables that β_3 is much larger than β_1 and β_2 for $h/a = 0.01$. This implies that the magnitude of Equation (5.101) is much larger than the magnitudes of Equations (5.99) and (5.100), and this point should be noted when performing the calculations for thin plates.

Vibration solutions for thick rectangular plates of other combinations of boundary conditions may be obtained from Liew, Xiang, and Kitipornchai (1993).

TABLE 5.39
Natural Frequencies $\bar{\omega} = \omega a^2\sqrt{\rho h/D}$ for Rectangular Plate with Edges $x = 0$, $x = a$, $y = b$ Simply Supported, and Edge $y = 0$ Clamped ($b/a = 0.5$)

h/a		1	2	3	4	5	6
	m	1	2	3	4	1	2
	$\beta_1 b/\pi$	1.22678	1.18086	1.13999	1.11063	2.24121	2.22162
0.01	$\beta_2 b/\pi$	1.41444	1.84020	2.40455	3.03234	2.34179	2.62416
	$\beta_3 b/\pi$	51.3284	51.3357	51.3478	51.3647	51.3282	51.3354
	$\bar{\omega}$	69.1986	94.3686	139.782	205.851	207.400	233.353
	m	1	2	3	1	4	2
	$\beta_1 b/\pi$	1.19144	1.13851	1.09383	2.15774	1.06382	2.13395
0.1	$\beta_2 b/\pi$	1.26438	1.65147	2.13189	1.75581	2.62706	2.00512
	$\beta_3 b/\pi$	5.13617	5.19282	5.27522	5.02147	5.37220	5.06548
	$\bar{\omega}$	59.4801	79.1951	112.678	151.182	156.813	167.125
	m	1	2	3	1	4	2
	$\beta_1 b/\pi$	1.12978	1.08145	1.04735	2.06342	1.02796	2.04940
0.2	$\beta_2 b/\pi$	1.01976	1.37891	1.79106	1.07233	2.20149	1.35295
	$\beta_3 b/\pi$	2.51916	2.59666	2.69208	2.10638	2.79040	2.16340
	$\bar{\omega}$	45.0569	59.1227	81.1493	99.7234	107.726	109.747

TABLE 5.40
Natural Frequencies $\bar{\omega} = \omega a^2 \sqrt{\rho h/D}$ for Rectangular Plate with Edges $x = 0$, $x = a$ Simply Supported and the Other Two Edges Clamped ($b/a = 0.5$)

h/a		1	2	3	4	5	6
0.01	m	1	2	3	4	1	2
	$\beta_1 b/\pi$	1.46956	1.38887	1.30586	1.24121	2.48439	2.45096
	$\beta_2 b/\pi$	1.62832	1.97911	2.48662	3.08181	2.57185	2.81753
	$\beta_3 b/\pi$	51.3284	51.3357	51.3478	51.3647	51.3281	51.3353
	$\bar{\omega}$	94.9657	115.392	155.714	217.884	252.397	275.275
0.1	m	1	2	3	4	1	2
	$\beta_1 b/\pi$	1.38423	1.28718	1.19643	1.13294	2.29989	2.25956
	$\beta_2 b/\pi$	1.37958	1.70925	2.15496	2.63545	1.80739	2.03951
	$\beta_3 b/\pi$	5.12373	5.18215	5.26663	5.36554	4.99160	5.03831
	$\bar{\omega}$	75.1962	90.0396	119.150	160.554	166.781	180.228
0.2	m	1	2	3	1	4	2
	$\beta_1 b/\pi$	1.26070	1.16625	1.09667	2.11210	1.05680	2.09145
	$\beta_2 b/\pi$	1.05050	1.38617	1.79064	1.06384	2.19998	1.34531
	$\beta_3 b/\pi$	2.48603	2.57473	2.67880	2.07114	2.78243	2.13326
	$\bar{\omega}$	52.1283	62.9729	82.9509	102.737	108.577	112.166

TABLE 5.41
Natural Frequencies $\bar{\omega} = \omega a^2 \sqrt{\rho h/D}$ for Rectangular Plate with Edges $x = 0$, $x = a$, $y = 0$ Simply Supported and Edge $y = b$ Free ($b/a = 0.5$)

h/a		1	2	3	4	5	6
0.01	m	1	2	1	3	2	3
	$\beta_1 b/\pi$	0.39732	0.42705	1.28597	0.42454	1.34337	1.38257
	$\beta_2 b/\pi$	0.81094	1.47662	1.46582	2.16146	1.94771	2.52728
	$\beta_3 b/\pi$	51.3285	51.3358	51.3284	51.3479	51.3357	51.3478
	$\bar{\omega}$	16.0971	46.6393	75.0554	95.7777	110.504	163.809
0.1	m	1	2	1	3	2	3
	$\beta_1 b/\pi$	0.38862	0.41312	1.27843	0.40267	1.32525	1.35933
	$\beta_2 b/\pi$	0.79341	1.41557	1.31713	2.01036	1.72404	2.19245
	$\beta_3 b/\pi$	5.15551	5.21849	5.13107	5.30857	5.17903	5.25061
	$\bar{\omega}$	15.4054	42.8870	66.3720	82.7190	92.9718	130.346
0.2	m	1	2	1	3	2	3
	$\beta_1 b/\pi$	0.38050	0.40254	1.27288	0.38951	1.31271	1.34432
	$\beta_2 b/\pi$	0.75651	1.29906	1.05299	1.78191	1.39527	1.78392
	$\beta_3 b/\pi$	2.60532	2.69637	2.48261	2.80251	2.53074	2.59962
	$\bar{\omega}$	14.1341	36.1646	52.8012	63.7960	69.9757	92.8079

TABLE 5.42
Natural Frequencies $\bar{\omega} = \omega a^2\sqrt{\rho h/D}$ for Rectangular Plate with Edges $x = 0$, $x = a$ Simply Supported and the Other Two Edges Free ($b/a = 0.5$)

h/a		1	2	3	4	5	6
0.01	m	1	1	2	2	3	1
	$\beta_1 b/\pi$	0.09552 i	0.66586	0.15711 i	0.79257	0.20119 i	1.55421
	$\beta_2 b/\pi$	0.70057	0.97092	1.40498	1.62001	2.11013	1.70457
	$\beta_3 b/\pi$	51.3285	51.3285	51.3358	51.3357	51.3479	51.3284
	$\bar{\omega}$	9.5078	27.3597	38.4774	64.2036	87.0926	105.035
0.1	m	1	1	2	2	3	1
	$\beta_1 b/\pi$	0.09846 i	0.64108	0.16981 i	0.76100	0.22553 i	1.54652
	$\beta_2 b/\pi$	0.69456	0.92435	1.36148	1.51302	1.98048	1.47096
	$\beta_3 b/\pi$	5.15639	5.15325	5.22163	5.21065	5.31444	5.10975
	$\bar{\omega}$	9.3306	24.9711	35.9987	56.5363	76.2163	89.5926
0.2	m	1	1	2	2	3	1
	$\beta_1 b/\pi$	0.10021 i	0.61198	0.17708 i	0.73390	0.23710 i	1.55290
	$\beta_2 b/\pi$	0.67913	0.84233	1.27194	1.33894	1.77602	1.09251
	$\beta_3 b/\pi$	2.61090	2.59349	2.71157	2.66269	2.82461	2.38612
	$\bar{\omega}$	8.9007	21.3271	31.0963	45.3418	59.6198	68.8138

TABLE 5.43
Natural Frequencies $\bar{\omega} = \omega a^2\sqrt{\rho h/D}$ for Rectangular Plate with Two Edges $x = 0$, $x = a$ Simply Supported, Edge $y = 0$ Clamped, and Edge $y = b$ Free ($b/a = 0.5$)

h/a		1	2	3	4	5	6
0.01	m	1	2	3	1	2	4
	$\beta_1 b/\pi$	0.57143	0.53200	0.49854	1.50727	1.52918	0.46938
	$\beta_2 b/\pi$	0.90888	1.51020	2.17710	1.66220	2.07912	2.86274
	$\beta_3 b/\pi$	51.3285	51.3358	51.3479	51.3284	51.3357	51.3648
	$\bar{\omega}$	22.7512	50.6057	98.4649	99.3823	131.485	166.117
0.1	m	1	2	1	3	2	4
	$\beta_1 b/\pi$	0.55420	0.49965	1.45142	0.45519	1.45721	0.41432
	$\beta_2 b/\pi$	0.87546	1.43572	1.41811	2.01644	1.77488	2.56597
	$\beta_3 b/\pi$	5.15427	5.21712	5.11836	5.30728	5.16686	5.41184
	$\bar{\omega}$	21.1870	45.5725	81.0357	84.0786	103.590	132.299
0.2	m	1	2	1	3	2	4
	$\beta_1 b/\pi$	0.53062	0.46545	1.36352	0.42077	1.37525	0.38214
	$\beta_2 b/\pi$	0.81088	1.30581	1.06956	1.78251	1.39766	2.21966
	$\beta_3 b/\pi$	2.59874	2.69180	2.45516	2.79976	2.50944	2.90608
	$\bar{\omega}$	18.4903	37.5487	57.8767	64.2947	73.0862	94.1986

5.5.3 Circular Plates

Mindlin and Deresiewicz (1954) showed that the governing equations of motion for thick circular plates given by Equations (5.88) to (5.90) may be recast into polar coordinates (r, θ) and three harmonic equations involving potentials Θ_1, Θ_2, Θ_3 defined as

$$\psi_r = (\sigma_1 - 1)\frac{\partial \Theta_1}{\partial \chi} + (\sigma_2 - 1)\frac{\partial \Theta_2}{\partial \chi} + \frac{1}{\chi}\frac{\partial \Theta_3}{\partial \theta} \tag{5.125}$$

$$\psi_\theta = (\sigma_1 - 1)\frac{1}{\chi}\frac{\partial \Theta_1}{\partial \theta} + (\sigma_2 - 1)\frac{1}{\chi}\frac{\partial \Theta_2}{\partial \theta} - \frac{\partial \Theta_3}{\partial \chi} \tag{5.126}$$

$$\bar{w} = \Theta_1 + \Theta_2 \tag{5.127}$$

where

$$\sigma_1, \sigma_2 = \frac{\left(\delta_2^2, \delta_1^2\right)}{\left[\frac{\tau^2\lambda^2}{12} - \frac{6(1-\nu)\kappa^2}{\tau^2}\right]} = \frac{2\left(\delta_2^2, \delta_1^2\right)}{\delta_3^2(1-\nu)} \tag{5.128}$$

$$\delta_1^2, \delta_2^2 = \frac{\lambda^2}{2}\left[\frac{\tau^2}{12} + \frac{\tau^2}{6(1-\nu)\kappa^2} \pm \sqrt{\left(\frac{\tau^2}{12} + \frac{\tau^2}{6(1-\nu)\kappa^2}\right)^2 + \frac{4}{\lambda^2}}\right] \tag{5.129}$$

$$\delta_3^2 = \frac{2}{(1-\nu)}\left[\frac{\tau^2\lambda^2}{12} - \frac{6(1-\nu)\kappa^2}{\tau^2}\right] \tag{5.130}$$

$$\bar{w} = \frac{w}{R}; \quad \chi = \frac{r}{R}; \quad \tau = \frac{h}{R}; \quad \lambda = \omega R^2\sqrt{\frac{\rho h}{D}} \tag{5.131}$$

in which w is the transverse deflection, r is the radial coordinate, R is the plate radius, h is the plate thickness, ω is the circular frequency, D is the flexural rigidity, ρ is the mass density per unit volume, and the subscripts r,θ denote the quantities in the radial and circumferential directions, respectively. Note that λ is the frequency parameter.

Based on the aforementioned three potentials, the governing equations of vibrating plates, in polar coordinates, can now be expressed as

$$\left(\nabla^2 + \delta_1^2\right)\Theta_1 = 0 \tag{5.132}$$

$$\left(\nabla^2 + \delta_2^2\right)\Theta_2 = 0 \tag{5.133}$$

$$\left(\nabla^2 + \delta_3^2\right)\Theta_3 = 0 \tag{5.134}$$

where the Laplacian operator $\nabla^2(\bullet) = \partial^2(\bullet)/\partial\chi^2 + (1/\chi)\partial(\bullet)/\partial\chi + (1/\chi^2)\partial^2(\bullet)/\partial\theta^2$. The solutions to Equations (5.132) to (5.134) are given by

$$\Theta_1 = A_1 J_n(\delta_1 \chi)\cos n\theta + B_1 Y_n(\delta_1 \chi)\cos n\theta \tag{5.135}$$

$$\Theta_2 = A_2 J_n(\delta_2 \chi)\cos n\theta + B_2 Y_n(\delta_2 \chi)\cos n\theta \tag{5.136}$$

$$\Theta_3 = A_3 J_n(\delta_3 \chi)\sin n\theta + B_3 Y_n(\delta_3 \chi)\sin n\theta \tag{5.137}$$

where A_i and B_i are arbitrary constants; $J_n(\bullet)$ and $Y_n(\bullet)$ are the Bessel functions of the first and second kinds of order n, respectively; and n corresponds to the number of nodal diameters.

The boundary conditions for circular Mindlin plates at the edge $r = R$ (i.e., for $\chi = 1$) are given by

$$M_{rr} = 0,\ \ M_{r\theta} = 0,\ \ Q_r = 0 \quad \text{for a free edge} \tag{5.138}$$

$$M_{rr} = 0,\ \ M_{r\theta} = 0, \bar{w} = 0 \quad \text{for a simply supported edge (soft type)} \tag{5.139}$$

$$M_{rr} = 0,\ \ \psi_\theta = 0,\ \ \bar{w} = 0 \quad \text{for a simply supported edge (hard type)} \tag{5.140}$$

$$\psi_r = 0,\ \ \psi_\theta = 0,\ \ \bar{w} = 0 \quad \text{for a clamped edge} \tag{5.141}$$

For solid circular plates with the previously mentioned boundary conditions, the constants B_i are set to zero in Equations (5.135) to (5.137) so as to avoid infinite displacements, slopes, and bending moments at $r = 0$. The frequency parameter λ can then be determined by substituting Equations (5.135) to (5.137) into Equations (5.132) to (5.134) to yield the determinant equation given by

$$\begin{vmatrix} C_{11} & C_{12} & C_{13} \\ C_{21} & C_{22} & C_{23} \\ C_{31} & C_{32} & C_{33} \end{vmatrix} = 0 \Rightarrow C_{11}(C_{22}C_{33} - C_{23}C_{32}) - C_{12}(C_{21}C_{33} - C_{23}C_{31}) + C_{13}(C_{21}C_{32} - C_{22}C_{31}) = 0 \tag{5.142}$$

- For a circular plate with a free edge, the elements of the determinant of Equation (5.142) are

$$C_{1i} = (\sigma_i - 1)\left[J_n''(\delta_i) + \nu J_n'(\delta_i) - \nu n^2 J_n(\delta_i)\right] \tag{5.143a}$$

$$C_{2i} = -2n(\sigma_i - 1)[J_n'(\delta_i) - J_n(\delta_i)] \tag{5.143b}$$

$$C_{3i} = \sigma_i J_n'(\delta_i) \tag{5.143c}$$

$$C_{13} = n(1-\nu)[J_n'(\delta_3) - J_n(\delta_3)] \tag{5.143d}$$

$$C_{23} = -J_n''(\delta_3) + J_n'(\delta_3) - n^2 J_n(\delta_3) \tag{5.143e}$$

$$C_{33} = nJ_n(\delta_3) \tag{5.143f}$$

- For a circular plate with a soft-type simply supported edge, the elements of the determinant are

$$C_{1i} = (\sigma_i - 1)\left[J_n''(\delta_i) + \nu J_n'(\delta_i) - \nu n^2 J_n(\delta_i)\right] \tag{5.144a}$$

$$C_{2i} = -2n(\sigma_i - 1)[J_n'(\delta_i) - J_n(\delta_i)] \tag{5.144b}$$

$$C_{3i} = \sigma_i J_n(\delta_i) \tag{5.144c}$$

$$C_{13} = n(1-\nu)[J_n'(\delta_3) - J_n(\delta_3)] \tag{5.144d}$$

$$C_{23} = -J_n''(\delta_3) + J_n'(\delta_3) - n^2 J_n(\delta_3) \tag{5.144e}$$

$$C_{33} = 0 \tag{5.144f}$$

- For a circular plate with a hard-type simply supported edge, the elements of the determinant are

$$C_{1i} = (\sigma_i - 1)\left[J_n''(\delta_i) + \nu J_n'(\delta_i) - \nu n^2 J_n(\delta_i)\right] \tag{5.145a}$$

$$C_{2i} = -2n(\sigma_i - 1)J_n(\delta_i) \tag{5.145b}$$

$$C_{3i} = J_n(\delta_i) \tag{5.145c}$$

$$C_{13} = n(1-\nu)[J_n'(\delta_3) - J_n(\delta_3)] \tag{5.145d}$$

$$C_{23} = J_n'(\delta_3) \tag{5.145e}$$

$$C_{33} = 0 \tag{5.145f}$$

- For a circular plate with a clamped edge, the elements of the determinant are

$$C_{1i} = (\sigma_i - 1)J_n'(\delta_i) \tag{5.146a}$$

$$C_{2i} = n(\sigma_i - 1)J_n(\delta_i) \tag{5.146b}$$

$$C_{3i} = J_n(\delta_i) \tag{5.146c}$$

$$C_{13} = nJ_n(\delta_3) \tag{5.146d}$$

$$C_{23} = J_n'(\delta_3) \tag{5.146e}$$

$$C_{33} = 0 \tag{5.146f}$$

where $i = 1,2$.

Tables 5.44 to 5.46 present some values of the natural frequency parameters $\lambda = \omega R^2/\sqrt{\rho h/D}$ for circular plates of radius R, thickness h, and with free, soft simply supported, and clamped edges, respectively. The Poisson ratio is taken as $\nu = 0.3$

TABLE 5.44
Frequency Parameter λ of Circular Plates with Free Edge

n	s	$\tau = 0$	$\tau = 0.1$	$\tau = 0.2$
0	1	9.003	8.868	8.505
	2	38.443	36.041	31.111
	3	87.750	76.676	59.645
	4	156.82	126.27	90.059
1	1	20.475	19.711	17.978
	2	59.812	54.257	44.434
	3	118.96	99.935	74.331
	4	197.87	152.75	105.03
2	0	5.358	5.278	5.114
	1	35.260	33.033	28.668
	2	84.366	73.875	57.722
	3	153.31	123.77	88.530
3	0	12.439	12.064	11.214
	1	53.008	48.227	39.960
	2	111.95	94.531	70.862
	3	190.69	147.99	102.27
4	0	21.835	20.801	18.816
	1	73.543	64.891	51.545
	2	142.43	115.96	83.801
	3	231.03	172.45	115.57

and the shear correction factor as $\kappa^2 = \pi^2/12$. The frequency parameters for classical thin plates are also given under the column heading $\tau = h/R = 0$ in the tables. In the tables, n refers to the number of nodal diameters, while s denotes the number of nodal circles excluding the boundary circle.

5.5.4 Annular Plates

Irie, Yamada, and Takagi (1982) derived the exact vibration solutions for thick annular plates with nine combinations of boundary conditions for the inner and outer edges. The boundary conditions are given in Equations (5.138) to (5.141). The potentials given in Equations (5.135) to (5.137) are valid for the annular plates. Unlike the circular plate solutions where the constants B_i are set to zero to eliminate singularities at the plate origin (i.e., $r = 0$ or $\chi = 0$), these constants are nonzero for annular plates. By substituting Equations (5.135) to (5.137) into Equations (5.125) to (5.127) and then into the appropriate boundary conditions for each of the nine annular plate cases, one obtains the characteristic equation that, upon solving, yields the natural frequencies of vibration of annular thick plates based on the Mindlin plate theory.

TABLE 5.45
Frequency Parameter λ of Circular Plates with Soft Simply Supported Edge

n	s	$\tau = 0$	$\tau = 0.1$	$\tau = 0.2$
0	1	4.935	4.894	4.777
	2	29.720	28.240	24.994
	3	74.156	65.942	52.514
	4	138.32	113.57	82.766
1	1	13.898	13.510	12.620
	2	48.479	44.691	37.537
	3	102.77	87.994	66.946
	4	176.80	139.27	97.873
2	0	25.613	24.313	21.687
	1	70.117	62.552	50.126
	2	134.30	110.66	80.950
	3	218.20	165.02	112.43
3	0	39.957	36.962	31.547
	1	94.549	81.526	62.675
	2	168.68	133.77	94.597
	3	262.49	190.77	126.53
4	0	56.842	51.158	41.908
	1	121.70	101.37	75.137
	2	205.85	157.20	107.94
	3	309.61	216.47	140.25

The elements of the determinantal equation are given in a paper by Irie, Yamada, and Takagi (1982).

Tables 5.47 to 5.55 present sample frequency parameters $\lambda = \omega a^2/\sqrt{\rho h/D}$, which were obtained by Irie, Yamada, and Takagi (1982), for the nine combinations of boundary conditions for annular plates. The annular plates have an outer radius a, inner radius b, thickness h, and with free, hard simply supported, and clamped edges, respectively. The Poisson ratio is taken as $\nu = 0.3$ and the shear correction factor as $\kappa^2 = \pi^2/12$. In the tables, n refers to the number of nodal diameters, while s denotes the number of nodal circles.

5.5.5 Sectorial Plates

Huang, McGee, and Leissa (1994) derived the exact vibration solutions for thick sectorial plates having a sector angle α, radius R, thickness h, simply supported radial edges, and a circular edge that may be clamped, simply supported, or free. The boundary conditions for the radial simply supported edges are given by

TABLE 5.46
Frequency Parameter λ of Circular Plates with Clamped Edge

n	*s*	τ = 0	τ = 0.1	τ = 0.2
0	1	10.216	9.941	9.240
	2	39.771	36.479	30.211
	3	89.104	75.664	56.682
	4	158.18	123.32	85.571
1	1	21.260	20.232	17.834
	2	60.829	53.890	42.409
	3	120.08	97.907	70.473
	4	199.05	148.70	100.12
2	0	34.877	32.406	27.214
	1	84.583	72.368	54.557
	2	153.82	120.55	83.937
	3	242.72	174.05	114.24
3	0	51.030	46.178	37.109
	1	115.02	91.712	66.667
	2	190.30	143.50	97.152
	3	289.18	199.36	128.02
4	0	69.666	61.272	47.340
	1	140.11	111.74	78.733
	2	229.52	116.69	110.16
	3	338.41	224.61	141.53

TABLE 5.47
Frequency Parameter λ of Annular Plates with Free Inner Edge and Free Outer Edge

		b/a = 0.3		*b/a* = 0.5		*b/a* = 0.7	
n	*s*	*h/a* = 0.1	*h/a* = 0.2	*h/a* = 0.1	*h/a* = 0.2	*h/a* = 0.1	*h/a* = 0.2
0	1	8.23	7.89	9.10	8.55	12.46	10.89
	2	46.63	39.57	81.03	64.01	188.95	131.06
1	1	17.02	15.13	15.76	13.77	19.12	15.27
	2	52.50	43.17	83.48	65.32	190.04	131.38
2	1	4.80	4.61	4.17	4.00	3.47	3.29
	2	30.77	26.63	28.05	23.64	31.83	24.19

TABLE 5.48
Frequency Parameter λ of Annular Plates with Free Inner Edge and Simply Supported Outer Edge

		$b/a = 0.3$		$b/a = 0.5$		$b/a = 0.7$	
n	s	$h/a = 0.1$	$h/a = 0.2$	$h/a = 0.1$	$h/a = 0.2$	$h/a = 0.1$	$h/a = 0.2$
0	1	4.63	4.53	5.03	4.91	6.81	6.49
	2	34.92	30.49	59.53	48.56	139.61	97.41
1	1	12.19	11.19	10.90	9.95	12.29	10.90
	2	41.45	34.80	62.28	50.22	140.90	98.04
2	1	23.07	20.71	20.92	18.56	22.01	18.86
	2	57.18	45.73	70.09	55.00	144.71	99.90

TABLE 5.49
Frequency Parameter λ of Annular Plates with Free Inner Edge and Clamped Outer Edge

		$b/a = 0.3$		$b/a = 0.5$		$b/a = 0.7$	
n	s	$h/a = 0.1$	$h/a = 0.2$	$h/a = 0.1$	$h/a = 0.2$	$h/a = 0.1$	$h/a = 0.2$
0	1	11.12	10.35	17.02	15.40	39.37	32.44
	2	46.25	36.77	77.24	55.27	165.42	99.31
1	1	18.12	15.87	20.48	17.94	40.85	33.35
	2	51.74	40.18	79.41	56.57	166.42	99.85
2	1	30.08	25.33	29.02	24.33	45.19	36.08
	2	66.24	49.74	85.76	60.44	169.38	101.49

TABLE 5.50
Frequency Parameter λ of Annular Plates with Simply Supported Inner Edge and Free Outer Edge

		$b/a = 0.3$		$b/a = 0.5$		$b/a = 0.7$	
n	s	$h/a = 0.1$	$h/a = 0.2$	$h/a = 0.1$	$h/a = 0.2$	$h/a = 0.1$	$h/a = 0.2$
0	1	3.40	3.33	4.09	4.01	6.10	5.86
	2	29.79	25.89	55.12	44.64	134.41	92.39
1	1	3.33	3.23	4.79	4.65	8.09	7.63
	2	32.29	27.87	56.87	45.85	135.50	92.99
2	1	5.96	5.74	7.79	7.42	12.79	11.79
	2	39.64	33.52	61.99	49.35	138.71	94.78

TABLE 5.51
Frequency Parameter λ of Annular Plates with Simply Supported Inner Edge and Simply Supported Outer Edge

		b/a = 0.3		b/a = 0.5		b/a = 0.7	
n	s	h/a = 0.1	h/a = 0.2	h/a = 0.1	h/a = 0.2	h/a = 0.1	h/a = 0.2
0	1	20.22	18.21	37.33	31.87	93.26	70.06
	2	71.71	56.08	127.17	90.64	280.58	162.31
1	1	22.30	19.98	38.86	33.04	94.27	70.70
	2	73.99	57.60	128.46	91.39	281.24	162.70
2	1	28.67	25.22	43.45	36.49	97.30	72.60
	2	80.78	62.11	132.29	93.64	283.19	163.90

TABLE 5.52
Frequency Parameter λ of Annular Plates with Simply Supported Inner Edge and Clamped Outer Edge

		b/a = 0.3		b/a = 0.5		b/a = 0.7	
n	s	h/a = 0.1	h/a = 0.2	h/a = 0.1	h/a = 0.2	h/a = 0.1	h/a = 0.2
0	1	31.01	25.68	55.09	41.62	124.94	80.34
	2	85.50	61.56	145.82	95.27	300.62	171.69
1	1	32.84	72.04	56.26	42.41	125.61	80.78
	2	87.52	62.87	146.89	95.93	301.17	172.06
2	1	38.64	31.32	59.84	44.85	127.64	82.13
	2	93.61	66.81	150.10	97.89	302.80	173.17

TABLE 5.53
Frequency Parameter λ of Annular Plates with Clamped Inner Edge and Free Outer Edge

		b/a = 0.3		b/a = 0.5		b/a = 0.7	
n	s	h/a = 0.1	h/a = 0.2	h/a = 0.1	h/a = 0.2	h/a = 0.1	h/a = 0.2
0	1	6.52	6.14	12.57	11.46	33.87	28.05
	2	37.89	29.76	69.58	49.27	156.63	93.37
1	1	6.31	5.79	12.71	11.43	34.16	28.13
	2	39.54	31.20	70.80	50.25	157.45	93.96
2	1	7.55	6.89	13.79	12.14	35.23	28.63
	2	44.83	35.68	74.48	53.15	159.91	95.71

TABLE 5.54
Frequency Parameter λ of Annular Plates with Clamped Inner Edge and Simply Supported Outer Edge

		$b/a = 0.3$		$b/a = 0.5$		$b/a = 0.7$	
n	s	$h/a = 0.1$	$h/a = 0.2$	$h/a = 0.1$	$h/a = 0.2$	$h/a = 0.1$	$h/a = 0.2$
0	1	27.38	22.44	51.22	38.36	119.99	76.76
	2	82.17	59.38	142.71	93.78	297.80	172.17
1	1	28.60	23.52	52.14	39.12	120.61	77.26
	2	83.71	60.62	143.65	94.45	298.34	172.49
2	1	32.89	27.32	55.05	41.51	122.15	78.78
	2	88.59	64.42	146.51	96.46	299.95	173.45

$$w(r,0) = 0, \quad w(r,\alpha) = 0 \tag{5.147a}$$

$$M_\theta(r,0) = 0, \quad M_\theta(r,\alpha) = 0 \tag{5.147b}$$

$$\psi_r(r,0) = 0, \quad \psi_r(r,\alpha) = 0 \tag{5.147c}$$

The potentials that satisfy the foregoing boundary conditions are given by

- For the case $\delta_1^2 > 0, \delta_2^2 < 0, \delta_3^2 < 0$

TABLE 5.55
Frequency Parameter λ of Annular Plates with Clamped Inner Edge and Clamped Outer Edge

		$b/a = 0.3$		$b/a = 0.5$		$b/a = 0.7$	
n	s	$h/a = 0.1$	$h/a = 0.2$	$h/a = 0.1$	$h/a = 0.2$	$h/a = 0.1$	$h/a = 0.2$
0	1	39.40	30.04	70.28	48.31	152.15	90.20
	2	95.59	64.23	159.78	97.39	314.09	172.31
1	1	40.37	30.77	70.90	48.73	152.49	90.41
	2	96.99	65.36	160.60	98.02	314.57	172.66
2	1	43.98	33.67	72.96	50.18	153.57	91.12
	2	101.43	68.81	163.08	99.90	316.03	173.70

$$\Theta_1 = [A_1 J_\mu(\delta_1\chi) + B_1 Y_\mu(\delta_1\chi)]\sin\frac{n\pi\theta}{\alpha} \quad (5.148a)$$

$$\Theta_2 = [A_2 I_\mu(\delta_2\chi) + B_2 K_\mu(\delta_2\chi)]\sin\frac{n\pi\theta}{\alpha} \quad (5.148b)$$

$$\Theta_3 = [A_3 I_\mu(\delta_3\chi) + B_3 K_\mu(\delta_3\chi)]\cos\frac{n\pi\theta}{\alpha} \quad (5.148c)$$

- For the case $\delta_1^2 > 0,\ \delta_2^2 > 0,\ \delta_3^2 < 0$

$$\Theta_1 = [A_1 J_\mu(\delta_1\chi) + B_1 Y_\mu(\delta_1\chi)]\sin\frac{n\pi\theta}{\alpha} \quad (5.149a)$$

$$\Theta_2 = [A_2 J_\mu(\delta_2\chi) + B_2 Y_\mu(\delta_2\chi)]\sin\frac{n\pi\theta}{\alpha} \quad (5.149b)$$

$$\Theta_3 = [A_3 J_\mu(\delta_3\chi) + B_3 Y_\mu(\delta_3\chi)]\cos\frac{n\pi\theta}{\alpha} \quad (5.149c)$$

where $J_\mu, Y_\mu, I_\mu, K_\mu$ are ordinary and modified Bessel functions of the first and second kinds, and A_i, B_i (i = 1,2,3) are arbitrary constants of integration. Note that μ is not, in general, an integer, and B_i (i = 1,2,3) are not necessarily set equal to zero.

By substituting Equations (5.148a) to (5.148c) or Equations (5.149a) to (5.149c) into Equations (5.125) to (5.127) and then into the appropriate boundary conditions for each of the three circular edge cases,

$$w(R,\theta) = 0,\ \psi_r(R,\theta) = 0,\ \psi_\theta(R,\theta) = 0 \ \text{ for clamped circular edge} \quad (5.150a)$$

$$w(R,\theta) = 0,\ M_r(R,\theta) = 0,\ \psi_\theta(R,\theta) = 0 \ \text{ for simply supported circular edge} \quad (5.150b)$$

$$M_r(R,\theta) = 0,\ M_{r\theta}(R,\theta) = 0,\ Q_r(R,\theta) = 0 \ \text{ for free circular edge} \quad (5.150c)$$

one obtains the characteristic equation that, upon solving, yields the natural frequencies of vibration of thick sectorial plates. The details for the characteristic equations are given in a paper by Huang, McGee, and Leissa (1994).

Tables 5.56 to 5.58 present sample frequency parameters $\lambda = \omega R^2/\sqrt{\rho h/D}$, which were obtained by Huang, McGee, and Leissa (1994), for the clamped, simply supported, and free circular edge for sectorial plates with simply supported radial edges. The Poisson ratio is taken as ν = 0.3 and the shear correction factor as $\kappa^2 = \pi^2/12$. In the tables, s denotes the number of nodal circles. The mode shapes have no radial nodal lines.

TABLE 5.56
Frequency Parameter λ of Sectorial Plates with Simply Supported Radial Edges and a Clamped Circular Edge

			$\lambda = \omega R^2/\sqrt{\rho h/D}$	
α	μ	s	$h/R = 0.1$	$h/R = 0.2$
60°	3	1	45.8329	36.8449
		2	91.5849	66.5856
		3	143.520	97.2196
		4	199.374	128.142
		5	257.924	159.012
120°	1.5	1	26.0100	22.3729
		2	63.0436	48.4416
		3	109.203	77.2641
		4	161.544	107.330
		5	217.563	137.828
180°	1	1	20.2500	17.8084
		2	53.8756	42.3801
		3	98.0100	70.5600
		4	148.840	100.200
		5	203.918	130.645
270°	0.6667	1	18.7907	16.2102
		2	50.5139	39.4485
		3	93.0316	66.7843
		4	142.641	95.9219
		5	196.836	125.931
330°	0.5455	1	18.7246	15.7963
		2	49.6543	38.3595
		3	91.4357	65.3261
		4	140.429	94.2448
		5	194.153	124.142

Source: Huang, McGee, and Leissa (1994).
Note: The Poisson ratio is taken as $\nu = 0.3$ and the shear correction factor as $\kappa^2 = \pi^2/12$. The variable s denotes the number of nodal circles. The mode shapes have no radial nodal lines.

TABLE 5.57
Frequency Parameter λ of Sectorial Plates with All Edges Simply Supported

			$\lambda R^2/\sqrt{\rho h/D}$	
α	μ	s	h/R = 0.1	h/R = 0.2
60°	3	1	37.4544	32.1489
		2	81.9025	63.0436
		3	134.096	94.8686
		4	190.992	126.788
		5	250.906	158.508
120°	1.5	1	18.8356	17.2225
		2	53.5824	43.9569
		3	99.4009	74.1321
		4	152.276	105.473
		5	209.670	136.890
180°	1	1	13.6161	12.7449
		2	44.7561	37.6996
		3	88.1721	67.0761
		4	139.476	98.0100
		5	195.720	129.277
270°	0.6667	1	12.3598	11.2021
		2	41.6562	34.6347
		3	83.2835	63.1562
		4	133.239	93.5760
		5	188.457	124.570
330°	0.5455	1	12.3463	10.8524
		2	40.9486	33.5701
		3	81.7788	61.6466
		4	131.054	91.8415
		5	185.753	122.730

Source: Huang, McGee, and Leissa (1994).

Note: The Poisson ratio is taken as $\nu = 0.3$ and the shear correction factor as $\kappa^2 = \pi^2/12$. The variable s denotes the number of nodal circles. The mode shapes have no radial nodal lines.

TABLE 5.58
Frequency Parameter λ of Sectorial Plates with Simply Supported Radial Edges and a Free Circular Edge

			$\lambda = \omega R^2/\sqrt{\rho h/D}$	
α	μ	s	$h/R = 0.1$	$h/R = 0.2$
60°	3	1	12.0645	11.3138
		2	48.2275	39.9601
		3	94.5309	70.8627
		4	147.992	102.271
		5	205.725	132.874
120°	1.5	1	2.6651	2.6195
		2	26.1080	23.2208
		3	63.9168	51.0925
		4	111.797	81.4903
		5	166.008	112.256
180°	1	1	Rigid body rotation	Rigid body rotation
		2	19.7109	17.9784
		3	54.2580	44.4342
		4	99.9360	74.3320
		5	152.752	105.034
270°	0.6667	1	2.0614	1.8052
		2	17.8576	15.8946
		3	50.6900	41.2080
		4	94.7738	70.3547
		5	146.388	100.719
330°	0.5455	1	2.5769	2.2498
		2	17.6299	15.2906
		3	49.7738	40.0081
		4	93.0921	68.7747
		5	144.052	98.9829

Source: Huang, McGee, and Leissa (1994).

Note: The Poisson ratio is taken as $\nu = 0.3$ and the shear correction factor as $\kappa^2 = \pi^2/12$. The variable s denotes the number of nodal circles. The mode shapes have no radial nodal lines.

5.6 VIBRATION OF THICK RECTANGULAR PLATES BASED ON 3-D ELASTICITY THEORY

In very thick plates, there are symmetric thickness vibration modes that even the Mindlin plate theory is unable to identify. Srinivas, Joga Rao, and Rao (1970) derived the exact vibration solutions for very thick rectangular plates based on the three-dimensional (3-D), linear, small-deformation theory of elasticity. This section is based on that work.

The basic equations of elasticity in terms of displacements for a thick rectangular plate executing simple harmonic motion with angular frequency ω are

$$\nabla^2 u + \frac{1}{1-2\nu}\frac{\partial e}{\partial x} + \rho\frac{\omega^2}{G}u = 0 \tag{5.151}$$

$$\nabla^2 v + \frac{1}{1-2\nu}\frac{\partial e}{\partial y} + \rho\frac{\omega^2}{G}v = 0 \tag{5.152}$$

$$\nabla^2 w + \frac{1}{1-2\nu}\frac{\partial e}{\partial z} + \rho\frac{\omega^2}{G}w = 0 \tag{5.153}$$

where u, v, and w are the displacements in the x-, y-, and z-directions, respectively; G is the shear modulus; ν is the Poisson ratio; ρ is the mass density; and

$$e = \frac{\partial u}{\partial x} + \frac{\partial v}{\partial y} + \frac{\partial w}{\partial z} \tag{5.154}$$

Referring to Figure 5.2, the edge conditions for the simply supported rectangular plate, of length a and width b, are

$$\text{at } x = 0 \quad \text{and} \quad x = \text{a: } \sigma_x = 0, \quad v = 0 \quad \text{and} \quad w = 0 \tag{5.155}$$

$$\text{at } y = 0 \quad \text{and} \quad y = b\text{: } \sigma_y = 0, \; u = 0 \quad \text{and} \quad w = 0 \tag{5.156}$$

The boundary conditions, given in Equations (5.155) and (5.156), can be satisfied by setting

$$\begin{bmatrix} u \\ v \\ w \end{bmatrix} = h\sum_{m=1}^{\infty}\sum_{n=1}^{\infty}\begin{bmatrix} \varphi(\bar{z})\cos m\pi\bar{x}\sin n\pi\bar{y} \\ \psi(\bar{z})\sin m\pi\bar{x}\cos n\pi\bar{y} \\ \chi(\bar{z})\sin m\pi\bar{x}\sin n\pi\bar{y} \end{bmatrix} \tag{5.157}$$

where h is the plate thickness and $\bar{z} = z/h, \bar{x} = x/a, \bar{y} = y/b$.

By substituting Equation (5.157) into Equations (5.151) to (5.153) and using the stress-displacement relationships and the stress-free surface condition at

$$z = 0 \text{ and } z = h\text{: } \sigma_x = 0, \; \tau_{xz} = 0 \text{ and } \tau_{yz} = 0 \tag{5.158}$$

one obtains the characteristic equation

$$\left[8g^2rs(r^2+g^2)^2(1-\cosh r\cosh s) + \left\{16g^4r^2s^2 + (r^2+g^2)^4\right\}\sinh r\sinh s\right]\sinh r = 0 \tag{5.159}$$

where

TABLE 5.59
Frequency Parameters $\lambda = \omega R^2/\sqrt{\rho h/D}$ for Simply Supported Rectangular Plates

$\frac{g}{\pi^2}$	Mindlin Plate Theory			3-D Elasticity Theory					
	I-A	II-A	III-A	I-A	I-S	II-S	II-A	III-A	III-S
0.18	0.68208	3.4126	3.9926	0.68893	1.3329	2.2171	3.4126	3.9310	5.4903
0.20	0.74312	3.4414	4.0720	0.75111	1.4050	2.3320	3.4414	4.0037	5.4795
0.26	0.91520	3.5264	4.2982	0.92678	1.6019	2.6407	3.5264	4.2099	5.4621
0.32	1.0735	3.6094	4.5098	1.0889	1.7772	2.9066	3.6094	4.4013	5.4635
0.50	1.4890	3.8476	5.0804	1.5158	2.2214	3.5306	3.8476	4.9086	5.5554

$$g = \sqrt{\left(\frac{m\pi h}{a}\right)^2 + \left(\frac{n\pi h}{b}\right)^2} \tag{5.160}$$

$$r = \sqrt{g^2 - \omega^2 \frac{\rho h^2}{G}} \tag{5.161}$$

$$s = \sqrt{g^2 - \omega^2 \frac{\rho h^2}{G} \frac{1-2\nu}{2(1-\nu)}} \tag{5.162}$$

The solution to this characteristic equation for each combination of (m, n) yields an infinite number of natural frequency values. A sample of the first six frequency values, extracted from a paper by Srinivas, Joga Rao, and Rao (1970), is given in Table 5.59. These values are compared with all three of the frequency values furnished by the Mindlin plate theory with $\kappa^2 = \pi^2/12$. The Poisson ratio ν is taken as 0.3. Note that A and S denote modes that are antisymmetric and symmetric about the mid-plane, respectively. The frequencies under column II-A of the 3-D elasticity theory are associated with thickness-twist modes. It can be seen that the first flexural frequencies of the Mindlin plate theory are only slightly lower than the first 3-D elasticity theory frequencies. However if we assume a Mindlin shear correction factor of $\kappa^2 = 0.88$, the Mindlin frequencies can be made almost equal to the 3-D elasticity theory results.

REFERENCES

Conway, H. D. 1960. Analogies between buckling and vibration of polygonal plates and membranes. *Canadian Aeronautical Journal* 6:263.

Gabrielson, T. B. 1999. Frequency constants for transverse vibration of annular disks. *J. Acoust. Soc. Am.* 105:3311–17.

Hashemi, S. H., and M. Arsanjani. 2005. Exact characteristic equations for some of classical boundary conditions of vibrating moderately thick rectangular plates. *International Journal of Solids and Structures* 42:819–53.

Huang, C. S., A. W. Leissa, and O. G. McGee. 1993. Exact analytical solutions for the vibrations of sectorial plates with simply supported edges. *Journal of Applied Mechanics* 60:478–83.

Huang, C. S., O. G. McGee, and A. W. Leissa. 1994. Exact analytical solutions for free vibrations of thick sectorial plates with simply supported radial edges. *International Journal of Solids and Structures* 31:1609–31.

Irie, T., G. Yamada, and K. Takagi. 1982. Natural frequencies of thick annular plates. *Journal of Applied Mechanics* 49:633–38.

Itao, K., and S. H. Crandall. 1979. Natural modes and natural frequencies of uniform, circular, free-edge plates. *Transactions of ASME, Journal of Applied Mechanics* 46:448–53.

Kirchhoff, G. 1850. Uber das gleichgwich und die bewegung einer elastischen scheibe. *J. Angew. Math.* 40:51–88.

Leissa, A. W. 1969. *Vibration of plates*. NASA SP-160. Washington, DC: U.S. Government Printing Office. Repr. Sewickley, PA: Acoustical Society of America, 1993.

Leissa, A. W. 1973. The free vibration of rectangular plates. *Journal of Sound and Vibration* 31 (3): 257–93.

Levy, M. 1899. Sur l'equilibrie elastique d'une plaque rectangulaire. *C. R. Acad. Sci.* 129:535–39.

Liew, K. M. 1993. On the use of pb-2 Rayleigh-Ritz method for free flexural vibration of triangular plates with curved internal supports. *Journal of Sound and Vibration* 165:329–40.

Liew, K. M., and K. Y. Lam. 1991. A set of orthogonal plate functions for flexural vibration of regular polygonal plates. *Transactions of ASME, Journal of Vibration and Acoustics* 113:182–86.

Liew, K. M., and M. K. Lim. 1993. Transverse vibration of trapezoidal plates of variable thickness: Symmetric trapezoids. *Journal of Sound and Vibration* 165:45–67.

Liew, K. M., C. M. Wang, Y. Xiang, and S. Kitipornchai. 1998. *Vibration of Mindlin plates: Programming the p-version Ritz method*. Oxford, UK: Elsevier Science.

Liew, K. M., Y. Xiang, and S. Kitipornchai. 1993. Transverse vibration of thick rectangular plates—I: Comprehensive sets of boundary conditions. *Computers and Structures* 49:1–29.

Liew, K. M., Y. Xiang, S. Kitipornchai, and C. M. Wang. 1993. Vibration of thick skew plates based on Mindlin shear deformation plate theory. *Journal of Sound and Vibration* 168:39–69.

Mindlin, R. D. 1951. Influence of rotary inertia and shear on flexural motions of isotropic, elastic plates. *Journal of Applied Mechanics* 18:31–38.

Mindlin, R. D., and H. Deresiewicz. 1954. Thickness-shear and flexural vibrations of a circular disk. *Journal of Applied Physics* 25:1329–32.

Mindlin, R. D., A. Schacknow, and H. Deresiewicz. 1956. Flexural vibration of rectangular plates. *Transactions of ASME, Journal of Applied Mechanics* 23:430–36.

Navier, C. L. M. H. 1823. Extrait des recherches sur la flexion des plans elastiques. *Bull. Sci. Soc. Philomarhique de Paris* 5:95–102.

Ng, F. L. 1974. Tabulation of methods for the numerical solution of the hollow waveguide problem. *IEEE Transactions on Microwave Theory and Techniques* MTT-22:322–29.

Pnueli, D. 1975. Lower bounds to the gravest and all higher frequencies of homogeneous vibrating plates of arbitrary shape. *Journal of Applied Mechanics* 42:815–20.

Ramakrishnan, R., and V. X. Kunukkasseril. 1973. Free vibration of annular sector plates. *Journal of Sound and Vibration* 30 (1): 127–29.

Reddy, J. N. 2007. *Theory and analysis of elastic plates and shells*. 2nd ed. Boca Raton, FL: CRC Press.

Schelkunoff, S. A. 1943. *Electromagnetic waves*. New York: Van Nostrand.

Shanmugam, N. E., and C. M. Wang. 2007. *Dynamics*. Vol. 2 of *Analysis and design of plated structures*. Abingdon, England: Woodhead Publishing.

Soedel, W. 2004. *Vibrations of shells and plates*. 3rd ed. Boca Raton, FL: CRC Press.

Srinivas, S., C. V. Joga Rao, and A. K. Rao. 1970. An exact analysis for vibration of simply supported homogeneous and laminated thick rectangular plates. *Journal of Sound and Vibration* 12 (2): 187–99.

Szilard, R. 1974. *Theory and analysis of plates*. Englewood Cliffs, NJ: Prentice-Hall.

Timoshenko, S. P., and S. Woinowsky-Krieger. 1959. *Theory of plates and shells*. New York: McGraw-Hill.

Ugural, A. C. 1981. *Stresses in plates and shells*. New York: McGraw-Hill.

Vogel, S. M., and D. W. Skinner. 1965. Natural frequencies of transversely vibrating uniform annular plates. *Journal of Applied Mechanics* 32:926–31.

Voigt, W. 1893. Bemerkungen zu dem Problem der transversalen Schwingungen rechteckiger Platten. *Nachr. Ges. Wiss* (Göttingen) 6:225–30.

Wang, C. M. 1994. Natural frequency formula for simply supported Mindlin plates. *Journal of Vibration and Acoustics, Transactions of the ASME* 116:536–40.

Wang, C. M., S. Kitipornchai, and J. N. Reddy. 2000. Relationship between vibration frequencies of Reddy and Kirchhoff polygonal plates with simply supported edges. *Trans. ASME, Journal of Vibration and Acoustics* 122 (1): 77–81.

Wang, C. Y. 2010. Exact solution of equilateral triangular waveguide. *Electronics Letters* 4:925–27.

Wang, C. Y., and C. M. Wang. 2005. Examination of the fundamental frequencies of annular plates with small core. *Journal of Sound and Vibration* 280:1116–24.

Wang, C. Y., C. M. Wang, and Z. Y. Tay. 2013. Forthcoming. Analogy of TE waveguide and vibrating plate with sliding edge condition and exact solutions. *Journal of Engineering Mechanics* 139.

Xing, Y. F., and B. Liu. 2009a. New exact solutions for free vibrations of rectangular thin plates by symplectic dual method. *Acta Mechanica Sinica* 25:265–70.

———. 2009b. Characteristic equations and closed-form solutions for free vibrations of rectangular Mindlin plates. *Acta Mechanica Solida Sinica* 22:125–36.

———. 2009c. Closed form solutions for free vibrations of rectangular Mindlin plates. *Acta Mechanica Sinica* 25:689–98.

6 Vibration of Plates with Complicating Effects

6.1 INTRODUCTION

In Chapter 5, plates with classical boundary conditions such as clamped, simply supported, sliding, or free are treated. In this chapter, we present exact vibration solutions for plates with complicating effects such as the presence of in-plane forces, internal spring supports, internal hinge, elastic foundation, and nonuniform thickness distribution. When these complicating effects are removed, the solutions reduce to those presented in Chapter 5.

6.2 PLATES WITH IN-PLANE FORCES

With in-plane forces, there are basically two types of plate problems with exact solutions. They are (a) rectangular plates with two parallel sides simply supported and (b) circular and annular plates with uniform normal stress.

6.2.1 Rectangular Plates with In-Plane Forces

Consider the rectangular plate shown in Figure 6.1. The edges at $\bar{y}=0$ and $\bar{y}=b$ are simply supported. The plate is under uniform compressive forces $\bar{N}_1,\bar{N}_2$ in the $\bar{x}$, and $\bar{y}$ -directions, respectively.

Similar to Section 5.2, the governing equation for a vibrating rectangular plate with in-plane forces is given by

$$D\nabla^4 w+\bar{N}_1\frac{\partial^2 w}{\partial \bar{x}^2}+\bar{N}_2\frac{\partial^2 w}{\partial \bar{y}^2}-\rho h\bar{\omega}^2 w=0 \tag{6.1}$$

Let $x=\bar{x}/a,\ y=\bar{y}/a$. Since the horizontal sides are simply supported, we set

$$w(x,y) = W(x)\sin(\alpha y) \tag{6.2}$$

where $\alpha = n\pi a/b$. In view of Equation (6.2), Equation (6.1) becomes

$$\frac{d^4W}{dx^4}+(N_1-2\alpha^2)\frac{d^2W}{dx^2}-(\omega^2+N_2\alpha^2-\alpha^4)W=0,\quad 0\le x\le 1 \tag{6.3}$$

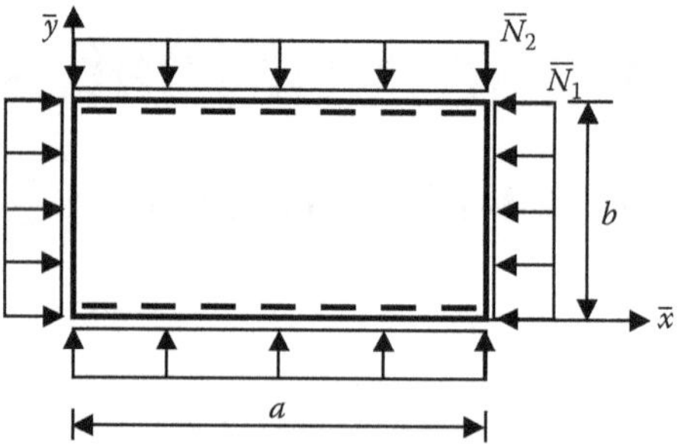

FIGURE 6.1 Vibration of rectangular plate with in-plane forces.

Here the nondimensional parameters are

$$N_1 = \bar{N}_1 a^2/D,\ N_2 = \bar{N}_2 a^2/D,\ \omega^2 = \rho h \bar{\omega}^2 a^4/D \tag{6.4}$$

Note that the frequency is normalized with respect to the width a instead of the height b as in Chapter 5.

The boundary conditions for a variety of edges are as follows:

- For a clamped vertical edge

$$W = 0,\quad \frac{dW}{dx} = 0 \tag{6.5}$$

- For a simply supported vertical edge

$$W = 0,\quad \frac{d^2W}{dx^2} - \alpha^2 \nu W = 0 \tag{6.6}$$

- For a sliding vertical edge

$$\frac{dW}{dx} = 0,\quad \frac{d^3W}{dx^3} - \alpha^2(2-\nu)\frac{dW}{dx} + N_1\frac{dW}{dx} = 0 \tag{6.7}$$

- For a free vertical edge

$$\frac{d^2W}{dx^2} - \alpha^2 \nu W = 0,\quad \frac{d^3W}{dx^3} - \alpha^2(2-\nu)\frac{dW}{dx} + N_1\frac{dW}{dx} = 0 \tag{6.8}$$

- For a spring-supported sliding edge with translational spring constant k_1

$$\frac{dW}{dx} = 0,\quad \frac{d^3W}{dx^3} - \alpha^2(2-\nu)\frac{dW}{dx} + N_1\frac{dW}{dx} = \mp\xi W \tag{6.9}$$

Here $\xi = k_1 a^3/D$, and the top and bottom signs are for the left and right edges, respectively.

- For a simply supported edge with torsional spring constant k_2

$$W = 0, \quad \frac{d^2W}{dx^2} - \alpha^2 \nu W = \pm \varsigma \frac{dW}{dx} \tag{6.10}$$

where $\varsigma = k_2 a/D$.

Other boundary conditions are possible.

6.2.1.1 Analogy with Beam Vibration

The vibration of a uniform beam with axial compressive force is given in Equation (4.16), i.e.,

$$\frac{d^4 w_b}{dx^4} + a_b \frac{d^2 w_b}{dx^2} - \omega_b^2 w_b = 0, \quad 0 \le x \le 1 \tag{6.11}$$

where the subscript b refers to the beam and

$$a_b = F'L^2/EI, \quad \omega_b^2 = \bar{\omega}_b^2 L^4 \rho_b / EI \tag{6.12}$$

Here F' is the axial compressive force, L is the length of the beam, ρ_b is the mass per length, and EI is the flexural rigidity. The boundary conditions are as follows:

- For a clamped end

$$w_b = 0, \quad \frac{dw_b}{dx} = 0 \tag{6.13}$$

- For a simply supported vertical edge

$$w_b = 0, \quad \frac{d^2 w_b}{dx^2} = 0 \tag{6.14}$$

- For a sliding vertical edge

$$\frac{dw_b}{dx} = 0, \quad \frac{d^3 w_b}{dx^3} + a_b \frac{dw_b}{dx} = 0 \tag{6.15}$$

- For a free vertical edge

$$\frac{d^2 w_b}{dx^2} = 0, \quad \frac{d^3 w_b}{dx^3} + a_b \frac{dw_b}{dx} = 0 \tag{6.16}$$

- For a spring-supported sliding end with translational spring constant k_{1b}

$$\frac{dw_b}{dx} = 0, \quad \frac{d^3 w_b}{dx^3} + a_b \frac{dw_b}{dx} = \mp \xi_b w_b \tag{6.17}$$

where $\xi_b = k_{1b} L^3 / EI$.

- For a simply supported end with torsional spring constant k_{2b}

$$w_b = 0, \quad \frac{d^2 w_b}{dx^2} = \pm \varsigma_b \frac{dw_b}{dx} \tag{6.18}$$

where $\varsigma_b = k_{2b} L / EI$.

By comparing Equations (6.3) and (6.11) and their respective boundary conditions, an analogy exists between a vibrating plate with two parallel simply supported sides and a vibrating beam. The boundary conditions need to be one of the following: clamped, simply supported, sliding, sliding with translational spring, and simply supported with torsional spring. Then

$$N_1 - 2\alpha^2 = a_b \tag{6.19}$$

$$\omega^2 + N_2 \alpha^2 - \alpha^4 = \omega_b^2 \tag{6.20}$$

$$\xi = \xi_b, \quad \varsigma = \varsigma_b \tag{6.21}$$

Given the plate parameters α and N_1,N_2, one can find the analogous beam compression force a_b from Equation (6.19), and from Chapter 4 the frequencies ω_b. Then plate frequencies ω can be obtained from Equation (6.20). Since an analogy exists, we shall not present the numerical results for these cases here.

6.2.1.2 Plates with Free Vertical Edge

If the plate has two horizontal simply supported edges and at least one free vertical edge, then the analogy of the previous section does not apply. We shall solve Equations (6.3) and (6.8) directly as follows. Let $W = e^{\lambda x}$ and

$$A = \alpha^2 - \frac{N_1}{2}, \quad B = \omega^2 + N_2 \alpha^2 - \alpha^4 \tag{6.22}$$

Equation (6.3) yields

$$\lambda = \pm \sqrt{A \pm \sqrt{A^2 + B}} \tag{6.23}$$

Let

$$\lambda_1 = \sqrt{A + \sqrt{A^2 + B}}, \quad \lambda_2 = \sqrt{\left|A - \sqrt{A^2 + B}\right|} \tag{6.24}$$

which in some cases may be complex. If $B > 0$, W can be expressed in a linear combination of $\cosh(\lambda_1 x), \sinh(\lambda_1 x), \cos(\lambda_2 x), \sin(\lambda_2 x)$. If $B < 0$, W can be expressed in a linear combination of $\cosh(\lambda_1 x), \sinh(\lambda_1 x), \cosh(\lambda_2 x), \sinh(\lambda_2 x)$.

We consider the following three cases: Case 1, with left side simply supported and right side free; Case 2, with left side clamped and right side free; and Case 3, with left side sliding and right side free. In all three cases, the solution that satisfies the left-side conditions is in the form

$$W = c_1 f(x) + c_2 g(x) \tag{6.25}$$

For nontrivial solutions, the right-side free-boundary conditions of Equation (6.8) give the exact characteristic equation

$$\left| \begin{matrix} f'' - \alpha^2 \nu f & g'' - \alpha^2 \nu g \\ f''' - [\alpha^2(2-\nu) - N_1]f' & g''' - [\alpha^2(2-\nu) - N_1]g' \end{matrix} \right|_{x=1} = 0 \tag{6.26}$$

The frequencies are obtained from Equation (6.26).

6.2.1.2.1 *Case 1: Left Side Simply Supported and Right Side Free*

The solution that satisfies the left-side boundary conditions is

$$f = \sinh(\lambda_1 x), \quad g = \begin{cases} \sin(\lambda_2 x) & B > 0 \\ \sinh(\lambda_2 x) & B < 0 \end{cases} \tag{6.27}$$

Usually $B > 0$, except at very low frequencies. Typical normalized frequencies ω of rectangular plates with equal in-plane forces, i.e., $N_1 = N_2 = N$, are given in Table 6.1. The number in parentheses is the vertical number of half waves n. The asterisk denotes that the plate has buckled. The frequencies corresponding to the zero-force $N = 0$ case agrees with those presented in Table 5.2, in which the frequencies should be multiplied by $(a/b)^2$ due to the different normalization. The number of internal horizontal nodes is $n - 1$.

Note that tensile force ($N < 0$) increases frequency, since the plate is stiffened, whereas compressive force ($N > 0$) reduces frequency. If the compressive force is large enough, buckling may occur. The fundamental mode is always at $n = 1$. Notice also that some frequencies are repeated, due to reflections about the simply supported edges.

TABLE 6.1
Normalized Frequency ω for Plate with Three Sides Simply Supported and One Side Free with In-Plane Forces, $N_1 = N_2 = N$

N	$a/b = 0.5$	$a/b = 1$	$a/b = 2$
	8.2676 (1)	13.808 (1)	46.067 (1)
	16.224 (2)	33.501 (1)	64.236 (1)
−10	25.287 (1)	46.067 (2)	99.994 (1)
	28.767 (3)	62.411 (2)	154.18 (1)
	33.501 (2)	68.002 (1)	164.08 (2)
	4.0337 (1)	11.685 (1)	41.197 (1)
	11.685 (2)	27.756 (1)	59.066 (1)
0	18.821 (1)	41.197 (2)	94.484 (1)
	24.010 (3)	59.066 (2)	148.51 (1)
	27.756 (2)	61.861 (1)	159.08 (2)
	* (1)	2.3732 (1)	35.643 (1)
	2.3732 (2)	20.192 (1)	53.360 (1)
10	7.6883 (1)	35.643 (2)	88.595 (1)
	17.975 (3)	53.360 (2)	142.59 (1)
	20.192 (2)	54.961 (1)	153.91 (2)

Note: The values in parentheses represent n, the vertical number of half waves. An asterisk denotes that the plate has buckled.

6.2.1.2.2 *Case 2: Left Side Clamped and Right Side Free*

The solution that satisfies the left-side boundary conditions is

$$f = \lambda_2 \sinh(\lambda_1 x) - \lambda_1 \begin{cases} \sin(\lambda_2 x) \\ \sinh(\lambda_2 x) \end{cases}, \quad g = \cosh(\lambda_1 x) - \begin{cases} \cos(\lambda_2 x) & B > 0 \\ \cosh(\lambda_2 x) & B < 0 \end{cases} \tag{6.28}$$

Table 6.2 shows the normalized frequencies obtained from solving Equation (6.26) with Equation (6.28) and $N_1 = N_2 = N$.

6.2.1.2.3 *Case 3: Left Side Sliding and Right Side Free*

The solution that satisfies the left-side boundary conditions is

$$f = \cosh(\lambda_1 x), \quad g = \begin{cases} \cos(\lambda_2 x) & B > 0 \\ \cosh(\lambda_2 x) & B < 0 \end{cases} \tag{6.29}$$

TABLE 6.2
Normalized Frequency ω for Plate with Opposite Sides Simply Supported and Other Two Sides Clamped and Free, $N_1 = N_2 = N$

N	$a/b = 0.5$	$a/b = 1$	$a/b = 2$
	9.7247 (1)	14.004 (1)	46.563 (1)
	17.170 (2)	38.569 (1)	68.089 (1)
−10	29.429 (3)	46.563 (2)	108.49 (1)
	30.922 (1)	68.089 (2)	162.84 (2)
	38.569 (2)	78.121 (1)	167.81 (1)
	5.7039 (1)	12.687 (1)	41.702 (1)
	12.687 (2)	33.065 (1)	63.015 (1)
0	24.694 (3)	41.702 (2)	103.16 (1)
	24.944 (1)	63.015 (2)	159.30 (2)
	33.065 (2)	72.398 (1)	162.37 (1)
	* (1)	4.4032 (1)	36.161 (1)
	4.4032 (2)	26.151 (1)	57.442 (1)
10	16.589 (1)	34.114 (2)	97.499 (1)
	18.704 (3)	57.442 (2)	154.14 (2)
	26.151 (2)	66.122 (1)	156.72 (1)

Note: The values in parentheses represent n, the vertical number of half waves. An asterisk denotes that the plate has buckled.

Table 6.3 shows the frequency results obtained by solving Equation (6.26) with Equation (6.29) and $N_1 = N_2 = N$.

We add that the frequencies of a plate with two opposite simply supported sides and two free sides can be constructed by combining the simply supported–free results (x-antisymmetric mode) and the sliding–free (x-symmetric mode) results. The method can be extended to unequal in-plane forces, although the tables are for $N_1 = N_2$.

6.2.2 Circular Plates with In-Plane Forces

There are no analogies for circular plates. Exact solutions are possible if the plate is under uniform in-plane forces. The governing equations are derived following Leissa (1969).

Figure 6.2 shows a circular plate with radius R under uniform compressive force $\bar{N}$. If the vibration frequency is $\bar{\omega}$, the deflection amplitude is governed by

$$D\nabla^4 w + \bar{N}\nabla^2 w - \rho h\bar{\omega}^2 w = 0 \tag{6.30}$$

TABLE 6.3
Normalized Frequency ω for Plate with Opposite Sides Simply Supported and the Other Sides Sliding and Free, $N_1 = N_2 = N$

N	$a/b = 0.5$	$a/b = 1$	$a/b = 2$
	5.5325 (1)	13.934 (1)	43.964 (1)
	13.934 (2)	21.731 (1)	51.665 (1)
−10	14.936 (1)	43.964 (2)	124.72 (1)
	22.861 (2)	48.432 (1)	162.23 (2)
	26.252 (3)	52.956 (2)	171.30 (2)
	2.4079 (1)	9.7362 (1)	39.188 (1)
	9.1814 (1)	17.685 (1)	47.967 (1)
0	9.7362 (2)	39.188 (2)	119.10 (1)
	17.685 (2)	42.384 (1)	157.26 (2)
	21.997 (3)	47.967 (2)	166.29 (2)
	* (1)	* (1)	33.730 (1)
	* (1)	9.7574 (1)	42.367 (1)
10	* (2)	33.730 (2)	68.712 (1)
	9.7574 (2)	35.167 (1)	113.18 (1)
	16.103 (3)	42.367 (2)	152.13 (2)

Note: The values in parentheses represent n, the vertical number of half waves. An asterisk denotes that the plate has buckled.

Normalize the radial coordinate by R and let

$$w = W(r)\cos(n\theta) \tag{6.31}$$

Equation (6.30) then becomes

$$\nabla^4 W + 2A\nabla^2 W - \omega^2 W = 0 \tag{6.32}$$

where

$$\nabla^2 = \frac{d^2}{dr^2} + \frac{1}{r}\frac{d}{dr} - \frac{n^2}{r^2}, \quad A = \frac{\bar{N}R^2}{2D}, \quad \omega^2 = \frac{\rho h R^4 \bar{\omega}^2}{D} \tag{6.33}$$

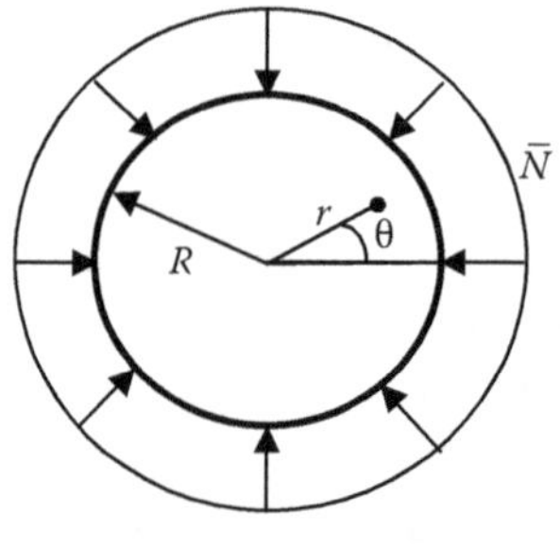

FIGURE 6.2 Circular plate under uniform compressive force.

Then Equation (6.32) can be factored into

$$\left(\nabla^2+\lambda_1^2\right)\left(\nabla^2-\lambda_2^2\right)W=0 \tag{6.34}$$

where

$$\lambda_1^2=\sqrt{A^2+\omega^2}+A,\quad \lambda_2^2=\sqrt{A^2+\omega^2}-A \tag{6.35}$$

The solution of W is a linear combination of the Bessel functions $J_n(\lambda_1 r)$, $Y_n(\lambda_1 r)$, $I_n(\lambda_2 r)$, $K_n(\lambda_2 r)$. For a full circular plate, n is an integer, and Y_n, K_n are not used.

The bending moment M_r and the effective shear force V_r are given by

$$M_r=-\frac{D}{R^2}\left[\frac{d^2W}{dr^2}+\frac{\nu}{r}\left(\frac{dW}{dr}-\frac{n^2}{r}W\right)\right] \tag{6.36}$$

$$V_r=-\frac{D}{R^3}\left[\frac{d^3W}{dr^3}+\frac{1}{r}\frac{d^2W}{dr^2}-\frac{1+(2-\nu)n^2}{r^2}\frac{dW}{dr}+\frac{(3-\nu)n^2}{r^3}W\right]-\frac{N}{R}\frac{dW}{dr} \tag{6.37}$$

If the edge is clamped, the boundary condition is

$$W=0,\quad \frac{dW}{dr}=0\quad \text{at}\quad r=1 \tag{6.38}$$

This yields the characteristic equation

$$\lambda_2 J_n(\lambda_1)I_n'(\lambda_2)-\lambda_1 J_n'(\lambda_1)I_n(\lambda_2)=0 \tag{6.39}$$

Table 6.4 shows the frequency results. The number of diametric nodes n is in parentheses. Our values for $A = 0$, i.e., no in-plane force, agrees with Table 5.10. The plate buckles for A larger than 7.3410.

TABLE 6.4
Frequency ω for a Clamped Circular Plate under Uniform Compressive In-Plane Force

A = 5	A = 0	A = −5	A = −10
5.8274 (0)	10.216 (0)	13.146 (0)	15.490 (0)
16.821 (1)	21.261 (1)	24.881 (1)	28.011 (1)
30.356 (2)	34.877 (2)	38.852 (2)	42.438 (2)
35.316 (0)	39.771 (0)	43.769 (0)	47.423 (0)
46.446 (3)	51.030 (3)	55.220 (3)	59.100 (3)

Note: The value of n (number of diametric nodes) is in parentheses.

TABLE 6.5
Frequency ω for a Simply Supported Circular Plate under Uniform Compressive In-Plane Force

$A = 2$	$A = 0$	$A = -5$	$A = -10$
1.0714 (0)	4.9351 (0)	9.0732 (0)	11.843 (0)
11.591 (1)	13.898 (1)	18.442 (1)	22.069 (1)
23.462 (2)	25.613 (2)	30.330 (2)	34.406 (2)
27.593 (0)	29.720 (0)	34.468 (0)	38.637 (0)
37.864 (3)	39.957 (3)	44.764 (3)	49.102 (3)

Note: The value of n (number of diametric nodes) is in parentheses.

If the plate is simply supported, the boundary conditions are

$$W = 0, \quad \frac{d^2W}{dr^2} + \nu \frac{dW}{dr} = 0 \quad \text{at} \quad r = 1 \tag{6.40}$$

Using the Bessel functions J_n, I_n and their properties, Equation (6.40) yields the characteristic equation

$$(1-\nu)[\lambda_1 I_n(\lambda_2)J_n'(\lambda_1) - \lambda_2 J_n(\lambda_1)I_n'(\lambda_2)] + (\lambda_1^2 + \lambda_2^2)J_n(\lambda_1)I_n(\lambda_2) = 0 \tag{6.41}$$

Table 6.5 shows the frequencies. The number of diametric nodes n is in parentheses. The simply supported plate buckles when the compressive force $A > 2.0989$.

If the edge is sliding, the boundary conditions are reduced to

$$\frac{dW}{dr} = 0, \quad \frac{d^3W}{dr^3} + \frac{d^2W}{dr^2} + n^2(3-\nu)W = 0 \quad \text{at} \quad r = 1 \tag{6.42}$$

The characteristic equation is

$$(1-\nu)n^2[\lambda_1 I_n(\lambda_2)J_n'(\lambda_1) - \lambda_2 J_n(\lambda_1)I_n'(\lambda_2)] + \lambda_1\lambda_2(\lambda_1^2 + \lambda_2^2)J_n'(\lambda_1)I_n'(\lambda_2) = 0 \tag{6.43}$$

Since the sliding case with in-plane force is somewhat rare, the numerical results are not presented here. The free plate with in-plane force is also rare, and is unstable for compressive forces. The frequency equations for annular plates can be similarly constructed.

6.3 PLATES WITH INTERNAL SPRING SUPPORT

We consider plates with no in-plane forces but with an internal translational spring support for vibration control. For rectangular plates, the internal line spring is perpendicular to two simply supported edges, as shown in Figure 6.3a. For circular plates, the spring support is on an internal concentric circle, as shown in Figure 6.3b.

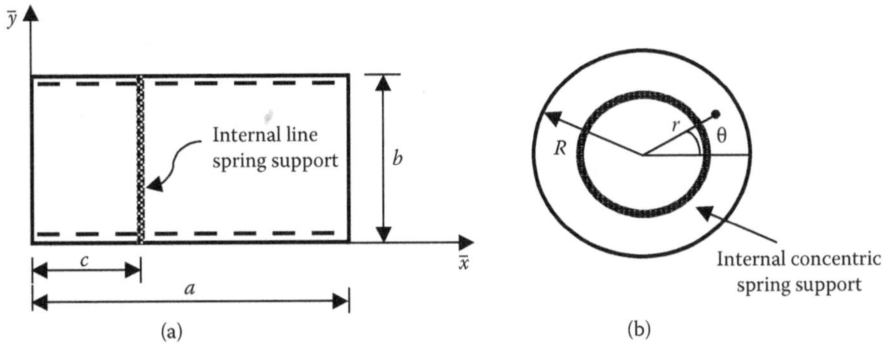

FIGURE 6.3 (a) Rectangular plate with two opposite sides simply supported and an internal line spring support, and (b) circular plate with an internal concentric spring support.

If the normalized spring constant ξ is zero, it is a plate without a support. If the spring constant is infinity, it is a plate with a rigid line support.

6.3.1 Rectangular Plates with Line Spring Support

Consider the rectangular plate shown in Figure 6.3a. The edges at $\bar{y}=0$ and $\bar{y}=b$ are simply supported, and there is an internal line spring support at $\bar{x}=c$. Our derivation mostly follows the work of Li (2003), who gave characteristic formulas but no numerical results.

Normalize all lengths by the width a as in Section 6.2.1, and let $\beta = c/a$. Separate the plate into two regions, where the subscript 1 denotes $0 \le x \le \beta$ and the subscript 2 denotes $\beta \le x \le 1$. By using Equation (6.2), the governing equations are

$$\frac{d^4W_i}{dx^4} - 2\alpha^2\frac{d^2W_i}{dx^2} - (\omega^2 - \alpha^4)W_i = 0 \quad i = 1,2 \tag{6.44}$$

The boundary conditions at the vertical edges could be any of the Equations (6.5–6.10). At the support $x = \beta$, we need continuity of displacement, slope, and moment, but vertical shear is affected by the spring.

$$W_1 = W_2 \tag{6.45}$$

$$\frac{dW_1}{dx} = \frac{dW_2}{dx} \tag{6.46}$$

$$\frac{d^2W_1}{dx^2} - \alpha^2\nu W_1 = \frac{d^2W_2}{dx^2} - \alpha^2\nu W_2 \tag{6.47}$$

$$\frac{d^3W_1}{dx^3} - \alpha^2(2-\nu)\frac{dW_1}{dx} - \xi W_1 = \frac{d^3W_2}{dx^3} - \alpha^2(2-\nu)\frac{dW_2}{dx} \tag{6.48}$$

Suppose

$$W_1 = c_1 f(x) + c_2 g(x), \quad W_2 = c_3 h(x) + c_4 k(x) \tag{6.49}$$

which satisfy the boundary conditions at $x = 0$ and $x = 1$, respectively. Then the continuity equations give the characteristic equation

$$\begin{vmatrix} f & g & -h & -k \\ \frac{df}{dx} & \frac{dg}{dx} & -\frac{dh}{dx} & -\frac{dk}{dx} \\ \frac{d^2f}{dx^2} & \frac{d^2g}{dx^2} & -\frac{d^2h}{dx^2} & -\frac{d^2k}{dx^2} \\ \frac{d^3f}{dx^3} - \xi f & \frac{d^3g}{dx^3} - \xi g & -\frac{d^3h}{dx^3} & -\frac{d^3k}{dx^3} \end{vmatrix}_{x=\beta} = 0 \tag{6.50}$$

The parameters for the problem include the aspect ratio a/b, the location of the line spring β, the normalized spring constant ξ, and the various combinations of left- and right-edge conditions (there are 10 classical combinations). Here we shall only present two cases where the support is at the midpoint: (a) where the vertical edges are either both simply supported or (b) where both are clamped.

6.3.1.1 Case 1: All Sides Simply Supported

Let

$$\lambda_1 = \sqrt{\omega + \alpha^2}, \quad \lambda_2 = \sqrt{|\omega - \alpha^2|} \tag{6.51}$$

The solutions that satisfy the left and right boundary conditions are, respectively,

$$f = \sinh(\lambda_1 x), \quad g = \begin{cases} \sin(\lambda_2 x) & \omega > \alpha^2 \\ \sinh(\lambda_2 x) & \omega < \alpha^2 \end{cases} \tag{6.52}$$

$$h = \sinh[\lambda_1 (x-1)], \quad k = \begin{cases} \sin[\lambda_2 (x-1)] & \omega > \alpha^2 \\ \sinh[\lambda_2 (x-1)] & \omega < \alpha^2 \end{cases} \tag{6.53}$$

For a given α, ξ, the frequency is obtained from Equation (6.50), and these are presented in Table 6.6 as well as the value of n (in parentheses). Some frequencies are duplicated for different aspect ratios, and some are duplicated for different modes. Also, if a node is right on the line support, the frequency will be independent of the spring constant ξ.

TABLE 6.6
Frequencies for the Simply Supported Plate with an Internal Line Spring Support at Mid-Span

ξ	$a/b = 0.5$	$a/b = 1$	$a/b = 2$
	12.337 (1)	19.739 (1)	49.348 (1)
	19.739 (2)	49.348 (1)	78.957 (1)
0	32.076 (3)	78.597 (2)	128.30 (1)
	41.946 (1)	98.696 (1)	167.78 (2)
	49.348 (2)	128.30 (2)	197.39 (1)
	23.007 (1)	27.736 (1)	53.112 (1)
	27.736 (2)	49.348 (1)	78.957 (1)
200	37.572 (3)	53.112 (2)	129.88 (1)
	41.946 (1)	78.957 (2)	168.95 (2)
	49.348 (2)	100.77 (1)	197.39 (1)
	41.946 (1)	49.348 (1)	94.586 (1)
	49.348 (2)	69.327 (1)	78.957 (1)
∞	61.685 (3)	78.957 (2)	197.39 (1)
	63.534 (1)	94.586 (2)	234.58 (1)
	78.957 (4)	128.30 (3)	260.70 (2)

Note: The value of n (number of vertical half waves) is in parentheses.

6.3.1.2 Case 2: Both Horizontal Sides Simply Supported and Both Vertical Sides Clamped

We take

$$f = \lambda_2 \sinh(\lambda_1 x) - \lambda_1 \begin{cases} \sin(\lambda_2 x) \\ \sinh(\lambda_2 x) \end{cases}, \quad g = \cosh(\lambda_1 x) - \begin{cases} \cos(\lambda_2 x) \\ \cosh(\lambda_2 x) \end{cases} \tag{6.54}$$

$$h = \lambda_2 \sinh[\lambda_1(x-1)] - \lambda_1 \begin{cases} \sin[\lambda_2(x-1)] \\ \sinh[\lambda_2(x-1)] \end{cases}, \quad k = \cosh[\lambda_1(x-1)] - \begin{cases} \cos[\lambda_2(x-1)] \\ \cosh[\lambda_2(x-1)] \end{cases}, \tag{6.55}$$

where the upper form is for $\omega > \alpha^2$ and the lower form for $\omega < \alpha^2$. Equation (6.50) gives the results in Table 6.7, where the value of n is in parentheses.

6.3.2 Circular Plates with Concentric Spring Support

We extend the work of Wang and Wang (2003). Consider a circular plate of radius R with internal elastic spring support on a concentric circle of radius βR. Normalize

TABLE 6.7
Frequencies for the Plate with Opposite Sides Simply Supported and Opposite Sides Clamped and with an Internal Line Spring Support at Mid-Span

ξ	$a/b = 0.5$	$a/b = 1$	$a/b = 2$
	23.816 (1)	28.951 (1)	54.743 (1)
	28.951 (2)	54.743 (2)	94.585 (1)
0	39.089 (3)	69.327 (1)	154.78 (1)
	54.743 (4)	94.585 (2)	170.35 (2)
	63.535 (1)	102.22 (3)	206.70 (2)
	32.434 (1)	36.293 (1)	58.822 (1)
	36.293 (2)	58.822 (2)	94.585 (1)
200	44.713 (3)	69.327 (1)	156.13 (1)
	58.822 (4)	94.585 (2)	171.63 (2)
	63.535 (1)	104.39 (3)	206.70 (2)
	63.535 (1)	69.327 (1)	94.585 (1)
	69.327 (2)	94.585 (2)	115.80 (1)
∞	79.525 (3)	95.263 (1)	206.70 (2)
	90.872 (1)	115.80 (2)	218.97 (2)
	94.585 (4)	140.20 (3)	234.59 (1)

Note: The value of n (number of vertical half waves) is in parentheses.

the radial coordinate by R. Let the subscript 1 denote the region $0 \le r \le \beta$ and the subscript 2 denote the region $\beta < r \le 1$. Similar to Section 6.2.2, the governing equation for the vibration of such a spring-supported circular plate is given by

$$\nabla^4 W_i - \omega^2 W_i = 0, \quad i = 1,2 \tag{6.56}$$

The solutions are

$$W_1 = c_1 J_n(\sqrt{\omega} r) + c_2 I_n(\sqrt{\omega} r) \tag{6.57}$$

$$W_2 = c_3 J_n(\sqrt{\omega} r) + c_4 I_n(\sqrt{\omega} r) + c_5 Y_n(\sqrt{\omega} r) + c_6 K_n(\sqrt{\omega} r) \tag{6.58}$$

Continuity at $r = \beta$ simplifies to

$$W_1 = W_2, \ \frac{dW_1}{dr} = \frac{dW_2}{dr}, \ \frac{d^2W_1}{dr^2} = \frac{d^2W_2}{dr^2} \tag{6.59}$$

$$\frac{d^3W_1}{dr^3} - \xi W_1 = \frac{d^3W_2}{dr^3} \tag{6.60}$$

Here, $\xi = k_1 R^3 / D$ is the normalized spring constant.

6.3.2.1 Case 1: Plate Is Simply Supported at the Edge

The boundary condition at $r = 1$ is

$$W_2 = 0, \quad \frac{d^2W_2}{dr^2} + \nu\frac{dW_2}{dr} = 0 \tag{6.61}$$

Let

$$S_1 = J_n(\sqrt{\omega}r), \quad S_2 = I_n(\sqrt{\omega}r), S_3 = Y_n(\sqrt{\omega}r), \quad S_4 = K_n(\sqrt{\omega}r) \tag{6.62}$$

From Equations (6.59) and (6.60), we can construct the submatrix

$$U = \begin{pmatrix} S_1 & S_2 & -S_1 & -S_2 & -S_3 & -S_4 \\ \frac{dS_1}{dr} & \frac{dS_2}{dr} & -\frac{dS_1}{dr} & -\frac{dS_2}{dr} & -\frac{dS_3}{dr} & -\frac{dS_4}{dr} \\ \frac{d^2S_1}{dr^2} & \frac{d^2S_2}{dr^2} & -\frac{d^2S_1}{dr^2} & -\frac{d^2S_2}{dr^2} & -\frac{d^2S_3}{dr^2} & -\frac{d^2S_4}{dr^2} \\ \frac{d^3S_1}{dr^3} - \xi S_1 & \frac{d^3S_2}{dr^3} - \xi S_2 & -\frac{d^3S_1}{dr^3} & -\frac{d^3S_2}{dr^3} & -\frac{d^3S_3}{dr^3} & -\frac{d^3S_4}{dr^3} \end{pmatrix}_{r=\beta} \tag{6.63}$$

and from Equation (6.61), we can construct the submatrix

$$V = \begin{pmatrix} 0 & 0 & S_1 & S_2 & S_3 & S_4 \\ 0 & 0 & \frac{d^2S_1}{dr^2} + \nu\frac{dS_1}{dr} & \frac{d^2S_2}{dr^2} + \nu\frac{dS_2}{dr} & \frac{d^2S_3}{dr^2} + \nu\frac{dS_3}{dr} & \frac{d^2S_4}{dr^2} + \nu\frac{dS_4}{dr} \end{pmatrix}_{r=1} \tag{6.64}$$

Then the frequencies are obtained from the determinant equation

$$\begin{vmatrix} U \\ V \end{vmatrix} = 0 \tag{6.65}$$

Table 6.8 shows the results. When $\xi = 0$ or the spring is absent, the frequencies are the same for all support locations.

6.3.2.2 Case 2: Plate Is Clamped at the Edge

Since $W = 0$ and $dW/dr = 0$ at $r = 1$, U is same as before, but

$$V = \begin{pmatrix} 0 & 0 & S_1 & S_2 & S_3 & S_4 \\ 0 & 0 & \frac{dS_1}{dr} & \frac{dS_2}{dr} & \frac{dS_3}{dr} & \frac{dS_4}{dr} \end{pmatrix}_{r=1} \tag{6.66}$$

Equation (6.65) yields the frequency results that are presented in Table 6.9.

TABLE 6.8
Frequencies for the Simply Supported Circular Plate with an Internal Ring Spring Support

ξ	β = 0.25	β = 0.5	β = 0.75
		4.9351 (0)	
		13.898 (1)	
0		25.613 (2)	
		29.720 (0)	
		39.957 (3)	
	14.994 (0)	17.754 (0)	10.680 (0)
	16.729 (1)	23.950 (1)	20.677 (1)
200	26.085 (2)	30.628 (0)	32.277 (2)
	35.544 (0)	31.259 (2)	35.766 (0)
	40.024 (3)	42.808 (3)	45.955 (3)
	21.089 (0)	27.800 (0)	15.458 (0)
	24.527 (1)	45.447 (1)	32.774 (1)
∞	31.029 (2)	52.594 (0)	54.535 (2)
	42.101 (3)	53.979 (2)	61.923 (0)
	72.492 (0)	63.071 (3)	80.627 (3)

Note: The value of *n* (number of diametric nodes) is in parentheses.

TABLE 6.9
Frequencies for the Clamped Circular Plate with an Internal Ring Spring Support

ξ	β = 0.25	β = 0.5	β = 0.75
		10.216 (0)	
		21.261 (1)	
0		34.877 (2)	
		39.771 (0)	
		51.030 (3)	
	21.047 (0)	18.805 (0)	11.795 (0)
	24.689 (1)	30.327 (1)	23.824 (1)
200	35.566 (2)	41.130 (2)	38.085 (2)
	43.043 (0)	42.607 (0)	43.105 (0)
	51.143 (3)	54.687 (3)	54.552 (3)
	32.573 (0)	30.628 (0)	15.974 (0)
	36.640 (1)	60.625 (1)	33.658 (1)
∞	43.308 (2)	74.979 (0)	55.780 (2)
	54.677 (3)	76.942 (2)	63.508 (0)
	89.030 (0)	86.983 (3)	82.293 (3)

Note: The value of *n* (number of diametric nodes) is in parentheses.

6.3.2.3 Case 3: Free Plate with Support

The boundary conditions at $r = 1$ simplify to

$$\frac{d^2W}{dr^2} + \nu\left(\frac{dW}{dr} - n^2W\right) \equiv L_1(W) = 0, \tag{6.67}$$

$$\frac{d^3W}{dr^3} - [1+\nu+(2-\nu)n^2]\frac{dW}{dr} + 3n^2W \equiv L_2(W) = 0 \tag{6.68}$$

where L_1, L_2 are differential operators as defined. We take

$$V = \begin{pmatrix} 0 & 0 & L_1(S_1) & L_1(S_2) & L_1(S_3) & L_1(S_4) \\ 0 & 0 & L_2(S_1) & L_2(S_2) & L_2(S_3) & L_2(S_4) \end{pmatrix}_{x=1} \tag{6.69}$$

Equation (6.65) furnishes the frequencies, which are listed in Table 6.10. Notice the changes in mode shapes, especially the fundamental mode of vibration. The frequencies depend on both the location and stiffness of the ring support.

TABLE 6.10
Frequencies for the Free Circular Plate with an Internal Ring Spring Support

ξ	$\beta = 0.25$	$\beta = 0.5$	$\beta = 0.75$
0		5.3584 (2) 9.0031 (0) 12.439 (3) 20.475 (1) 21.835 (4)	
200	5.4961 (2) 12.448 (3) 19.362 (0) 21.836 (4) 23.649 (1)	6.5636 (0) 7.9133 (2) 17.923 (0) 22.068 (4) 28.841 (1)	8.3897 (0) 16.775 (1) 17.127 (0) 17.135 (2) 19.876 (3)
∞	4.5339 (0) 30.410 (0) 35.057 (1) 43.403 (2) 56.713 (3)	6.9296 (0) 11.790 (2) 16.602 (3) 24.369 (4) 31.343 (0)	8.5270 (0) 20.067 (1) 30.200 (2) 31.177 (0) 38.835 (3)

Note: The value of n (number of diametric nodes) is in parentheses.

6.4 PLATES WITH INTERNAL ROTATIONAL HINGE

An internal rotational hinge models a partial crack. If the rotational spring constant is infinite, the plate is continuous. If the rotational spring constant is zero, it is a swivel for doors and panel sections. Exact solutions exist for rectangular plates with two parallel sides simply supported and for circular or annular plates.

6.4.1 RECTANGULAR PLATES WITH INTERNAL ROTATIONAL HINGE

Similar to Figure 6.3a, the edges at $\bar{y}=0$ and $\bar{y}=b$ are simply supported, and there is an internal rotational spring at $\bar{x}=c$. Normalize all lengths by the width a as in Section 6.2.1, and let $\beta = c/a$. Separate the plate into two regions, where the subscript 1 denotes $0 \le x \le \beta$ and the subscript 2 denotes $\beta \le x \le 1$. The governing equations of motion are given by

$$\frac{d^4W_i}{dx^4} - 2\alpha^2\frac{d^2W_i}{dx^2} - (\omega^2 - \alpha^4)W_i = 0, \quad \text{where } i = 1,2 \tag{6.70}$$

At the hinge, we have continuity of displacement, bending moment, and shear force, i.e.,

$$W_1 = W_2, \qquad \frac{d^2W_1}{dx^2} - \alpha^2\nu W_1 = \frac{d^2W_2}{dx^2} - \alpha^2\nu W_2$$

$$\frac{d^3W_1}{dx^3} - \alpha^2(2-\nu)\frac{dW_1}{dx} = \frac{d^3W_2}{dx^3} - \alpha^2(2-\nu)\frac{dW_2}{dx} \tag{6.71}$$

Also, the bending moment is proportional to the difference in slope

$$\frac{d^2W_1}{dx^2} - \alpha^2\nu W_1 = \varsigma\left(\frac{dW_2}{dx} - \frac{dW_1}{dx}\right) \tag{6.72}$$

where ς is the normalized rotational spring constant defined after Equation (6.10).

Suppose

$$W_1 = c_1 f(x) + c_2 g(x), \quad W_2 = c_3 h(x) + c_4 k(x) \tag{6.73}$$

which satisfy the boundary conditions at $x = 0$ and $x = 1$, respectively. Then Equations (6.71) and (6.72) give the characteristic equation

$$\begin{vmatrix} f & g & -h & -k \\ \dfrac{d^2f}{dx^2} & \dfrac{d^2g}{dx^2} & -\dfrac{d^2h}{dx^2} & -\dfrac{d^2k}{dx^2} \\ \dfrac{d^3f}{dx^3} - \alpha^2(2-\nu)\dfrac{df}{dx} & \dfrac{d^3g}{dx^3} - \alpha^2(2-\nu)\dfrac{dg}{dx} & -\dfrac{d^3h}{dx^3} + \alpha^2(2-\nu)\dfrac{dh}{dx} & -\dfrac{d^3k}{dx^3} + \alpha^2(2-\nu)\dfrac{dk}{dx} \\ \dfrac{d^2f}{dx^2} + \varsigma\dfrac{df}{dx} - \alpha^2\nu f & \dfrac{d^2g}{dx^2} + \varsigma\dfrac{dg}{dx} - \alpha^2\nu g & -\varsigma\dfrac{dh}{dx} & -\varsigma\dfrac{dk}{dx} \end{vmatrix}_{x=\beta} = 0 \tag{6.74}$$

We shall only present some cases where the hinge is at the midpoint and where both vertical edges are either simply supported or clamped.

6.4.1.1 Case 1: All Sides Simply Supported

We use Equations (6.51) to (6.53) and substitute them into Equation (6.73). The results are given in Table 6.11. The $\varsigma = \infty$ case is the simply supported plate without the rotational spring. Note that the frequency is sensitive to the spring constant at small ς.

6.4.1.2 Case 2: Two Parallel Sides Simply Supported, with a Midline Internal Rotational Spring Parallel to the Other Two Clamped Sides

We use Equations (6.54), (6.55) and (6.73) to obtain the frequencies presented in Table 6.12.

6.4.2 Circular Plates with Concentric Internal Rotational Hinge

We extend the work of Wang (2002). As in Section 6.3.2, the plate is separated into two regions. The general solutions are given by Equations (6.57) and (6.58). At the joint $r = \beta$, the displacement, bending moment, and shear force are continuous, i.e.,

$$W_1 = W_2, \quad \frac{d^2W_1}{dr^2} + \frac{\nu}{r}\left(\frac{dW_1}{dr} - \frac{n^2}{r}W_1\right) = \frac{d^2W_2}{dr^2} + \frac{\nu}{r}\left(\frac{dW_2}{dr} - \frac{n^2}{r}W_2\right) \tag{6.75}$$

$$\begin{aligned} &\frac{d^3W_1}{dr^3} + \frac{1}{r}\frac{d^2W_1}{dr^2} - \frac{1+(2-\nu)n^2}{r^2}\frac{dW_1}{dr} + \frac{(3-\nu)n^2}{r^3}W_1 \\ &= \frac{d^3W_2}{dr^3} + \frac{1}{r}\frac{d^2W_2}{dr^2} - \frac{1+(2-\nu)n^2}{r^2}\frac{dW_2}{dr} + \frac{(3-\nu)n^2}{r^3}W_2 \end{aligned} \tag{6.76}$$

Also, the bending moment is proportional to the difference in slopes, i.e.,

$$\frac{d^2W_1}{dr^2} + \frac{\nu}{r}\left(\frac{dW_1}{dr} - \frac{n^2}{r}W_1\right) = \varsigma\left(\frac{dW_2}{dr} - \frac{dW_1}{dr}\right) \tag{6.77}$$

Define S_i as in Equation (6.62) and the operators

$$Z_1 = \frac{d^2}{dr^2} + \frac{\nu}{\beta}\frac{d}{dr}, \quad Z_2 = \frac{d^3}{dr^3} + \frac{1}{\beta}\frac{d^2}{dr^2} - \frac{1+(2-\nu)n^2}{\beta^2}\frac{d}{dr}, \quad Z_3 = \varsigma\frac{d}{dr} - \frac{\nu n^2}{\beta^2} \tag{6.78}$$

TABLE 6.11
Frequencies for the Rectangular Plate with All Sides Simply Supported with a Midline Rotational Spring

ς	a/b = 0.5	a/b = 1	a/b = 2
	6.8805 (1)	16.135 (1)	46.738 (1)
	16.135 (2)	46.738 (2)	78.957 (1)
0	29.208 (3)	49.348 (1)	111.03 (1)
	41.946 (1)	75.283 (1)	164.79 (2)
	46.738 (4)	78.957 (2)	197.39 (1)
	10.982 (1)	18.598 (1)	48.243 (1)
	18.598 (2)	48.243 (2)	78.957 (1)
5	31.005 (3)	49.348 (1)	119.97 (1)
	41.946 (1)	78.957 (2)	166.08 (2)
	48.243 (4)	88.714 (1)	197.39 (1)
	12.337 (1)	19.739 (1)	49.348 (1)
	19.739 (2)	49.348 (1)	78.957 (1)
∞	32.076 (3)	78.957 (2)	128.30 (1)
	41.946 (1)	98.696 (1)	167.78 (2)
	49.348 (2)	128.30 (2)	197.39 (1)

Note: Number of vertical half waves n is in parenthesis.

TABLE 6.12
Frequencies for the Plate with Two Parallel Sides Simply Supported and Two Sides Clamped with a Midline Rotational Spring

ς	a/b = 0.5	a/b = 1	a/b = 2
	16.194 (1)	22.815 (1)	50.749 (1)
	22.815 (2)	50.749 (2)	94.585 (1)
0	34.280 (3)	69.327 (1)	132.26 (1)
	63.535 (1)	94.585 (2)	166.81 (2)
	69.327 (2)	98.777 (3)	206.70 (2)
	21.529 (1)	26.862 (1)	53.012 (1)
	26.862 (2)	53.012 (2)	94.585 (1)
5	37.223 (3)	69.327 (1)	143.59 (1)
	63.535 (1)	94.585 (2)	168.33 (2)
	69.237 (2)	100.46 (3)	206.70 (2)
	23.816 (1)	28.951 (1)	54.743 (1)
	28.951 (2)	54.743 (2)	94.585 (1)
∞	39.089 (3)	69.327 (1)	154.78 (1)
	54.743 (4)	94.585 (2)	170.35 (2)
	63.535 (1)	102.22 (3)	206.70 (2)

Note: Number of vertical half waves n is in parenthesis.

Then Equations (6.75) to (6.77) lead to the submatrix

$$U = \begin{pmatrix} S_1 & S_2 & -S_1 & -S_2 & -S_3 & -S_4 \\ Z_1S_1 & Z_1S_2 & -Z_1S_1 & -Z_1S_2 & -Z_1S_3 & -Z_1S_4 \\ Z_2S_1 & Z_2S_2 & -Z_1S_1 & -Z_2S_2 & -Z_2S_3 & -Z_2S_4 \\ (Z_1+Z_3)S_1 & (Z_1+Z_3)S_2 & -\varsigma\dfrac{dS_1}{dr} & -\varsigma\dfrac{dS_2}{dr} & -\varsigma\dfrac{dS_3}{dr} & -\varsigma\dfrac{dS_4}{dr} \end{pmatrix}_{r=\beta} \tag{6.79}$$

6.4.2.1 Case 1: Plate Is Simply Supported at the Edge

The V matrix is given by Equation (6.64), and with Equation (6.79), we solve Equation (6.65) to obtain the frequency results. In Table 6.13, we observe that the frequencies are very sensitive to small ς. For large ς, the frequencies rapidly approach the no-hinge values presented in Chapter 5.

6.4.2.2 Case 2: Plate Is Clamped at the Edge

We use Equation (6.66) and Equation (6.79) to obtain the frequency results presented in Table 6.14.

TABLE 6.13
Frequencies for Simply Supported Circular Plate with an Internal Ring Hinge

ς	$\beta = 0.25$	$\beta = 0.5$	$\beta = 0.75$
0	4.1409 (0) 12.865 (1) 25.613 (2) 27.383 (0) 39.910 (3)	3.3722 (0) 9.2188 (1) 22.126 (2) 24.345 (0) 38.242 (3)	3.3233 (0) 8.1628 (1) 13.974 (0) 17.887 (2) 30.955 (3)
1	4.3887 (0) 13.163 (1) 25.613 (2) 28.019 (0) 39.916 (3)	4.0846 (0) 10.793 (1) 23.058 (2) 26.282 (0) 38.628 (3)	4.4390 (0) 10.911 (1) 19.626 (0) 20.465 (2) 33.261 (3)
∞		4.9351 (0) 13.898 (1) 25.613 (2) 29.720 (0) 39.957 (3)	

Note: The value of n (number of diametric nodes) is in parentheses.

TABLE 6.14
Frequencies for Clamped Circular Plate with an Internal Ring Hinge

ς	β = 0.25	β = 0.5	β = 0.75
	9.0247 (0)	9.8593 (0)	8.3916 (0)
	18.974 (1)	15.999 (1)	21.236 (1)
0	34.827 (2)	28.239 (0)	33.444 (2)
	37.973 (0)	28.795 (2)	34.945 (0)
	50.992 (3)	46.831 (3)	46.123 (3)
	9.3728 (0)	10.014 (0)	9.0873 (0)
	19.606 (1)	17.628 (1)	21.245 (1)
1	34.836 (2)	30.303 (2)	33.917 (2)
	38.465 (0)	31.791 (0)	36.617 (0)
	50.997 (3)	46.831 (3)	47.445 (3)
		10.216 (0)	
		21.261 (1)	
∞		34.877 (2)	
		39.771 (0)	
		51.030 (3)	

Note: The value of n (number of diametric nodes) is in parentheses.

6.4.2.3 Case 3: Plate Is Free at the Edge

We use Equation (6.69) and Equation (6.79) to obtain the results presented in Table 6.15.

6.5 PLATES WITH PARTIAL ELASTIC FOUNDATION

An elastic foundation model supports over such areas as soils or rubber material. We shall consider only the Winkler foundation, where the restoring force is proportional to the displacement. The equation of motion for a plate with a Winkler foundation, as shown in Figure 6.4, is given by

$$D\nabla^4 w - kw - \rho h\bar{\omega}^2 w = 0 \tag{6.80}$$

where k is the foundation stiffness.

By normalizing the lengths by a characteristic length a, Equation (6.80) becomes

$$\nabla^4 w - (\omega^2 - \eta)w = 0 \tag{6.81}$$

where ω is defined as in Equation (6.4), and $\eta = ka^4/D$ is the normalized stiffness.

TABLE 6.15
Frequencies for Free Circular Plate with an Internal Ring Hinge

ς	β = 0.25	β = 0.5	β = 0.75
0	5.2954 (2)	5.2739 (2)	5.3494 (2)
	7.2987 (0)	6.4490 (0)	7.8007 (0)
	12.417 (3)	12.347 (3)	12.436 (3)
	18.120 (1)	13.099 (1)	14.614 (1)
	21.830 (4)	21.762 (4)	21.835 (4)
1	5.3066 (2)	5.3004 (2)	5.3536 (2)
	7.8125 (0)	7.6828 (0)	8.7739 (0)
	12.420 (3)	12.368 (3)	12.437 (3)
	18.773 (1)	15.505 (1)	18.109 (1)
	21.830 (4)	21.775 (4)	21.835 (4)
∞		5.3584 (2)	
		9.0031 (0)	
		12.439 (3)	
		20.475 (1)	
		21.835 (4)	

Note: The value of n (number of diametric nodes) is in parentheses.

6.5.1 Plates with Full Foundation

Let w_0 denote the solution without a foundation with frequency ω_0. The governing equation of motion is given by

$$\nabla^4 w_0 - \omega_0^2 w_0 = 0 \tag{6.82}$$

By comparing Equations (6.81) and (6.82), we find that the frequency of a plate with a full foundation (see Figure 6.4a) is related to the frequency without foundation (Leissa 1969) as follows:

$$\omega = \sqrt{\omega_0^2 + \eta} \tag{6.83}$$

But for partial foundations, the relation in Equation (6.83) does not hold.

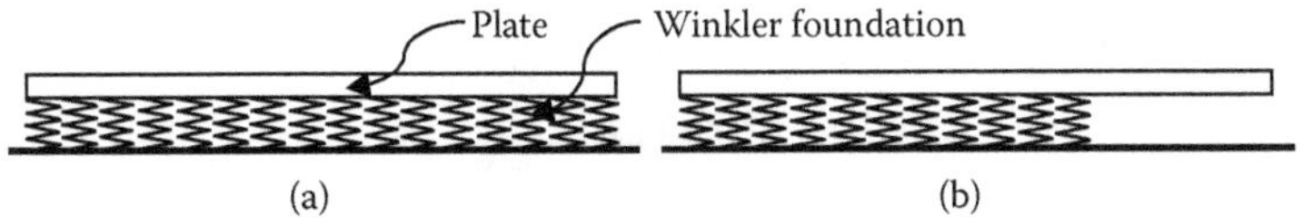

FIGURE 6.4 Plate with (a) full Winkler foundation and (b) partial Winkler foundation.

6.5.2 Rectangular Plates with Partial Foundation

Consider the simply supported, rectangular plate, where there is a foundation supporting region 1 for which $0 \le x \le \beta$ and no foundation for $\beta \le x \le 1$. Here, x is the normalized length, and $\beta = c/a$. Since all the edges are simply supported, we use Equation (6.31) to obtain

$$\frac{d^4W_1}{dx^4} - 2\alpha^2\frac{d^2W_1}{dx^2} + (\alpha^4 - \omega^2 + \eta)W_1 = 0 \tag{6.84}$$

$$\frac{d^4W_2}{dx^4} - 2\alpha^2\frac{d^2W_2}{dx^2} + (\alpha^4 - \omega^2)W_2 = 0 \tag{6.85}$$

At the joint, the plate is completely continuous

$$W_1 = W_2, \quad \frac{dW_1}{dx} = \frac{dW_2}{dx}, \quad \frac{d^2W_1}{dx^2} = \frac{d^2W_2}{dx^2}, \quad \frac{d^3W_1}{dx^3} = \frac{d^3W_2}{dx^3} \tag{6.86}$$

Of course, W_1 also satisfies the boundary conditions at the left edge and W_2 at the right edge.

The general solution to Equation (6.85) depends on whether $\omega > \alpha^2$, $\omega = \alpha^2$, or $\omega < \alpha^2$. The general solution to Equation (6.84) depends on the relative magnitudes of $\omega, \sqrt{\eta}$, and $\sqrt{\alpha^4 + \eta}$. The independent solutions are products of hyperbolic and circular functions. The characteristic determinant is still exact, but we shall not generate numerical solutions for this section.

6.5.3 Circular Plates with Partial Foundation

Figure 6.5 shows a circular plate of radius R with a concentric region of radius βR supported by a partial elastic Winkler foundation.

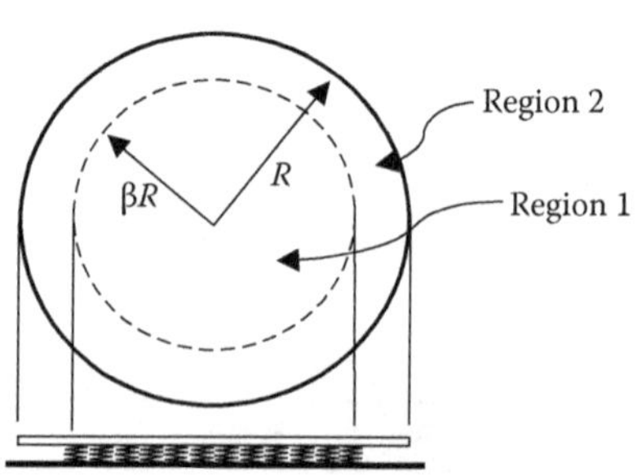

FIGURE 6.5 Circular plate with partial Winkler foundation.

Here, we extend the work of Wang (2005). Let subscript 1 denote the inner region and 2 denote the outer region. The equations of motion are

$$\nabla^4 w_1 - (\omega^2 - \eta) w_1 = 0 \tag{6.87}$$

$$\nabla^4 w_2 - \omega^2 w_2 = 0 \tag{6.88}$$

By using Equation (6.31), the general solution to Equation (6.88) is

$$W_2 = c_1 S_1 + c_2 S_2 + c_3 S_3 + c_4 S_4 \tag{6.89}$$

where S_i is defined as in Equation (6.62). Let

$$\hat{\omega} = \left|\omega^2 - \eta\right|^{1/2} \tag{6.90}$$

The bounded general solution to Equation (6.87) is

$$W_1 = c_5 S_5 + c_6 S_6 = \begin{cases} c_5 J_n(\sqrt{\hat{\omega}} r) + c_6 I_n(\sqrt{\hat{\omega}} r), & \omega > \sqrt{\eta} \\ c_5 r^n + c_6 r^{n+2}, & \omega = \sqrt{\eta} \\ c_5 \operatorname{Re}\left[J_n(\sqrt{i\hat{\omega}} r)\right] + c_6 \operatorname{Im}\left[J_n(\sqrt{i\hat{\omega}} r)\right], & \omega < \sqrt{\eta} \end{cases} \tag{6.91}$$

Continuity relations as in Equations (6.86) give the matrix

$$U = \begin{pmatrix} S_5 & S_6 & -S_1 & -S_2 & -S_3 & -S_4 \\ \dfrac{dS_5}{dr} & \dfrac{dS_6}{dr} & -\dfrac{dS_1}{dr} & -\dfrac{dS_2}{dr} & -\dfrac{dS_3}{dr} & -\dfrac{dS_4}{dr} \\ \dfrac{d^2S_5}{dr^2} & \dfrac{d^2S_6}{dr^2} & -\dfrac{d^2S_1}{dr^2} & -\dfrac{d^2S_2}{dr^2} & -\dfrac{d^2S_3}{dr^2} & -\dfrac{d^2S_4}{dr^2} \\ \dfrac{d^3S_5}{dr^3} & \dfrac{d^3S_6}{dr^3} & -\dfrac{d^3S_1}{dr^3} & -\dfrac{d^3S_2}{dr^3} & -\dfrac{d^3S_3}{dr^3} & -\dfrac{d^3S_4}{dr^3} \end{pmatrix}_{r=\beta} \tag{6.92}$$

The outer-edge boundary conditions then give a submatrix V, and the characteristic determinant is given by Equation (6.65).

TABLE 6.16
Frequencies for Simply Supported Circular Plate with a Concentric Elastic Foundation

η	β = 0.25	β = 0.5	β = 0.75
0		4.9351 (0) 13.898 (1) 25.613 (2) 29.720 (0) 39.957 (3)	
100	6.6081 (0) 14.025 (1) 25.625 (2) 30.298 (0) 39.958 (3)	9.2282 (0) 15.132 (1) 26.001 (2) 30.512 (0) 40.092 (3)	10.855 (0) 16.706 (1) 27.090 (2) 31.004 (0) 40.820 (3)
10,000	17.998 (0) 19.068 (1) 26.507 (2) 40.064 (3) 56.853 (4)	33.948 (0) 35.212 (1) 39.250 (2) 47.332 (3) 60.337 (4)	76.284 (0) 77.008 (1) 79.237 (2) 83.476 (3) 90.741 (4)
∞	25.980 (0) 27.474 (1) 32.755 (2) 42.935 (3) 57.836 (4)	59.845 (0) 61.012 (1) 64.654 (2) 71.126 (3) 86.821 (4)	243.42 (0) 244.34 (1) 247.13 (2) 251.83 (3) 258.52 (4)

Note: The value of n (number of diametric nodes) is in parentheses.

6.5.3.1 Case 1: Plate Is Simply Supported at the Edge

The submatrix V is given by Equation (6.64). Table 6.16 gives the results. When the stiffness is zero, it is a homogeneous simply supported plate, as considered in Chapter 5. When the stiffness is infinity, it is the clamped, simply supported annular plate. However, the approach to the inner-edge clamped case is very slow, as indicated by the $\eta = 10{,}000$ entries.

6.5.3.2 Case 2: Plate Is Clamped at the Edge

The V submatrix is given in Equation (6.66). The results are shown in Table 6.17.

6.5.3.3 Case 3: Plate Is Free at the Edge

The V submatrix is given in Equation (6.69). The results are shown in Table 6.18. Notice the switching of modes, especially the fundamental mode, as the stiffness is increased. For the $\eta = 0$ case, we have omitted the zero frequency associated with rigid body motion.

TABLE 6.17
Frequencies for Clamped Circular Plate with Concentric Elastic Foundation

η	$\beta = 0.25$	$\beta = 0.5$	$\beta = 0.75$
		10.216 (0)	
		21.261 (1)	
0		34.877 (2)	
		39.771 (3)	
		51.030 (4)	
	11.546 (0)	13.507 (0)	14.251 (0)
	21.410 (1)	22.472 (1)	23.407 (1)
100	34.894 (2)	35.345 (2)	36.164 (2)
	40.191 (0)	40.311 (0)	40.884 (0)
	51.032 (3)	51.217 (3)	51.864 (3)
	27.337 (0)	49.846 (0)	96.398 (0)
	28.144 (1)	50.912 (1)	109.36 (1)
10,000	36.223 (2)	54.359 (2)	122.09 (2)
	51.214 (3)	62.037 (3)	125.48 (0)
	69.511 (0)	75.189 (4)	140.25 (3)
	39.442 (0)	89.251 (0)	357.79 (0)
	40.836 (1)	90.230 (1)	358.51 (1)
∞	45.876 (2)	93.321 (2)	360.69 (2)
	56.016 (3)	98.928 (3)	364.38 (3)
	71.482 (4)	107.57 (4)	369.67 (4)

Note: The value of n (number of diametric nodes) is in parentheses.

6.6 STEPPED PLATES

A stepped plate describes a plate consisting of regions of different constant thicknesses. Although the analysis can be generalized to plates with regions of different materials, here we consider only a plate with the same material properties but two different thicknesses. A representative reference is Xiang and Wang (2002).

6.6.1 Stepped Rectangular Plates

Consider a rectangular plate with two horizontal edges simply supported, as shown in Figure 6.6. Assume that the only difference between the two regions is the thickness. Let the left side be of thickness h_1 and the right side h_2 and

$$\gamma = h_2/h_1 \leq 1 \tag{6.93}$$

TABLE 6.18
Frequencies for Free Circular Plate with Concentric Elastic Foundation

η	β = 0.25	β = 0.5	β = 0.75
		5.3584 (2)	
		9.0031 (0)	
0		12.439 (3)	
		20.475 (1)	
		21.835 (4)	
	0.6185 (1)	2.3709 (1)	5.3250 (1)
	2.1896 (0)	3.8348 (0)	6.1988 (0)
100	5.3616 (2)	5.5416 (2)	6.9671 (2)
	10.155 (0)	11.898 (0)	12.421 (0)
	12.439 (3)	12.469 (3)	12.999 (3)
	3.5427 (1)	7.8578 (0)	20.924 (0)
	4.3148 (0)	7.8724 (1)	21.295 (1)
10,000	5.6000 (2)	9.1565 (2)	22.681 (2)
	12.452 (3)	13.889 (3)	25.775 (3)
	21.836 (0)	22.314 (4)	31.346 (4)
	5.8384 (0)	13.024 (0)	53.624 (0)
	5.6101 (1)	13.290 (1)	54.230 (1)
∞	7.1027 (2)	14.704 (2)	56.131 (2)
	12.855 (3)	18.562 (3)	59.555 (3)
	21.921 (4)	25.596 (4)	64.802 (4)

Note: The value of n (number of diametric nodes) is in parentheses.

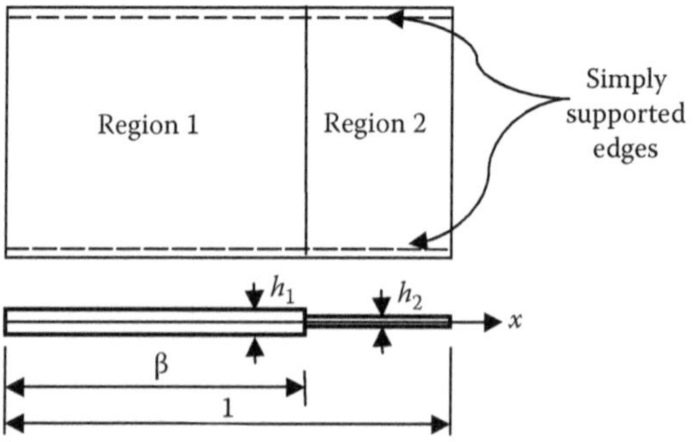

FIGURE 6.6 Rectangular plate with two piecewise constant-thickness distributions.

The governing equations for the vibration of the considered stepped rectangular plate are

$$D_1\nabla^4 w_1 - \rho_1 h_1 \bar{\omega}^2 w_1 = 0 \tag{6.94}$$

$$D_2\nabla^4 w_2 - \rho_2 h_2 \bar{\omega}^2 w_2 = 0 \tag{6.95}$$

Since $D \sim h^3$, normalize with the properties of the left region to obtain

$$\nabla^4 w_1 - \omega^2 w_1 = 0 \tag{6.96}$$

$$\nabla^4 w_2 - \frac{\omega^2}{\gamma^2} w_2 = 0 \tag{6.97}$$

In view of Equation (6.2), one can write the governing equations of motion as

$$\frac{d^4 W_1}{dx^4} - 2\alpha^2 \frac{d^2 W_1}{dx^2} - (\omega^2 - \alpha^4) W_1 = 0 \tag{6.98}$$

$$\frac{d^4 W_2}{dx^4} - 2\alpha^2 \frac{d^2 W_2}{dx^2} - \left(\frac{\omega^2}{\gamma^2} - \alpha^4\right) W_2 = 0 \tag{6.99}$$

At the junction $x = \beta$, we require continuity of displacement, slope, bending moment, and shear force, i.e.,

$$W_1 = W_2 \tag{6.100}$$

$$\frac{dW_1}{dx} = \frac{dW_2}{dx} \tag{6.101}$$

$$\frac{d^2 W_1}{dx^2} - \alpha^2 \nu W_1 = \gamma^3 \left(\frac{d^2 W_2}{dx^2} - \alpha^2 \nu W_2 \right) \tag{6.102}$$

$$\frac{d^3 W_1}{dx^3} - \alpha^2 (2-\nu) \frac{dW_1}{dx} = \gamma^3 \left(\frac{d^3 W_2}{dx^3} - \alpha^2 (2-\nu) \frac{dW_2}{dx} \right) \tag{6.103}$$

Let

$$W_1 = c_1 f(x) + c_2 g(x), \quad W_2 = c_3 h(x) + c_4 k(x) \tag{6.104}$$

which satisfy the boundary conditions at $x = 0$ and $x = 1$, respectively. Define the operators

$$L_3 = \frac{d^2}{dx^2} - \alpha^2 \nu, \quad L_4 = \frac{d^3}{dx^3} - \alpha^2 (2-\nu) \frac{d}{dx} \tag{6.105}$$

then the continuity equations give the following characteristic equation

$$\begin{vmatrix} f & g & -h & -k \\ \frac{df}{dx} & \frac{dg}{dx} & -\frac{dh}{dx} & -\frac{dk}{dx} \\ L_3 f & L_3 g & -\gamma^3 L_3 h & -\gamma^3 L_3 k \\ L_4 f & L_4 g & -\gamma^3 L_4 h & -\gamma^3 L_4 k \end{vmatrix}_{x=\beta} = 0 \tag{6.106}$$

We present some special cases in the following subsections.

6.6.1.1 Case 1: Plate Is Simply Supported on All Sides

Let

$$\lambda_1 = \sqrt{\omega + \alpha^2}, \quad \lambda_2 = \sqrt{|\omega - \alpha^2|}, \quad \lambda_3 = \sqrt{(\omega / \gamma) + \alpha^2}, \quad \lambda_4 = \sqrt{|(\omega / \gamma) - \alpha^2|} \tag{6.107}$$

The solutions that satisfy the left and right boundary conditions are

$$f = \sinh(\lambda_1 x), \quad g = \begin{cases} \sin(\lambda_2 x) & \omega > \alpha^2 \\ \sinh(\lambda_2 x) & \omega < \alpha^2 \end{cases} \tag{6.108}$$

$$h = \sinh[\lambda_3 (x-1)], \quad k = \begin{cases} \sin[\lambda_4 (x-1)] & \omega / \gamma > \alpha^2 \\ \sinh[\lambda_4 (x-1)] & \omega / \gamma < \alpha^2 \end{cases} \tag{6.109}$$

Then Equation (6.106) gives the vibration results presented in Table 6.19 for the 2:1 rectangular plate $a/b = 2$ and $\beta = c/a$.

6.6.1.2 Case 2: Plate Is Simply Supported on Opposite Sides and Clamped on Opposite Sides

We take

$$f = \lambda_2 \sinh(\lambda_1 x) - \lambda_1 \begin{cases} \sin(\lambda_2 x) \\ \sinh(\lambda_2 x) \end{cases}, \quad g = \cosh(\lambda_1 x) - \begin{cases} \cos(\lambda_2 x) \\ \cosh(\lambda_2 x) \end{cases} \tag{6.110}$$

$$h = \lambda_4 \sinh[\lambda_3 (x-1)] - \lambda_3 \begin{cases} \sin[\lambda_4 (x-1)] \\ \sinh[\lambda_4 (x-1)] \end{cases}, k = \cosh[\lambda_3 (x-1)] - \begin{cases} \cos[\lambda_4 (x-1)] \\ \cosh[\lambda_4 (x-1)] \end{cases}, \tag{6.111}$$

TABLE 6.19
Frequencies for the 2:1 Stepped Rectangular Plate with All Edges Simply Supported

	$\beta = 0.25$	$\beta = 0.5$	$\beta = 0.75$
	15.381 (1)	22.687 (1)	40.031 (1)
	29.971 (1)	43.309 (1)	61.348 (1)
$\gamma = 0.25$	44.486 (2)	51.370 (2)	81.048 (1)
	49.757 (1)	60.466 (1)	90.688 (2)
	59.118 (2)	84.598 (2)	131.73 (1)
	28.620 (1)	35.447 (1)	48.848 (1)
	48.583 (1)	55.536 (1)	67.797 (1)
$\gamma = 0.5$	74.442 (1)	92.404 (1)	110.53 (1)
	87.997 (2)	98.973 (2)	138.62 (2)
	113.56 (2)	134.32 (1)	161.95 (1)
	39.291 (1)	42.724 (1)	46.589 (1)
	63.665 (1)	68.730 (1)	73.424 (1)
$\gamma = 0.75$	103.07 (1)	110.58 (1)	119.67 (1)
	129.29 (2)	138.76 (2)	158.67 (2)
	158.45 (2)	171.35 (1)	184.45 (1)

Note: The value of n (number of diametric nodes) is in parentheses.

where the top and bottom forms are similarly differentiated in Equations (6.108) and (6.109). The results are shown in Table 6.20.

6.6.2 Stepped Circular Plates

Consider Figure 6.7, where the inner region of the circular plate has a higher density. The governing equations are Equations (6.98) and (6.99). Similar to Section 6.3.2, the solutions are given by

$$W_1 = c_1 J_n(\sqrt{\omega}r) + c_2 I_n(\sqrt{\omega}r) = c_1 S_1 + c_2 S_2 \tag{6.112}$$

$$W_2 = c_3 J_n(\sqrt{\omega/\gamma}r) + c_4 I_n(\sqrt{\omega/\gamma}r) + c_5 Y_n(\sqrt{\omega/\gamma}r) + c_6 K_n(\sqrt{\omega/\gamma}r)$$
$$= c_3 S_3 + c_4 S_4 + c_5 S_5 + c_6 S_6 \tag{6.113}$$

Continuity at $r = \beta$ yields

$$W_1 = W_2, \quad \frac{dW_1}{dr} = \frac{dW_2}{dr} \tag{6.114}$$

$$L_5 W_1 = \gamma^3 L_5 W_2, \quad L_6 W_1 = \gamma^3 L_6 W_2 \tag{6.115}$$

TABLE 6.20
Frequencies for the 2:1 Stepped Rectangular Plate with Two Opposite Edges Simply Supported and the Other Two Opposite Edges Clamped

	$\beta = 0.25$	$\beta = 0.5$	$\beta = 0.75$
	34.339 (1)	27.357 (1)	42.606 (1)
	45.332 (2)	47.914 (1)	73.559 (1)
$\gamma = 0.25$	58.271 (1)	54.320 (2)	100.80 (1)
	61.902 (2)	71.048 (1)	108.86 (2)
	81.294 (1)	92.546 (2)	150.92 (1)
	31.948 (1)	39.906 (1)	47.194 (1)
	57.989 (1)	65.513 (1)	79.591 (1)
$\gamma = 0.5$	89.479 (2)	103.40 (2)	133.09 (1)
	89.956 (1)	110.40 (1)	148.41 (2)
	118.34 (2)	151.93 (2)	182.82 (2)
	43.724 (1)	47.468 (1)	51.670 (1)
	76.186 (1)	82.208 (1)	87.835 (1)
$\gamma = 0.75$	123.80 (1)	133.07 (1)	143.94 (1)
	131.13 (2)	142.31 (2)	163.24 (2)
	164.90 (2)	182.04 (2)	194.15 (2)

Note: The value of n (number of vertical half waves) is in parentheses.

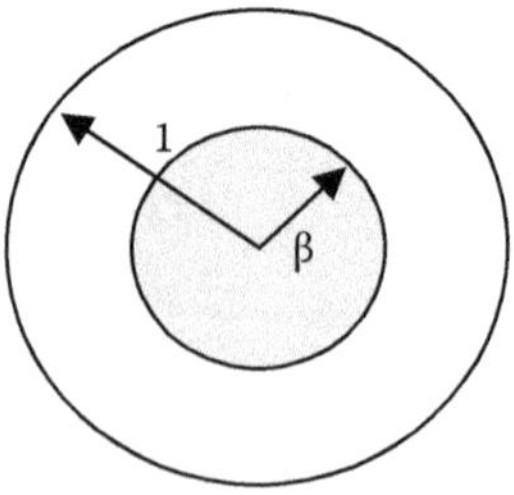

FIGURE 6.7 Stepped circular plate.

where

$$L_5 = \frac{d^2}{dr^2} + \frac{\nu}{r}\left(\frac{d}{dr} - \frac{n^2}{r}\right), \quad L_6 = \frac{d^3}{dr^3} + \frac{1}{r}\frac{d^2}{dr^2} - \frac{1+(2-\nu)n^2}{r^2}\frac{d}{dr} + \frac{(3-\nu)n^2}{r^3} \tag{6.116}$$

From Equations (6.114) and (6.115), one can construct the matrix

$$U = \begin{pmatrix} S_1 & S_2 & -S_3 & -S_4 & -S_5 & -S_6 \\ \frac{dS_1}{dr} & \frac{dS_2}{dr} & -\frac{dS_3}{dr} & -\frac{dS_4}{dr} & -\frac{dS_5}{dr} & -\frac{dS_6}{dr} \\ L_5S_1 & L_5S_2 & -\gamma^3 L_5S_3 & -\gamma^3 L_5S_4 & -\gamma^3 L_5S_5 & -\gamma^3 L_5S_6 \\ L_6S_1 & L_6S_2 & -\gamma^3 S_3 & -\gamma^3 S_4 & -\gamma^3 S_5 & -\gamma^3 S_6 \end{pmatrix}_{r=\beta} \tag{6.117}$$

We shall illustrate with the following three cases.

6.6.2.1 Case 1: Circular Plate with Simply Supported Edge

The submatrix is

$$V = \begin{pmatrix} 0 & 0 & S_3 & S_4 & S_5 & S_6 \\ 0 & 0 & \frac{d^2S_3}{dr^2} + \nu\frac{dS_3}{dr} & \frac{d^2S_4}{dr^2} + \nu\frac{dS_4}{dr} & \frac{d^2S_5}{dr^2} + \nu\frac{dS_5}{dr} & \frac{d^2S_6}{dr^2} + \nu\frac{dS_6}{dr} \end{pmatrix}_{r=1} \tag{6.118}$$

Then Equation (6.65) gives the frequencies as shown in Table 6.21.

6.6.2.2 Case 2: Circular Plate with Clamped Edge

The submatrix is

$$V = \begin{pmatrix} 0 & 0 & S_3 & S_4 & S_5 & S_6 \\ 0 & 0 & \frac{dS_3}{dr} & \frac{dS_4}{dr} & \frac{dS_5}{dr} & \frac{dS_6}{dr} \end{pmatrix}_{r=1} \tag{6.119}$$

Then Equation (6.65) gives the frequencies in Table 6.22.

6.6.2.3 Case 3: Circular Plate with Free Edge

The submatrix is

$$V = \begin{pmatrix} 0 & 0 & L_1S_3 & L_1S_4 & L_1S_5 & L_1S_6 \\ 0 & 0 & L_2S_3 & L_2S_4 & L_2S_5 & L_2S_6 \end{pmatrix}_{r=1} \tag{6.120}$$

TABLE 6.21
Frequencies for Simply Supported Stepped Circular Plate

	$\beta = 0.25$	$\beta = 0.5$	$\beta = 0.75$
	1.1437 (0)	1.3461 (0)	2.5383 (0)
	3.3927 (1)	3.3234 (1)	5.0863 (1)
$\gamma = 0.25$	7.9779 (0)	12.361 (2)	11.060 (2)
	7.9786 (2)	14.464 (0)	16.177 (0)
	11.739 (1)	16.529 (3)	21.031 (3)
	2.5287 (0)	3.0155 (0)	4.1634 (0)
	6.9563 (1)	7.4651 (1)	10.373 (1)
$\gamma = 0.5$	14.407 (2)	15.913 (2)	17.741 (2)
	16.357 (0)	20.661 (0)	22.103 (0)
	24.804 (1)	25.311 (3)	27.583 (3)
	3.8058 (0)	4.1622 (0)	4.6394 (0)
	10.470 (1)	11.080 (1)	12.768 (1)
$\gamma = 0.75$	19.942 (2)	20.622 (2)	22.713 (2)
	23.442 (0)	24.873 (0)	27.262 (0)
	37.199 (1)	41.689 (1)	43.866 (1)

Note: The value of n (number of diametric nodes) is in parentheses.

TABLE 6.22
Frequencies for Clamped Stepped Circular Plate

	$\beta = 0.25$	$\beta = 0.5$	$\beta = 0.75$
	2.2507 (0)	2.6273 (0)	4.4420 (0)
	5.1137 (1)	5.1629 (1)	8.5725 (1)
$\gamma = 0.25$	11.124 (2)	15.779 (2)	14.330 (2)
	11.209 (0)	20.156 (0)	19.106 (0)
	13.898 (3)	22.422 (3)	24.114 (3)
	5.1088 (0)	5.7929 (0)	6.5751 (0)
	10.642 (1)	11.723 (1)	15.464 (1)
$\gamma = 0.5$	19.780 (2)	20.989 (2)	24.933 (2)
	22.371 (0)	26.826 (0)	29.895 (0)
	26.863 (3)	32.405 (3)	35.880 (3)
	7.7816 (0)	8.1032 (0)	8.2558 (0)
	16.052 (1)	17.161 (1)	18.544 (1)
$\gamma = 0.75$	27.200 (2)	28.085 (2)	30.844 (2)
	31.535 (0)	33.174 (0)	36.179 (0)
	38.971 (3)	41.355 (3)	44.999 (3)

Note: The value of n (number of diametric nodes) is in parentheses.

TABLE 6.23
Frequencies for Free Stepped Circular Plate

	$\beta = 0.25$	$\beta = 0.5$	$\beta = 0.75$
	1.7350 (2)	3.3127 (2)	6.5795 (2)
	2.2814 (0)	3.7298 (0)	9.8710 (0)
$\gamma = 0.25$	3.2050 (3)	4.4835 (3)	11.362 (3)
	4.9775 (1)	5.6396 (1)	14.174 (1)
	5.4785 (4)	6.3162 (4)	14.237 (4)
	3.1327 (2)	4.4781 (2)	5.7593 (2)
	4.8367 (0)	6.7306 (0)	9.6940 (0)
$\gamma = 0.5$	6.3335 (3)	7.7202 (3)	11.669 (3)
	10.293 (1)	11.920 (4)	17.933 (4)
	10.941 (4)	12.030 (1)	19.807 (1)
	4.2484 (2)	4.8290 (2)	5.3685 (2)
	7.0919 (0)	8.1514 (0)	9.1773 (0)
$\gamma = 0.75$	9.3891 (3)	10.057 (3)	11.636 (3)
	15.488 (1)	16.890 (4)	19.375 (4)
	16.389 (4)	17.048 (1)	20.318 (1)

Note: The value of n (number of diametric nodes) is in parentheses.

where L_1, L_2 are defined in Equations (6.67) and (6.68). The results are shown in Table 6.23.

6.7 VARIABLE-THICKNESS PLATES

We consider plates whose thicknesses vary continuously. The solutions have implications for plates made of functionally graded materials. So far, there are no exact solutions for variable-thickness rectangular plates or circular plates. Annular plates with power law thickness or rigidity have been presented by Lenox and Conway (1980) and Wang, Wang, and Chen (2012). Figure 6.8 shows the two kinds of variable-thickness plates considered.

The dynamic equation in polar coordinates for a plate with radial variable property is given by (Leissa 1969)

$$D\nabla^4 w + \frac{dD}{d\bar{r}}\left(2\frac{\partial^3 w}{\partial \bar{r}^3} + \frac{2+\nu}{\bar{r}}\frac{\partial^2 w}{\partial \bar{r}^2} - \frac{1}{\bar{r}^2}\frac{\partial w}{\partial \bar{r}} + \frac{2}{\bar{r}^2}\frac{\partial^3 w}{\partial \bar{r}\,\partial\theta^2} - \frac{3}{\bar{r}^3}\frac{\partial^2 w}{\partial\theta^2}\right)$$

$$+\frac{d^2 D}{d\bar{r}^2}\left(\frac{\partial^2 w}{\partial \bar{r}^2} + \frac{\nu}{\bar{r}}\frac{\partial w}{\partial \bar{r}} + \frac{\nu}{\bar{r}^2}\frac{\partial^2 w}{\partial\theta^2}\right) = -\rho h\frac{\partial^2 w}{\partial \bar{t}^2} \tag{6.121}$$

We normalize all lengths by the outer radius R and drop the over-bars. Let

$$w = W(r)\cos(n\theta)\cos(\bar{\omega}\bar{t}) \tag{6.122}$$

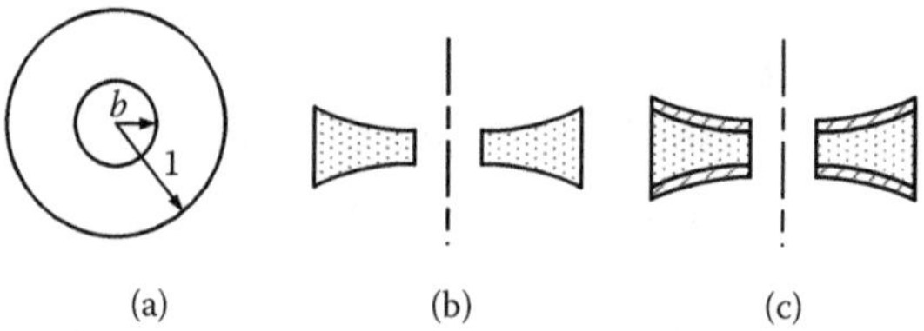

FIGURE 6.8 (a) Nonuniform thickness annular plate, (b) constant density with parabolic thickness, and (c) parabolic sandwich plate.

where $\bar{\omega}$ is the frequency. Let D_0 be the maximum rigidity; let ρh_0 be the maximum mass per area; and define normalized parameters ϕ, ψ, ω

$$D = D_0\phi(r), \quad \rho h = \rho h_0 \psi(r), \quad \omega = \bar{\omega} R^2 \sqrt{\rho h_0 / D_0} \tag{6.123}$$

Then Equation (6.121) becomes

$$\phi\left(\frac{d^2}{dr^2} + \frac{1}{r}\frac{d}{dr} - \frac{n^2}{r^2}\right)^2 W + \frac{d\phi}{dr}\left(2\frac{d^3}{dr^3} + \frac{2+\nu}{r}\frac{d^2}{dr^2} - \frac{1+2n^2}{r^2}\frac{d}{dr} + \frac{3n^2}{r^2}\right)W$$
$$+ \frac{d^2\phi}{dr^2}\left(\frac{d^2}{dr^2} + \frac{\nu}{r}\frac{d}{dr} - \frac{\nu n^2}{r^2}\right)W - \omega^2 \psi W = 0 \tag{6.124}$$

Note that if

$$\phi = r^{\alpha}, \quad \psi = r^{\alpha - 4} \tag{6.125}$$

then Equation (6.124) yields solutions in the form

$$W = r^{\lambda} \tag{6.126}$$

where λ satisfies the algebraic equation

$$(\lambda^2 - n^2)[(\lambda-2)^2 - n^2] + \alpha[\lambda(\lambda-1)(2\lambda - 2 + \nu) - (1+2n^2)\lambda + 3n^2]$$
$$+ \alpha(\alpha-1)[\lambda(\lambda-1) + \nu(\lambda - n^2)] - \omega^2 = 0 \tag{6.127}$$

Since a singularity exists at $r = 0$, only annular plates are considered. Let the inner radius be βR. Although α can be any exponent, we shall only present some physically meaningful cases.

6.7.1 Case 1: Constant Density with Parabolic Thickness

For this case, $\alpha = 6$. Let

$$b_1 = 14 + 2n^2 - 6\nu, \quad b_2 = 4[9(1-\nu)^2 - 8n^2(1-3\nu) + \omega^2], \; d = b_1 - \sqrt{b_2} \qquad (6.128)$$

The roots of Equation (6.127) are

$$\lambda_{1,2} = -2 \pm \sqrt{(b_1 + \sqrt{b_2})/2}, \; \lambda_{3,4} = -2 \pm \sqrt{(b_1 - \sqrt{b_2})/2} \qquad (6.129)$$

The general solution is

$$W = c_1 r^{\lambda_1} + c_2 r^{\lambda_2} + c_3 \begin{cases} r^{\lambda_3} \\ \cos[\sqrt{|d|/2}\ln r]/r^2 \end{cases} + c_4 \begin{cases} r^{\lambda_4} \\ \sin[\sqrt{|d|/2}\ln r]/r^2 \end{cases} \qquad (6.130)$$

where the top form is used if $d > 0$ and the bottom form if $d < 0$. Using the inner- and outer-edge conditions, Equation (6.130) gives the characteristic equation for frequency. Table 6.24 shows the results for both edges clamped (C-C), both edges simply supported (S-S), or both edges free (F-F). Notice the mode changes for the F-F case.

TABLE 6.24
Frequencies for the Annular Plate with Parabolic Thickness

	$\beta = 0.25$	$\beta = 0.5$	$\beta = 0.75$
C-C	15.666 (0)	50.126 (0)	273.73 (0)
	16.489 (1)	50.768 (1)	274.30 (1)
	19.054 (2)	52.760 (2)	276.01 (2)
	23.544 (3)	56.269 (3)	278.92 (3)
	30.082 (4)	61.513 (4)	283.08 (4)
S-S	5.7272 (0)	19.244 (0)	117.73 (0)
	7.2019 (1)	20.732 (1)	118.82 (1)
	12.205 (2)	24.889 (2)	122.06 (2)
	18.555 (3)	31.195 (3)	127.45 (3)
	23.571 (0)	39.346 (4)	134.93 (4)
F-F	2.6553 (2)	2.6550 (2)	2.6545 (2)
	4.5145 (0)	5.9863 (0)	7.3801 (3)
	7.1587 (1)	7.3980 (3)	12.016 (0)
	7.4035 (3)	9.5876 (1)	14.076 (4)
	12.197 (2)	14.145 (4)	19.181 (1)

Note: The value of n (number of diametric nodes) is in parentheses.

TABLE 6.25
Frequencies for a Sandwich Annular Plate

	β = 0.25	β = 0.5	β = 0.75
	13.034 (0)	47.906 (0)	271.66 (0)
	13.782 (1)	48.513 (1)	272.22 (1)
C-C	16.179 (2)	50.403 (2)	273.91 (2)
	20.511 (3)	53.768 (3)	276.78 (3)
	26.940 (4)	58.853 (4)	280.89 (4)
	5.2168 (0)	20.465 (0)	119.13 (0)
	6.7998 (1)	21.649 (1)	120.17 (1)
S-S	10.698 (2)	25.099 (2)	123.26 (2)
	16.322 (3)	30.627 (3)	128.41 (3)
	23.690 (4)	38.104 (4)	135.60 (4)
	2.6558 (2)	2.6552 (2)	2.6545 (2)
	3.1405 (0)	5.1889 (0)	7.3801 (3)
F-F	5.1541 (1)	7.4043 (3)	11.668 (0)
	6.7380 (2)	8.3575 (1)	14.079 (4)
	7.4160 (3)	14.168 (4)	18.630 (1)

Note: The value of n (number of diametric nodes) is in parentheses.

6.7.2 Case 2: Parabolic Sandwich Plate

Consider a sandwich plate with high-density surface laminates separated by a web or foam of negligible density. For this case, $\alpha = 4$. From Equation (6.127), the roots are

$$\lambda_{1,2} = -1 \pm \sqrt{(b_1 + \sqrt{b_2})/2}, \; \lambda_{3,4} = -1 \pm \sqrt{(b_1 - \sqrt{b_2})/2} \tag{6.131}$$

where

$$b_1 = 6 + 2n^2 - 4\nu, \quad b_2 = 16[(1-\nu)^2 - n^2(1-2\nu)] + 4\omega^2 \tag{6.132}$$

Then the solution is given by Equation (6.130) with r^2 replaced by r. Table 6.25 shows the results.

6.8 DISCUSSION

We have described the methods for obtaining exact characteristic equations for some basic complicating factors and presented some specific samples. Of course, there are many other boundary conditions and combinations, such as "vibration of a stepped plate with in-plane force and rotational springs on the boundary." No doubt, the reader can extend the method presented in this chapter and generate exact frequencies to suit particular needs.

REFERENCES

Leissa, A. W. 1969. *Vibration of plates*. NASA SP-160. Washington, DC: U.S. Government Printing Office. Repr. Sewickley, PA: Acoustical Society of America, 1993.

Lenox, T. A., and H. D. Conway. 1980. An exact closed form solution for the flexural vibration of a thin annular plate having a parabolic thickness variation. *J. Sound Vibr*. 68:231–39.

Li, Q. S. 2003. An exact approach for free vibration analysis of rectangular plates with line-concentrated mass and elastic line support. *Int. J. Mech. Sci*. 45:669–85.

Wang, C. Y. 2002. Fundamental frequency of a circular plate weakened along a concentric circle. *Z. Angew. Math. Mech*. 82:70–72.

———. 2005. Fundamental frequency of a circular plate supported by a partial elastic foundation. *J. Sound Vibr*. 285:1203–9.

Wang, C. Y., and C. M. Wang. 2003. Fundamental frequencies of circular plates with internal elastic ring support. *J. Sound Vibr*. 263:1071–78.

Wang, C. Y., C. M. Wang, and W. Q. Chen. 2012. Exact closed form solutions for free vibration of non-uniform annular plates. *IES J. Part A: Civil and Structural Engineering* 5 (1): 50–55.

Xiang, Y., and C. M. Wang. 2002. Exact buckling and vibration solutions for stepped rectangular plates. *J. Sound Vibr*. 250: 503–17.

7 Vibration of Nonisotropic Plates

7.1 INTRODUCTION

In the previous Chapters 5 and 6, we have assumed the plate material to be isotropic, which means that the material properties at a point are the same in all directions. However, certain materials have properties that are not independent of the direction. These materials are said to be anisotropic. Examples of anisotropic materials are two-way reinforced concrete slabs, plywood, and fiber-reinforced plastics. Structural anisotropy is also introduced by means of ribs or corrugations. Consequently, to obtain a reasonable agreement between analysis and the actual behavior, it is necessary to consider the anisotropy of such plates in the calculations.

In this chapter, we consider the vibration problems of nonisotropic plates such as orthotropic, sandwich, laminated, and functionally graded plates, and we provide the exact vibration solutions for some plate shapes and boundary conditions.

7.2 ORTHOTROPIC PLATES

If a plate has different elastic properties in two orthogonal directions, it is called an *orthotropic plate* (i.e., orthogonally isotropic). In practice, two forms of orthotropy may be identified: material orthotropy and shape orthotropy. A plywood sheet is orthotropic because of different elastic properties in two perpendicular directions, whereas a voided concrete slab is orthotropic because cross sections in the two orthogonal directions are essentially different.

7.2.1 Governing Vibration Equation

The stress-strain relationships for an orthotropic material for plates are given by (Szilard 1974)

$$\sigma_{xx} = \frac{E_x}{(1-\nu_x\nu_y)}[\varepsilon_x + \nu_y\varepsilon_y] \tag{7.1}$$

$$\sigma_{yy} = \frac{E_y}{(1-\nu_x\nu_y)}[\varepsilon_y + \nu_x\varepsilon_x] \tag{7.2}$$

$$\sigma_{xy} = G\gamma_{xy} = 2G\varepsilon_{xy} \tag{7.3}$$

where σ_{xx}, σ_{yy} are the normal stresses in the x- and y-directions, respectively; $\varepsilon_{xx}, \varepsilon_{yy}$ are the normal strains in the x- and y-directions; σ_{xy} is the shearing stress; γ_{xy} is the corresponding shearing strain; E_x, E_y are the moduli of elasticity in the x- and y-directions; ν_x, ν_y are the Poisson ratios in the x- and y-directions; and $G \approx \sqrt{E_x E_y} \Big/ \left[2(1+\sqrt{\nu_x \nu_y})\right]$.

Based on Betti's reciprocal theorem, we have

$$E_x \nu_y = E_y \nu_x \quad \text{or} \quad D_x \nu_y = D_y \nu_x \tag{7.4}$$

and then Equations (7.1) and (7.2) can be written as

$$\sigma_{xx} = \bar{E}_x \varepsilon_x + \breve{E} \varepsilon_y \tag{7.5}$$

$$\sigma_{yy} = \bar{E}_y \varepsilon_y + \breve{E} \varepsilon_x \tag{7.6}$$

where $\bar{E}$ and $\breve{E}$ are defined in Equation (7.17).

Thus, for an orthotropic plate, there are four material constants (E_x, E_y, ν_x, ν_y) as opposed to two material constants (E, ν) for an isotropic plate.

As in the isotropic classical thin-plate theory, which is based on Kirchhoff's assumptions, the strain-displacement relations for orthotropic plates are given by

$$\varepsilon_x = -z \frac{\partial^2 w}{\partial x^2} \tag{7.7a}$$

$$\varepsilon_y = -z \frac{\partial^2 w}{\partial y^2} \tag{7.7b}$$

$$\varepsilon_{xy} = -z \frac{\partial^2 w}{\partial x \partial y} \tag{7.7c}$$

where w is the transverse displacement, and z is the coordinate measured from the mid-plane of the plate.

The bending moments and twisting moments are given by

$$M_{xx} = \int_{-h/2}^{h/2} z \sigma_{xx} dz \tag{7.8a}$$

$$M_{yy} = \int_{-h/2}^{h/2} z \sigma_{yy} dz \tag{7.8b}$$

$$M_{yx} = \int_{-h/2}^{h/2} z \sigma_{xy} dz = -M_{yx} \tag{7.8c}$$

where h is the plate thickness.

By substituting Equations (7.7a), (7.7b), (7.7c), (7.5), (7.6), and (7.3) into Equations (7.8a) to (7.8c), we have the following moment curvature relationships

$$M_{xx} = -\left(D_x \frac{\partial^2 w}{\partial x^2} + D_1 \frac{\partial^2 w}{\partial y^2} \right) \tag{7.9a}$$

$$M_{yy} = -\left(D_y \frac{\partial^2 w}{\partial y^2} + D_1 \frac{\partial^2 w}{\partial x^2} \right) \tag{7.9b}$$

$$M_{yx} = -2D_{xy} \frac{\partial^2 w}{\partial x \partial y} = -M_{xy} \tag{7.9c}$$

where

$$D_x = \frac{\bar{E}_x h^3}{12},\ D_y = \frac{\bar{E}_y h^3}{12},\ D_1 = \nu_y D_x = \nu_x D_y,\ 2D_{xy} = \left(1 - \sqrt{\nu_x \nu_y}\right)\sqrt{D_x D_y} \tag{7.10}$$

D_x,D_y are the flexural rigidities of the orthotropic plate, and $2D_{xy}$ is the torsional rigidity. For an orthotropic plate of uniform thickness, the torsional rigidity can be written as $D_{xy} = Gh^3/12$.

The equations of free harmonic motion of classical thin plates are given by

$$Q_x = \frac{\partial M_{xx}}{\partial x} + \frac{\partial M_{yx}}{\partial y} \tag{7.11}$$

$$Q_y = \frac{\partial M_{yy}}{\partial y} + \frac{\partial M_{yx}}{\partial x} \tag{7.12}$$

$$\frac{\partial Q_x}{\partial x} + \frac{\partial Q_y}{\partial y} = \rho h \omega^2 w \tag{7.13}$$

By substituting Equations (7.9a), (7.9b), and (7.9c) into Equations (7.11) and (7.12), we obtain

$$Q_x = -\frac{\partial}{\partial x}\left(D_x \frac{\partial^2 w}{\partial x^2} + H \frac{\partial^2 w}{\partial y^2} \right) \tag{7.14a}$$

$$Q_y = -\frac{\partial}{\partial y}\left(D_y \frac{\partial^2 w}{\partial y^2} + H \frac{\partial^2 w}{\partial x^2} \right) \tag{7.14b}$$

where

$$H = D_1 + 2D_{xy} = \frac{1}{2}(\nu_y D_x + \nu_x D_y) + 2D_{xy} \tag{7.15}$$

Note that H is the effective torsional rigidity of the orthotropic plate.

The substitution of Equations (7.14a) and (7.14b) into Equation (7.13) furnishes the governing plate equation for the vibrating orthotropic plate

$$D_x \frac{\partial^4 w}{\partial x^4} + 2H \frac{\partial^4 w}{\partial x^2 \partial y^2} + D_y \frac{\partial^4 w}{\partial y^4} - \rho h \omega^2 w = 0 \tag{7.16}$$

Note that for the isotropic plate case,

$$\begin{aligned} &\bar{E}_x = \bar{E}_y = \frac{E}{1-\nu^2}, \breve{E} = \frac{\nu E}{1-\nu^2}, \ G = \frac{E}{2(1+\nu)}, \ D_x = D_y = \frac{Eh^3}{12(1-\nu^2)} = D, \\ &H = D_1 + 2D_{xy} = D \end{aligned} \tag{7.17}$$

and Equation (7.16) reduces to Equation (5.3) in Chapter 5.

7.2.2 Principal Rigidities for Special Orthotropic Plates

7.2.2.1 Corrugated Plates

Corrugated plates and plates with stiffeners are frequently treated mathematically as orthotropic plates. Certainly, these stiffened plates have varying rigidities in the directions perpendicular and parallel to the stiffeners. Often these plates are modeled by equivalent orthotropic plates with elastic properties equal to the average properties of the various plate components.

Consider the corrugated plate having a sinusoidal form of corrugation as shown in Figure 7.1. The estimated principal rigidities are given by (Timoshenko and Woinowsky-Krieger 1959; Szilard 1974)

$$D_x = \frac{s}{\mu} \frac{Eh^3}{12(1-\nu^2)}, \ D_y = EI, \ H = 2D_{xy} = \frac{\mu}{s} \frac{Eh^3}{12(1+\nu)}, \ D_1 = 0 \tag{7.18}$$

where E is the Young's modulus of the plate material, ν is the Poisson ratio of the plate material, s is the half wavelength of the corrugated wave, $\bar{H}$ is the amplitude of the corrugated wave, h is the thickness of the corrugated plate, and

$$\mu = s\left(1 + \frac{\pi^2 \bar{H}^2}{4s^2}\right), \ I = \frac{\bar{H}^2 h}{2}\left(1 - \frac{0.81}{1 + 2.5\left(\frac{\bar{H}}{2s}\right)^2}\right) \tag{7.19}$$

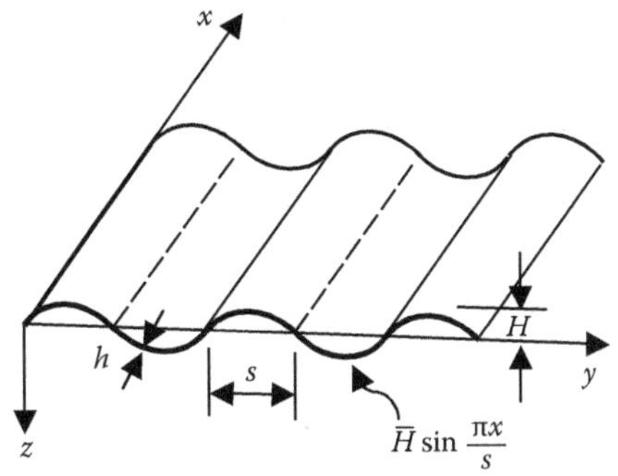

FIGURE 7.1 Corrugated plate.

Consequently, the net bending stiffness of the corrugated plate and the orthotropic model are made equal. It is evident that this procedure could yield poor results in certain local areas and yet give a good description of the overall plate stiffness. The precise meaning of the expression "equivalent orthotropic plate" becomes vague when a stiffened plate is modeled by an orthotropic plate. The orthotropic model can be structured in such a manner that a certain quantity, such as stress or deflection, in the orthotropic plate matches the corresponding quantity in the stiffened plate. However, it is not guaranteed that stresses will match if deflections are made to match and vice versa. Of course, the ideal situation is to closely match all the plate variables. It has been observed that a satisfactory matching of the overall plate behavior can be obtained if $s/a \ll 1$ and $s/b \ll 1$, where s is the distance between stiffeners and a and b are the overall plate dimensions.

7.2.2.2 Plate Reinforced by Equidistant Ribs/Stiffeners

Consider a plate reinforced by equidistant ribs/stiffeners in one direction, as shown in Figure 7.2. For such a plate, the principal rigidities may be approximated by (Timoshenko and Woinowsky-Krieger 1959)

$$D_x = H = \frac{Eh^3}{12(1-\nu^2)}, \quad D_y = \frac{Eh^3}{12(1-\nu^2)} + \frac{E_s I_s}{s} \tag{7.20}$$

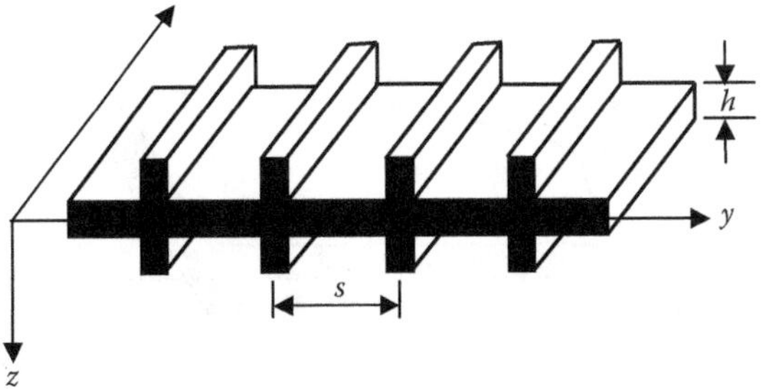

FIGURE 7.2 Plate with ribs/stiffeners.

where E and E_s are the Young's moduli of plate and stiffeners, respectively, ν is the Poisson ratio of the plate, s is the spacing between stiffeners, and I_s is the moment of inertia of a stiffener taken with respect to the middle plane of the plate.

If the plate is reinforced by two perpendicular sets of equidistant stiffeners, again assumed to be symmetric with respect to the middle surface of the plate, the orthotropic elastic constants are approximated by (Timoshenko and Woinowsky-Krieger 1959)

$$D_x = \frac{Eh^3}{12(1-\nu^2)} + \frac{E_s I_{s1}}{s_1}, \quad D_y = \frac{Eh^3}{12(1-\nu^2)} + \frac{E_s I_{s2}}{s_2}, \quad H = \frac{Eh^3}{12(1-\nu^2)} \tag{7.21}$$

in which I_{s1} and s_1 are, respectively, the second moment of area and the spacing of the stiffeners that run parallel to the y-axis, and I_{s2} and s_2 are, respectively, the second moment of area and the spacing of the stiffeners that run parallel to the x-axis.

7.2.2.3 Steel-Reinforced Concrete Slabs

For a slab with two-way reinforcement in the x- and y-directions, as shown in Figure 7.3, the principal rigidities may be approximated by (Timoshenko and Woinowsky-Krieger 1959; Reddy 2007)

$$D_x = \frac{E_c}{1-\nu_c^2}\left[I_{cx} + (n-1)I_{sx}\right], \quad D_y = \frac{E_c}{1-\nu_c^2}\left[I_{cy} + (n-1)I_{sy}\right], \quad D_1 = \nu_c\sqrt{D_x D_y},$$
$$D_{xy} = \frac{1-\nu_c}{2}\sqrt{D_x D_y}, \quad H = \sqrt{D_x D_y} \tag{7.22}$$

in which E_s is the Young's modulus of steel; E_c is the Young's modulus of concrete; ν_c is the Poisson ratio of concrete; $n = E_s/E_c$; and I_{cx} and I_{cy} are the moment of areas of the slab material and I_{sx} and I_{sy} are the moment of areas of the reinforcement bars, both taken about the neutral axis in the section x and section y, respectively. Note that $\nu_c = \breve{E}/\sqrt{\bar{E}_x\bar{E}_y}$ is assumed.

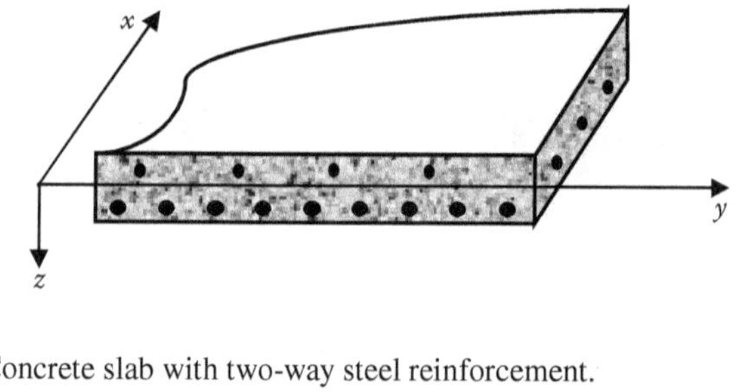

FIGURE 7.3 Concrete slab with two-way steel reinforcement.

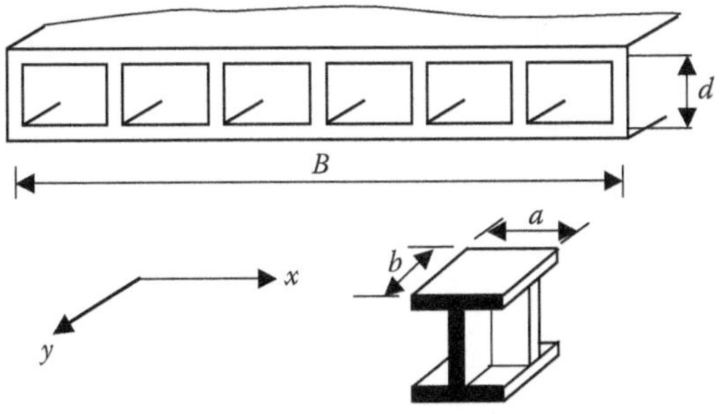

FIGURE 7.4 Multicell slab with transverse diaphragm and portion of a multicell slab.

7.2.2.4 Multicell Slab with Transverse Diaphragm

When closely spaced transverse diaphragms are incorporated into a multicell slab (as shown in Figure 7.4), local bending of the cell walls is largely prevented. The principal rigidities can be estimated from (Cope and Clark 1984)

$$D_x = \frac{EI_x}{b},\ D_y = \frac{EI_y}{a},\quad D_1 = \nu\sqrt{D_x D_y},\quad D_{xy} = \frac{EB^2 d^2}{2(1+\nu)B\sum \frac{ds}{t}} \tag{7.23}$$

where B is the total width of the slab between the outer web centerlines, d is the depth of the section between flange centerlines, ds/t is the length-to-thickness ratio of a rectangle making up the perimeter if the flange and outer webs are all of the same thickness t, and $\Sigma\, ds/t = 2(B+d)/t$. Note that the flange width is to be increased by $1/(1-\nu^2)$ when determining I_x and I_y about the centroid of the cross section.

7.2.2.5 Voided Slabs

For a slab with one-way circular voids as shown in Figure 7.5, the principal rigidities can be estimated from (Cope and Clark 1984)

$$\begin{aligned} D_x &= \frac{Eh^3}{12(1-\nu^2)}\left[1-\left(\frac{d}{h}\right)^4\right],\quad D_y = \frac{EI_y}{s(1-\nu^2)},\quad D_1 = \nu D_x, \\ D_{xy} &= \frac{Eh^3}{24(1+\nu)}\left[1-0.85\left(\frac{d}{h}\right)^4\right] \end{aligned} \tag{7.24}$$

The formulae in Equation 7.24 are calibrated for slabs with $0.47 < d/h < 0.81$.

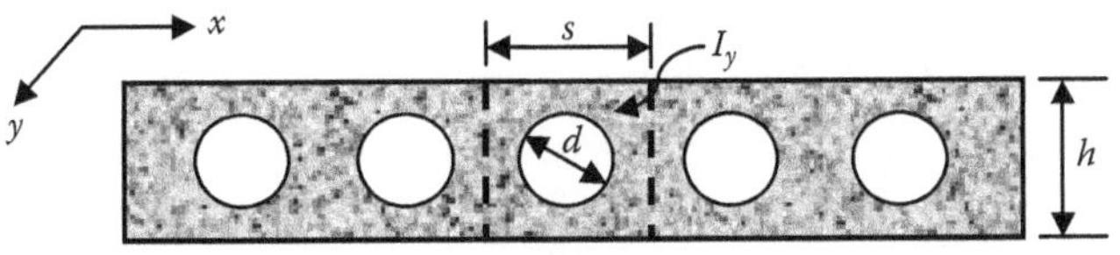

FIGURE 7.5 One-way circular voided slab.

7.2.3 Simply Supported Rectangular Orthotropic Plates

Exact vibration solutions may be obtained for simply supported rectangular orthotropic plates of length a and width b. The following deflection function w satisfies the simply supported boundary conditions

$$w = A_{mn} \sin\frac{m\pi x}{a} \sin\frac{n\pi y}{b} \tag{7.25}$$

By substituting Equation (7.25) into Equation (7.16), the natural frequency of vibration of the simply supported rectangular orthotropic plate is given by

$$\omega_{mn} = \frac{\pi^2}{\sqrt{\rho h}} \sqrt{\left[D_x\left(\frac{m}{a}\right)^4 + 2H\left(\frac{mn}{ba}\right)^2 + D_y\left(\frac{n}{b}\right)^4 \right]} \tag{7.26}$$

Sample frequency values $\bar{\omega}_{11} = \omega_{11}a^2\sqrt{\rho h/H}$ for isotropic plates $D_x/H = D_y/H = 1$ and orthotropic plates with various combinations of D_x/H and D_y/H ratios are given in Table 7.1.

7.2.4 Rectangular Orthotropic Plates with Two Parallel Sides Simply Supported

Exact vibration solutions are also possible for rectangular plates with two parallel sides simply supported, while the other two sides can take any combination of clamped, simply supported, and free boundary conditions using the Levy (1899) approach. The partial differential equation for such plates may be converted into an ordinary differential equation, since the two simply supported parallel edges (say, parallel to the x-axis) allow the mode shape to be made separable in the form

$$w(x, y) = W_n(x) \sin\frac{n\pi y}{b} \tag{7.27}$$

TABLE 7.1
Frequency $\bar{\omega}_{11} = \omega_{11}a^2\sqrt{\rho h/H}$ of Orthotropic Rectangular Plates with Simply Supported Edges

a/b	$D_x/H = 1/2$, $D_y/H = 1$	$D_x/H = 1$, $D_y/H = 1$ (isotropic case)	$D_x/H = 1$, $D_y/H = 2$	$D_x/H = 2$, $D_y/H = 1$
0.5	10.173	12.337	12.581	15.799
1.0	18.464	19.739	22.069	22.069
2.0	48.852	49.348	63.196	50.325

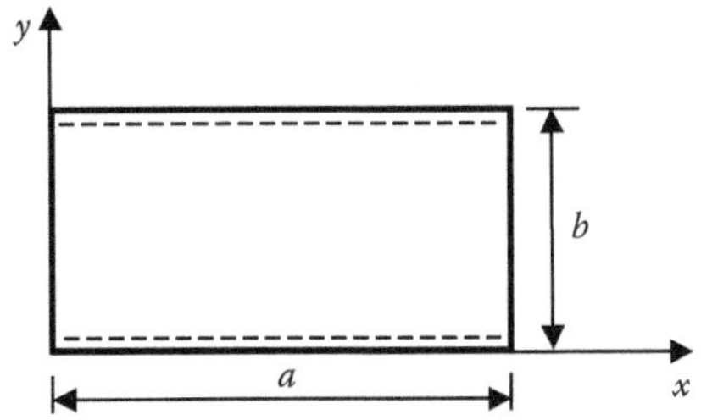

FIGURE 7.6 Rectangular plate with simply supported edges parallel to x-axis and the other two edges parallel to y-axis may be clamped, simply supported, or free.

where it satisfies the simply supported boundary conditions $w = 0$ and $M_{yy} = 0$ at $y = 0$ edge and $y = b$ edge (see Figure 7.6).

By substituting Equation (7.27) into Equation (7.16), we obtain the following fourth-order ordinary differential equation:

$$D_x \frac{d^4 W_n}{dx^4} - 2H\left(\frac{n\pi}{b}\right)^2 \frac{d^2 W_n}{dx^2} + \left[D_y\left(\frac{n\pi}{b}\right)^4 - \rho h\omega^2\right] W_n = 0 \tag{7.28}$$

The form of the solution to Equation (7.27) depends on the nature of the roots λ of the equation

$$\lambda^4 - 2\frac{H}{D_x}\left(\frac{n\pi}{b}\right)^2 \lambda^2 + \left[\frac{D_y}{D_x}\left(\frac{n\pi}{b}\right)^4 - \omega^2 \frac{\rho h}{D_x}\right] = 0 \tag{7.29}$$

There are two distinct cases. Case 1 is when $\omega^2 \geq (n\pi/b)^4 D_y/(\rho h)$, and Case 2 is when $\omega^2 < (n\pi/b)^4 D_y/(\rho h)$. The general solution for Equation (7.28) is given by (Voigt 1893)

$$W_n(x) = A_1 \cosh\lambda_1 x + A_2 \sinh\lambda_1 x + A_3 \cos\lambda_2 x + A_4 \sin\lambda_2 x \quad \text{for Case 1} \tag{7.30}$$

$$W_n(x) = A_1 \cosh\lambda_1 x + A_2 \sinh\lambda_1 x + A_3 \cosh\lambda_2 x + A_4 \sinh\lambda_2 x \quad \text{for Case 2} \tag{7.31}$$

where $A_i, i = 1,2,3,4$ are the integration constants and

$$(\lambda_1)^2 = \frac{H}{D_x}\left(\frac{n\pi}{b}\right)^2 + \sqrt{\left[\left(\frac{H}{D_x}\right)^2 - \frac{D_y}{D_x}\right]\left(\frac{n\pi}{b}\right)^4 + \omega^2 \frac{\rho h}{D_x}} \tag{7.32}$$

$$(\lambda_2)^2 = \left|\frac{H}{D_x}\left(\frac{n\pi}{b}\right)^2 - \sqrt{\left[\left(\frac{H}{D_x}\right)^2 - \frac{D_y}{D_x}\right]\left(\frac{n\pi}{b}\right)^4 + \omega^2 \frac{\rho h}{D_x}}\right| \tag{7.33}$$

In a similar manner as in Chapter 5, the characteristic equations for the various combinations of orthotropic rectangular plates with two parallel edges (i.e., $y = 0$, and $y = b$) simply supported are given in the following discussion. Note that the boundary conditions for a clamped, a simply supported, and a free edge at $x = 0$, and $x = a$ for the edges are given by

$$W_n = 0, \quad \frac{dW_n}{dx} = 0 \quad \text{for a clamped edge} \tag{7.34a}$$

$$W_n = 0, \quad D_x \frac{d^2W_n}{dx^2} - D_1\left(\frac{n\pi}{b}\right)^2 W_n = 0 \quad \text{for a simply supported edge} \tag{7.34b}$$

$$D_x \frac{d^2W_n}{dx^2} - D_1\left(\frac{n\pi}{b}\right)^2 W_n = 0, \quad D_x \frac{d^3W_n}{dx^3} - (D_1 + 4D_{xy})\left(\frac{n\pi}{b}\right)^2 \frac{dW_n}{dx} = 0 \quad \text{for a free edge} \tag{7.34c}$$

7.2.4.1 Two Parallel Edges (i.e., $y = 0$ and $y = b$) Simply Supported, with Simply Supported Edge $x = 0$ and Free Edge $x = a$ (designated as SSSF plates)

$$\lambda_2\Omega_1\bar{\Omega}_2 \sinh(\lambda_1 a)\cos(\lambda_2 a) - \lambda_1\Omega_2\bar{\Omega}_1 \cosh(\lambda_1 a)\sin(\lambda_2 a) = 0 \quad \text{for Case 1} \tag{7.35a}$$

$$\lambda_2\Omega_1\bar{\Omega}_3 \sinh(\lambda_1 a)\cosh(\lambda_2 a) - \lambda_1\Omega_3\bar{\Omega}_1 \cosh(\lambda_1 a)\sinh(\lambda_2 a) = 0 \quad \text{for Case 2} \tag{7.35b}$$

where

$$\Omega_1 = \lambda_1^2 - \frac{D_1}{D_x}\left(\frac{n\pi}{b}\right)^2, \quad \bar{\Omega}_1 = \lambda_1^2 - \frac{D_1 + 4D_{xy}}{D_x}\left(\frac{n\pi}{b}\right)^2 \tag{7.35c}$$

$$\Omega_2 = \lambda_2^2 + \frac{D_1}{D_x}\left(\frac{n\pi}{b}\right)^2, \quad \bar{\Omega}_2 = \lambda_2^2 + \frac{D_1 + 4D_{xy}}{D_x}\left(\frac{n\pi}{b}\right)^2 \tag{7.35d}$$

$$\Omega_3 = \lambda_2^2 - \frac{D_1}{D_x}\left(\frac{n\pi}{b}\right)^2, \quad \bar{\Omega}_3 = \lambda_2^2 - \frac{D_1 + 4D_{xy}}{D_x}\left(\frac{n\pi}{b}\right)^2 \tag{7.35e}$$

7.2.4.2 Two Parallel Edges (i.e., $y = 0$, and $y = b$) Simply Supported, with Clamped Edge $x = 0$ and Free Edge $x = a$ (designated as SCSF plates)

$$\Omega_1\bar{\Omega}_1 + \Omega_2\bar{\Omega}_2 + (\Omega_1\bar{\Omega}_2 + \Omega_2\bar{\Omega}_1)\cosh(\lambda_1 a)\cos(\lambda_2 a)$$
$$+ \frac{1}{\lambda_1\lambda_2}\left(\lambda_2^2\Omega_1\bar{\Omega}_2 - \lambda_1^2\Omega_2\bar{\Omega}_1\right)\sinh(\lambda_1 a)\sin(\lambda_2 a) = 0 \quad \text{for Case 1} \tag{7.36a}$$

$$\Omega_1\bar{\Omega}_1+\Omega_3\bar{\Omega}_3-\left(\Omega_1\bar{\Omega}_3+\Omega_3\bar{\Omega}_1\right)\cosh(\lambda_1 a)\cosh(\lambda_2 a)$$
$$+\frac{1}{\lambda_1\lambda_2}\left(\lambda_2^2\Omega_1\bar{\Omega}_3+\lambda_1^2\Omega_3\bar{\Omega}_1\right)\sinh(\lambda_1 a)\sinh(\lambda_2 a)=0 \quad \text{for Case 2} \qquad (7.36b)$$

7.2.4.3 Two Parallel Edges (i.e., $y = 0$ and $y = b$) Simply Supported, with Clamped Edges $x = 0$ and $x = a$ (designated as SCSC plates)

$$2\lambda_1\lambda_2\left[1-\cosh(\lambda_1 a)\cos(\lambda_2 a)\right]+\left(\lambda_1^2-\lambda_2^2\right)\sinh(\lambda_1 a)\sin(\lambda_2 a)=0 \quad \text{for Case 1} \qquad (7.37a)$$

$$2\lambda_1\lambda_2\left[1-\cosh(\lambda_1 a)\cosh(\lambda_2 a)\right]+\left(\lambda_1^2+\lambda_2^2\right)\sinh(\lambda_1 a)\sinh(\lambda_2 a)=0 \quad \text{for Case 2} \qquad (7.37b)$$

7.2.4.4 Two Parallel Edges (i.e., $y = 0$ and $y = b$) Simply Supported, with Clamped Edge $x = 0$ and Simply Supported Edge $x = a$ (designated as SCSS plates)

$$\lambda_1\cosh(\lambda_1 a)\sin(\lambda_2 a)-\lambda_2\sinh(\lambda_1 a)\cos(\lambda_2 a)=0 \quad \text{for Case 1} \qquad (7.38a)$$

$$\lambda_1\cosh(\lambda_1 a)\sinh(\lambda_2 a)-\lambda_2\sinh(\lambda_1 a)\cosh(\lambda_2 a)=0 \quad \text{for Case 2} \qquad (7.38b)$$

7.2.4.5 Two Parallel Edges (i.e., $y = 0$ and $y = b$) Simply Supported, with Free Edges $x = 0$ and $x = a$ (designated as SFSF plates)

$$2\left[1-\cosh(\lambda_1 a)\cos(\lambda_2 a)\right]+\left(\mu_0-\frac{1}{\mu_0}\right)\sinh(\lambda_1 a)\sin(\lambda_2 a)=0 \quad \text{for Case 1} \qquad (7.39a)$$

$$2\left[1-\cosh(\lambda_1 a)\cosh(\lambda_2 a)\right]+\left(\mu_0+\frac{1}{\mu_0}\right)\sinh(\lambda_1 a)\sinh(\lambda_2 a)=0 \quad \text{for Case 2} \qquad (7.39b)$$

where

$$\mu_0=\frac{\lambda_1\Omega_2\bar{\Omega}_1}{\lambda_2\Omega_1\bar{\Omega}_2} \quad \text{for Case 1 and} \quad \mu_0=\frac{\lambda_1\Omega_3\bar{\Omega}_1}{\lambda_2\Omega_1\bar{\Omega}_3} \quad \text{for Case 2} \qquad (7.39c)$$

Sample vibration results for orthotropic plates are given in Tables 7.2 to 7.6, where $m = n = 1$. When $D_x/H = D_y/H = 1$, the plate is isotropic, and the frequency values reduce to those given in Tables 5.2 to 5.6.

7.2.5 Rectangular Orthotropic Thick Plates

The exact solutions for free vibration of rectangular Mindlin plates with all edges simply supported were first obtained by Mindlin and Deresiewicz (1955). The

TABLE 7.2
Frequency $\bar{\omega}_{11} = \omega_{11}a^2\sqrt{\rho h/H}$ of Orthotropic SSSF Plates with $\nu_{xy} = \sqrt{\nu_x \nu_y} = 0.3$

a/b	$D_x/H = 1/2$, $D_y/H = 1$	$D_x/H = 1$, $D_y/H = 1$ (isotropic case)	$D_x/H = 1$, $D_y/H = 2$	$D_x/H = 2$, $D_y/H = 1$
0.5	4.168	4.034	4.511	3.799
1.0	11.868	11.685	14.957	11.347
2.0	41.467	41.197	56.441	40.704

TABLE 7.3
Frequency $\bar{\omega}_{11} = \omega_{11}a^2\sqrt{\rho h/H}$ of Orthotropic SCSF Plates with $\nu_{xy} = \sqrt{\nu_x \nu_y} = 0.3$

a/b	$D_x/H = 1/2$, $D_y/H = 1$	$D_x/H = 1$, $D_y/H = 1$ (isotropic case)	$D_x/H = 1$, $D_y/H = 2$	$D_x/H = 2$, $D_y/H = 1$
0.5	5.273	5.704	5.945	6.474
1.0	12.583	12.687	15.641	12.777
2.0	41.840	41.702	56.735	41.346

TABLE 7.4
Frequency $\bar{\omega}_{11} = \omega_{11}a^2\sqrt{\rho h/H}$ of Orthotropic SCSC Plates

a/b	$D_x/H = 1/2$, $D_y/H = 1$	$D_x/H = 1$, $D_y/H = 1$ (isotropic case)	$D_x/H = 1$, $D_y/H = 2$	$D_x/H = 2$, $D_y/H = 1$
0.5	17.797	23.816	23.943	32.678
1.0	24.200	28.951	30.587	36.606
2.0	52.221	54.743	67.493	59.258

derivation of the characteristic equations for other boundary conditions was recognized as too complex. With the recent development of symbolic manipulations, it becomes possible to derive the exact characteristic equations of orthotropic Mindlin plates as shown by Liu and Xing (2011).

Consider a thick rectangular plate of length a, width b, and uniform thickness h, oriented so that its undeformed middle surface contains the x- and y-axes

TABLE 7.5
Frequency $\bar{\omega}_{11} = \omega_{11}a^2\sqrt{\rho h/H}$ of Orthotropic SCSS Plates

a/b	$D_x/H = 1/2$, $D_y/H = 1$	$D_x/H = 1$, $D_y/H = 1$ (isotropic case)	$D_x/H = 1$, $D_y/H = 2$	$D_x/H = 2$, $D_y/H = 1$
0.5	13.466	17.332	17.507	23.200
1.0	20.928	23.646	25.623	28.255
2.0	50.360	51.674	65.029	54.044

TABLE 7.6
Frequency $\bar{\omega}_{11} = \omega_{11}a^2\sqrt{\rho h/H}$ of Orthotropic SFSF Plates (obtained from Equation 7.39b) with $\nu_{xy} = \sqrt{\nu_x \nu_y} = 0.3$

a/b	$D_x/H = 1/2$, $D_y/H = 1$	$D_x/H = 1$, $D_y/H = 1$ (isotropic case)	$D_x/H = 1$, $D_y/H = 2$	$D_x/H = 2$, $D_y/H = 1$
0.5	2.395	2.378	3.358	2.366
1.0	9.703	9.631	13.589	9.549
2.0	39.127	38.945	54.950	38.665

of a Cartesian coordinate system (x, y, z), as shown in Figure 7.7. According to the Mindlin plate theory, the displacements along the x-, y-, and z-directions are assumed to be

$$u = -z\psi_x(x,y,z,t), \quad v = -z\psi_y(x,y,z,t), \quad w = w(x,y,z,t) \tag{7.40}$$

where t is the time coordinate, w is the deflection, and ψ_x and ψ_y are the angles of rotations of a normal line due to plate bending with respect to y- and x-coordinates, respectively.

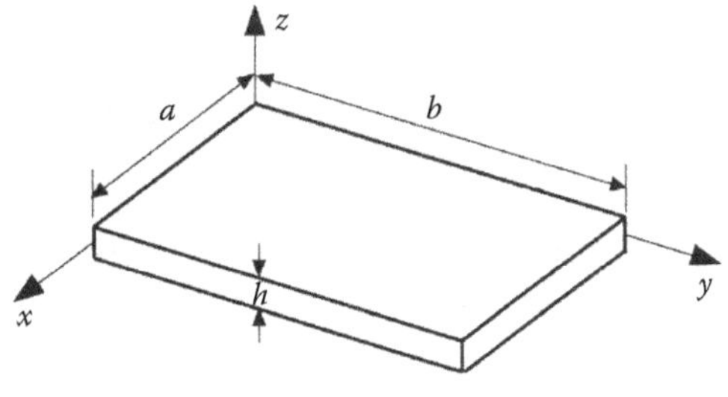

FIGURE 7.7 Mindlin plate and coordinates.

The relations between the stress resultants and the displacements for orthotropic Mindlin plates are given by

$$M_x = -\left(D_{11}\frac{\partial \psi_x}{\partial x} + D_{12}\frac{\partial \psi_y}{\partial y} \right), \quad M_y = -\left(D_{22}\frac{\partial \psi_y}{\partial y} + D_{21}\frac{\partial \psi_x}{\partial x} \right) \tag{7.41a}$$

$$M_{xy} = -D_{66}\left(\frac{\partial \psi_x}{\partial y} + \frac{\partial \psi_y}{\partial x} \right) \tag{7.41b}$$

$$Q_y = C_{44}\left(\frac{\partial w}{\partial y} - \psi_y \right), \quad Q_x = C_{55}\left(\frac{\partial w}{\partial x} - \psi_x \right) \tag{7.41c}$$

where the bending and shear rigidities are defined as

$$\begin{aligned} D_{11} &= \frac{E_x h^3}{12(1-\nu_x\nu_y)}, \quad D_{12} = \frac{\nu_y E_x h^3}{12(1-\nu_x\nu_y)}, \quad D_{22} = \frac{E_y h^3}{12(1-\nu_x\nu_y)}, \\ D_{21} &= \frac{\nu_x E_y h^3}{12(1-\nu_x\nu_y)}, \quad D_{66} = \frac{G_{xy} h^3}{12}, \quad C_{44} = \kappa G_{yz} h, \quad C_{55} = \kappa G_{zx} h \end{aligned} \tag{7.42}$$

and κ is the shear correction factor. In view of Betti's principle, the product $\nu_y E_x = \nu_x E_y$, and therefore $D_{12} = D_{21}$. The foregoing bending and shear rigidities are given for one layer. The formulations of rigidities for laminates can be found in textbooks or research papers (see, for example, the paper by Liew [1996]).

The governing equations of motion in terms of displacements are given by

$$D_1\frac{\partial^2 \psi_x}{\partial x^2} + \frac{\partial^2 \psi_x}{\partial y^2} + D_3\frac{\partial^2 \psi_y}{\partial x \partial y} + C_1\left(\frac{\partial w}{\partial x} - \psi_x \right) + \gamma^4 \psi_x = 0 \tag{7.43a}$$

$$D_2\frac{\partial^2 \psi_y}{\partial y^2} + \frac{\partial^2 \psi_y}{\partial x^2} + D_3\frac{\partial^2 \psi_x}{\partial x \partial y} + C_2\left(\frac{\partial w}{\partial y} - \psi_y \right) + \gamma^4 \psi_y = 0 \tag{7.43b}$$

$$C_1\left(\frac{\partial^2 w}{\partial x^2} - \frac{\partial \psi_x}{\partial x} \right) + C_2\left(\frac{\partial^2 w}{\partial y^2} - \frac{\partial \psi_y}{\partial y} \right) + \beta^4 w = 0 \tag{7.43c}$$

where the normalized material parameters and frequency parameters are defined as

$$D_1 = \frac{D_{11}}{D_{66}}, \quad D_2 = \frac{D_{22}}{D_{66}}, \quad D_3 = \frac{D_{12}+D_{66}}{D_{66}}, \quad C_1 = \frac{C_{55}}{D_{66}}, \quad C_2 = \frac{C_{44}}{D_{66}} \tag{7.44a}$$

$$\gamma^4 = \frac{\rho h^3 \omega^2}{12 D_{66}}, \quad \beta^4 = \frac{\rho h \omega^2}{D_{66}} \tag{7.44b}$$

As the governing differential equations (7.43a) to (7.43c) are in total of the sixth order, there are three boundary conditions on each edge. The boundary conditions are given as follows:

1. Simply supported edge (S),

$$w = 0, \quad \psi_s = 0$$
$$M_n = 0 \Rightarrow \frac{\partial \psi_n}{\partial n} + v_n \frac{\partial \psi_s}{\partial s} = 0 \Rightarrow \frac{\partial \psi_n}{\partial n} = 0 \tag{7.45}$$

2. Clamped edge (C),

$$w = 0, \quad \psi_s = 0, \quad \psi_n = 0 \tag{7.46}$$

3. Free edge (F), where stress resultants, M_n, M_{ns}, and Q_n are assigned zeros values, i.e.,

$$\frac{\partial \psi_n}{\partial n} + v_n \frac{\partial \psi_s}{\partial s} = 0, \quad \frac{\partial \psi_n}{\partial s} + \frac{\partial \psi_s}{\partial n} = 0, \quad \frac{\partial w}{\partial n} - \psi_n = 0 \tag{7.47}$$

Derived in the following discussion are exact solutions for Equations (7.43a) to (7.43c) for rectangular plates with one pair of opposite edges being simply supported, while the other pair of opposite edges can take any combination of the three types of boundary conditions.

By eliminating ψ_y from Equation (7.43a) by using Equation (7.43c), one gets

$$\left[1 - \frac{1}{C_1}\left(D_1 \frac{\partial^2}{\partial x^2} - D_3 \frac{C_1}{C_2}\frac{\partial^2}{\partial x^2} + \frac{\partial^2}{\partial y^2} + \gamma^4\right)\right]\psi_x = \left[1 + \frac{D_3}{C_1}\left(\frac{\partial^2}{\partial y^2} + \frac{C_1}{C_2}\frac{\partial^2}{\partial x^2} + \frac{\beta^4}{C_2}\right)\right]\frac{\partial w}{\partial x} \tag{7.48a}$$

Similarly, we can obtain

$$\left[1 - \frac{1}{C_2}\left(D_2 \frac{\partial^2}{\partial y^2} - D_3 \frac{C_2}{C_1}\frac{\partial^2}{\partial y^2} + \frac{\partial^2}{\partial x^2} + \gamma^4\right)\right]\psi_y = \left[1 + \frac{D_3}{C_2}\left(\frac{\partial^2}{\partial x^2} + \frac{C_2}{C_1}\frac{\partial^2}{\partial y^2} + \frac{\beta^4}{C_1}\right)\right]\frac{\partial w}{\partial y} \tag{7.48b}$$

The differentiation of Equation (7.43c) with respect to both x and y results in

$$\left(C_1 \frac{\partial^2}{\partial x^2} + C_2 \frac{\partial^2}{\partial y^2} + \beta^4\right)\frac{\partial w}{\partial x \partial y} = C_1 \frac{\partial^2}{\partial x^2}\frac{\partial \psi_x}{\partial y} + C_2 \frac{\partial^2}{\partial y^2}\frac{\partial \psi_y}{\partial x} \tag{7.48c}$$

Equations (7.48a) to (7.48c) may also be written as

$$\left.\begin{aligned} &L_1\frac{\partial\psi_x}{\partial y}=K_1\frac{\partial w}{\partial x\,\partial y},\quad L_2\frac{\partial\psi_y}{\partial x}=K_2\frac{\partial w}{\partial x\,\partial y}\\ &K_3\frac{\partial w}{\partial x\,\partial y}=C_1\frac{\partial^2}{\partial x^2}\frac{\partial\psi_x}{\partial y}+C_2\frac{\partial^2}{\partial y^2}\frac{\partial\psi_y}{\partial x}\end{aligned}\right\}\qquad(7.49\text{a,b,c})$$

where

$$L_1=1-\frac{1}{C_1}\left(D_1\frac{\partial^2}{\partial x^2}-D_3\frac{C_1}{C_2}\frac{\partial^2}{\partial x^2}+\frac{\partial^2}{\partial y^2}+\gamma^4\right)\qquad(7.50\text{a})$$

$$L_2=1-\frac{1}{C_2}\left(D_2\frac{\partial^2}{\partial y^2}-D_3\frac{C_2}{C_1}\frac{\partial^2}{\partial y^2}+\frac{\partial^2}{\partial x^2}+\gamma^4\right)\qquad(7.50\text{b})$$

$$K_1=1+\frac{D_3}{C_1}\left(\frac{\partial^2}{\partial y^2}+\frac{C_1}{C_2}\frac{\partial^2}{\partial x^2}+\frac{\beta^4}{C_2}\right)\qquad(7.50\text{c})$$

$$K_2=1+\frac{D_3}{C_2}\left(\frac{\partial^2}{\partial x^2}+\frac{C_2}{C_1}\frac{\partial^2}{\partial y^2}+\frac{\beta^4}{C_1}\right)\qquad(7.50\text{d})$$

$$K_3=C_1\frac{\partial^2}{\partial x^2}+C_2\frac{\partial^2}{\partial y^2}+\beta^4\qquad(7.50\text{e})$$

The elimination of both ψ_x and ψ_y from Equations (7.49a) to (7.49c) results in

$$\left(L_1L_2K_3-C_1K_1L_2\frac{\partial^2}{\partial x^2}-C_2L_1K_2\frac{\partial^2}{\partial y^2}\right)\frac{\partial w}{\partial x\,\partial y}=0\qquad(7.51\text{a})$$

Similarly, we can obtain

$$\left(L_1L_2K_3-C_1K_1L_2\frac{\partial^2}{\partial x^2}-C_2L_1K_2\frac{\partial^2}{\partial y^2}\right)\frac{\partial\psi_x}{\partial y}=0\qquad(7.51\text{b})$$

$$\left(L_1L_2K_3-C_1K_1L_2\frac{\partial^2}{\partial x^2}-C_2L_1K_2\frac{\partial^2}{\partial y^2}\right)\frac{\partial\psi_y}{\partial x}=0\qquad(7.51\text{c})$$

Assume the pair of opposite edges of $x = 0$ and a as simply supported. By using the separation-of-variables solution $w = e^{\mu x}e^{\lambda y}$—where $\mu = i\alpha$, $i=\sqrt{-1}$, $\alpha = m\pi/a$, and m is the number of half waves in the x-direction—Equation (7.51a) may be expressed as

$$\bar{a}\lambda^6 + \bar{b}\lambda^4 + \bar{c}\lambda^2 + \bar{d} = 0 \tag{7.52}$$

where

$$\bar{a} = \frac{D_2}{C_1} \tag{7.53a}$$

$$\bar{b} = \frac{1}{C_1C_2}\left[C_2(1+D_2)\gamma^4 + \left(\beta^4 - C_1C_2\right)D_2\right] - \frac{1}{C_1C_2}\left[C_2\left(D_1D_2 + 1 - D_3^2\right) + C_1D_2\right]\alpha^2 \tag{7.53b}$$

$$\bar{c} = \left[C_2(\gamma^8 - \beta^4) - C_1C_2\gamma^4 + (D_2+1)\beta^4\gamma^4 - C_1D_2\beta^4\right]\frac{1}{C_1C_2} + \left[C_2D_1 + C_1\left(1 + D_1D_2 - D_3^2\right)\right]\frac{\alpha^4}{C_1C_2} - \left[C_2(1+D_1) + C_1(1+D_2)\right]\frac{\alpha^2\gamma^4}{C_1C_2} - \left[\left(1 - D_3^2 + D_1D_2\right)\beta^4 - 2C_1C_2(D_3+1)\right]\frac{\alpha^2}{C_1C_2} \tag{7.53c}$$

$$\bar{d} = \left[C_1C_2 + \left(\gamma^4 - C_2 - C_1\right)\gamma^4\right]\frac{\beta^4}{C_1C_2} + \left[C_1(1+D_1)\gamma^4 + D_1\left(\beta^4 - C_1C_2\right)\right]\frac{\alpha^4}{C_1C_2} - \left[C_1\left(\gamma^8 - \beta^4\right) + (1+D_1)\beta^4\gamma^4 - C_2\left(C_1\gamma^4 + D_1\beta^4\right)\right]\frac{\alpha^2}{C_1C_2} - \frac{D_1\alpha^6}{C_2} \tag{7.53d}$$

Let

$$\lambda^2 = s - \frac{\bar{b}}{3\bar{a}} \tag{7.54}$$

The substitution of Equation (7.54) into Equation (7.52) yields

$$s^3 + ps + q = 0 \tag{7.55}$$

where

$$p = \frac{1}{\bar{a}}\left(\bar{c} - \frac{\bar{b}^2}{3\bar{a}}\right), \quad q = \frac{1}{\bar{a}}\left(\bar{d} + \frac{2\bar{b}^3}{27\bar{a}^2} - \frac{\bar{b}\bar{c}}{3\bar{a}}\right) \tag{7.56}$$

The roots of Equation (7.55) are given by

$$s_1 = \Delta_1 + \Delta_2, \quad s_2 = \varpi\Delta_1 + \varpi^2\Delta_2, \quad s_3 = \varpi^2\Delta_1 + \varpi\Delta_2 \tag{7.57a,b,c}$$

where

$$\Delta_1 = \sqrt[3]{-\frac{q}{2} + \sqrt{\left(\frac{q}{2}\right)^2 + \left(\frac{p}{3}\right)^3}}, \quad \Delta_2 = \sqrt[3]{-\frac{q}{2} - \sqrt{\left(\frac{q}{2}\right)^2 + \left(\frac{p}{3}\right)^3}} \tag{7.58}$$

$$\varpi = \frac{-1 + i\sqrt{3}}{2}, \quad \varpi^2 = \frac{-1 - i\sqrt{3}}{2} \tag{7.59}$$

So the roots of Equation (7.52) can be expressed as

$$\lambda_{1,2} = \pm i\beta_1, \quad \lambda_{3,4} = \pm i\beta_2, \quad \lambda_{5,6} = \pm i\beta_3 \tag{7.60}$$

where

$$\beta_j = \sqrt{\frac{\bar{b}}{3\bar{a}} - s_j} \quad j = 1, 2, 3 \tag{7.61}$$

From the foregoing derivations, one cannot directly determine whether β_j, $j =$ 1,2,3 is real, imaginary, or complex. We shall assume that they are real or imaginary but not complex, the correctness of which can be verified through calculations. The eigenfunctions w, ψ_x, and ψ_y (in a separation of variable form) can then be expressed in terms of the eigenvalues by three potentials W_j, $j = 1,2,3$, i.e.,

$$\psi_x = \sum_{j=1}^{3} g_j \frac{\partial W_j}{\partial x}, \quad \psi_y = \sum_{j=1}^{3} h_j \frac{\partial W_j}{\partial y}, \quad w = \sum_{j=1}^{3} W_j \tag{7.62a,b,c}$$

where

$$W_j(x, y) = (A_j \sin\beta_j y + B_j \cos\beta_j y)\sin\alpha x \tag{7.63}$$

The substitution of Equations (7.62) into Equations (7.49a,b) leads to

$$g_j = \left[1 - \frac{D_3}{C_1}\left(\beta_j^2 + \frac{C_1}{C_2}\alpha^2 - \frac{\beta^4}{C_2}\right)\right]\left[1 + \frac{1}{C_1}\left(D_1\alpha^2 - D_3\frac{C_1}{C_2}\alpha^2 + \beta_j^2 - \gamma^4\right)\right]^{-1} \tag{7.64a}$$

$$h_j = \left[1 - \frac{D_3}{C_2}\left(\alpha^2 + \frac{C_2}{C_1}\beta_j^2 - \frac{\beta^4}{C_1}\right)\right]\left[1 + \frac{1}{C_2}\left(D_2\beta_j^2 - D_3\frac{C_2}{C_1}\beta_j^2 + \alpha^2 - \gamma^4\right)\right]^{-1} \tag{7.64b}$$

where $j = 1, 2, 3$.

For plates with simply supported edges at ($x = 0, a$) while the edges at ($y = 0, b$) have arbitrary boundary conditions, the exact eigenequations and eigenfunctions can be obtained by substituting Equations (7.62a) to (7.62c) into the boundary conditions. The eigenequations and coefficients of eigenfunctions for the cases with $y = 0, b$ being S-C and S-F are listed as follows:

Case S-C

$$h_2\beta_2 \sin(b\beta_1)\tan(b\beta_3)(g_1 - g_3) - h_1\beta_1 \cos(b\beta_1)\tan(b\beta_2)\tan(b\beta_3)(g_2 - g_3) - h_3\beta_3 \sin(b\beta_1)\tan(b\beta_2)(g_1 - g_2) = 0 \tag{7.65}$$

$$\begin{aligned} A_1 &= \sin(b\beta_2)\sin(b\beta_3)(g_2 - g_3) \\ A_2 &= -\sin(b\beta_1)\sin(b\beta_3)(g_1 - g_3) \\ A_3 &= \sin(b\beta_1)\sin(b\beta_2)(g_1 - g_2) \\ B_1 &= 0, \quad B_2 = 0, \quad B_3 = 0 \end{aligned} \tag{7.66}$$

Case S-F

$$d_2 \cos(b\beta_1)\tan(b\beta_2)(e_1 f_3 - e_3 f_1) - d_1 \sin(b\beta_1)(e_2 f_3 - e_3 f_2) - d_3 \cos(b\beta_1)\tan(b\beta_3)(e_1 f_2 - e_2 f_1) = 0 \tag{7.67}$$

$$\begin{aligned} A_1 &= d_2 e_3 \cos(b\beta_3)\sin(b\beta_2) - d_3 e_2 \cos(b\beta_2)\sin(b\beta_3) \\ A_2 &= d_3 e_1 \cos(b\beta_1)\sin(b\beta_3) - d_1 e_3 \cos(b\beta_3)\sin(b\beta_1) \\ A_3 &= d_1 e_2 \cos(b\beta_2)\sin(b\beta_1) - d_2 e_1 \cos(b\beta_1)\sin(b\beta_2) \\ B_1 &= 0, \quad B_2 = 0, \quad B_3 = 0 \end{aligned} \tag{7.68}$$

where

$$d_j = g_j v_y \alpha^2 + h_j \beta_j^2, \quad e_j = (g_j + h_j)\alpha\beta_j, \quad f_j = (1 - h_j)\beta_j \tag{7.69}$$

and $j = 1,2,3$.

For rectangular plates with two simply supported edges $x = 0$ and $x = a$, whereas the other two edges $y = 0$ and $y = b$ are both clamped (C-C) or both free (F-F), the characteristic equation and coefficients of mode shape can be obtained in a similar form as the two foregoing cases after some simplifications, i.e., by moving the origin of the y coordinate to the center of the side $x = 0$, as shown in Figure 7.8.

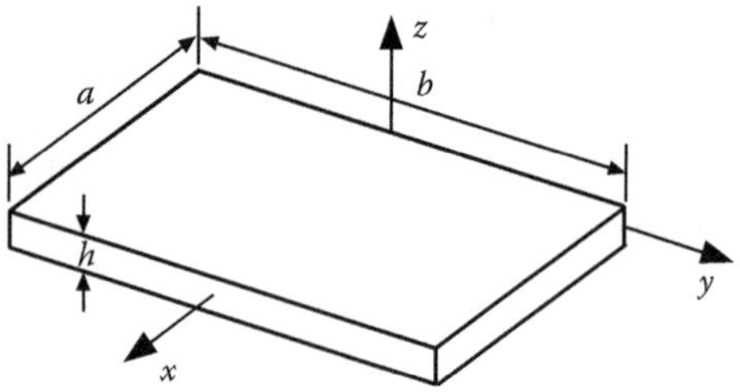

FIGURE 7.8 Mindlin plate and coordinates.

The mode shapes of w that are antisymmetric with respect to the x-axis can be written as

$$\psi_x = \sum_{j=1}^{3} g_j \frac{\partial \hat{W}_j}{\partial x}, \quad \psi_y = \sum_{j=1}^{3} h_j \frac{\partial \hat{W}_j}{\partial y}, \quad w = \sum_{j=1}^{3} \hat{W}_j \tag{7.70}$$

$$\hat{W}_j(x,y) = A_j \sin(\beta_j y)\sin(\alpha x) \tag{7.71}$$

The mode shapes of w that are symmetric with respect to the x-axis can be written as

$$\psi_x = \sum_{j=1}^{3} g_j \frac{\partial \tilde{W}_j}{\partial x}, \quad \psi_y = \sum_{j=1}^{3} h_j \frac{\partial \tilde{W}_j}{\partial y}, \quad w = \sum_{j=1}^{3} \tilde{W}_j \tag{7.72}$$

$$\tilde{W}_j(x,y) = B_j \cos(\beta_j y)\sin(\alpha x) \tag{7.73}$$

Consider a rectangular plate with two simply supported edges $x = 0$ and $x = a$, whereas the two edges $y = -b/2$ and $y = b/2$ are both clamped. The characteristic equation and coefficients of mode shape for the symmetric case are given by

$$\begin{aligned} &h_1\beta_1 \sin(\hat{b}\beta_1)(g_2 - g_3) - h_2\beta_2 \cos(\hat{b}\beta_1)\tan(\hat{b}\beta_2)(g_1 - g_3) \\ &+ h_3\beta_3 \cos(\hat{b}\beta_1)\tan(\hat{b}\beta_3)(g_1 - g_2) = 0 \end{aligned} \tag{7.74}$$

$$\begin{aligned} B_1 &= \cos(\hat{b}\beta_2)\cos(\hat{b}\beta_3)(g_2 - g_3) \\ B_2 &= -\cos(\hat{b}\beta_1)\cos(\hat{b}\beta_3)(g_1 - g_3) \\ B_3 &= \cos(\hat{b}\beta_1)\cos(\hat{b}\beta_2)(g_1 - g_2) \\ A_1 &= 0, \quad A_2 = 0, \quad A_3 = 0 \end{aligned} \tag{7.75}$$

where $\hat{b} = b/2$

Consider a rectangular plate with two simply supported edges $x = 0$ and $x = a$, while the other two edges $y = -b/2$ and $y = b/2$ are free. The characteristic equation and coefficients of mode shape for the symmetric case are given by

$$\begin{aligned} & d_2 \cot(\hat{b}\beta_2)\sin(\hat{b}\beta_1)(e_1 f_3 - e_3 f_1) - d_1 \cos(\hat{b}\beta_1)(e_2 f_3 - e_3 f_2) \\ & - d_3 \cot(\hat{b}\beta_3)\sin(\hat{b}\beta_1)(e_1 f_2 - e_2 f_1) = 0 \end{aligned} \tag{7.76}$$

$$\begin{aligned} B_1 &= d_2 e_3 \cos(\hat{b}\beta_2)\sin(\hat{b}\beta_3) - d_3 e_2 \cos(\hat{b}\beta_3)\sin(\hat{b}\beta_2) \\ B_2 &= d_3 e_1 \cos(\hat{b}\beta_3)\sin(\hat{b}\beta_1) - d_1 e_3 \cos(\hat{b}\beta_1)\sin(\hat{b}\beta_3) \\ B_3 &= d_1 e_2 \cos(\hat{b}\beta_1)\sin(\hat{b}\beta_2) - d_2 e_1 \cos(\hat{b}\beta_2)\sin(\hat{b}\beta_1) \\ A_1 &= 0, \quad A_2 = 0, \quad A_3 = 0 \end{aligned} \tag{7.77}$$

The d_j, e_j, f_j, $j = 1,2,3$ in Equations (7.76) and (7.77) are given by Equation (7.69). For antisymmetric cases of Equations (7.74) to (7.77), one only needs to change $\sin(b\beta_j)$, $\cos(b\beta_j)$, $\tan(b\beta_j)$, and $\cot(b\beta_j)$ by $\cos(b\beta_j)$, $\sin(b\beta_j)$, $\cot(b\beta_j)$, and $\tan(b\beta_j)$, respectively.

For a rectangular plate with two edges $x = 0$ and $x = a$ simply supported, whereas the edge $y = -b/2$ is clamped and the edge $y = b/2$ is free, the same procedure as in Equations (5.122) to (5.124) can be used.

Consider thick symmetric three-ply laminates with layers of equal thickness and stacking sequence (0°, 90°, 0°). The material properties for all layers of the laminates are identical: $E_x/E_y = 40$, $G_{23} = E_y/2$, $G_{12} = G_{31} = 3E_y/5$, $\nu_x = 1/4$, and $\nu_y = 0.00625$. The exact frequency parameters Ω

$$\Omega = \frac{\omega b^2}{\pi^2}\sqrt{\frac{\rho h}{D_0}} \tag{7.78}$$

where h is the total thickness and

$$D_0 = \frac{E_y h^3}{12(1 - \nu_x \nu_y)} \tag{7.79}$$

are tabulated in Tables 7.7 to 7.12 for square plates with various combinations of boundary conditions and h/b = 0.05, 0.10, and 0.2. The shear correction factor κ is taken as $\pi^2/12$. The eigenvalues $\alpha a/\pi$ and $\beta_j b/\pi$, j = 1,2,3 are also included in Tables 7.7 to 7.12, and they can be used to obtain the mode shapes. The assumption that β_j cannot be complex, but is either real or imaginary, is verified. It may be seen that β_2 and β_3 are imaginary in general, while β_1 is usually real, but it may become imaginary, as can be seen from Table 7.9.

TABLE 7.7
Frequency Parameters $\Omega = (\omega b^2/\pi^2)\sqrt{\rho h/D_0}$ for Three-Ply Laminated (0°, 90°, 0°) Square SSSS Plates

		Mode Sequence Number							
h/b		**1**	**2**	**3**	**4**	**5**	**6**	**7**	**8**
0.050	$\alpha a/\pi$	1	1	1	2	2	1	2	2
	$\beta_1 b/\pi$	1	2	3	1	2	4	3	4
	Ω	6.161	8.899	15.085	19.572	20.872	23.970	24.506	31.094
0.100	$\alpha a/\pi$	1	1	1	2	2	2	1	3
	$\beta_1 b/\pi$	1	2	3	1	2	3	4	1
	Ω	5.218	7.773	12.844	13.313	14.606	17.915	19.292	21.562
0.200	$\alpha a/\pi$	1	1	2	2	1	2	3	3
	$\beta_1 b/\pi$	1	2	1	2	3	3	1	2
	Ω	3.652	5.759	7.578	8.809	9.025	11.224	11.505	12.363

TABLE 7.8
Frequency Parameters $\Omega = (\omega b^2/\pi^2)\sqrt{\rho h/D_0}$ for Three-Ply Laminated (0°, 90°, 0°) Square SCSC Plates

		Mode Sequence Number							
h/b		**1**	**2**	**3**	**4**	**5**	**6**	**7**	**8**
0.05	$\alpha a/\pi$	1	1	1	2	2	2	1	2
	$\beta_1 b/\pi$	1.425	2.440	3.424	1.341	2.400	3.400	4.396	4.379
	$\beta_2 b/\pi$	21.540i	21.533i	21.516i	2.100i	25.587i	3.503i	21.481i	25.540i
	$\beta_3 b/\pi$	1.716i	2.532i	3.302i	25.593i	2.800i	25.571i	3.989i	4.148i
	Ω	6.907	11.230	18.578	19.833	21.984	26.789	28.058	34.307
0.10	$\alpha a/\pi$	1	1	2	1	2	2	1	3
	$\beta_1 b/\pi$	1.408	2.368	1.368	3.306	2.350	3.295	4.240	1.345
	$\beta_2 b/\pi$	12.791i	12.762i	18.800i	12.693i	18.779i	18.731i	12.572i	1.771i
	$\beta_3 b/\pi$	1.548i	2.183i	1.684i	2.663i	2.274i	2.730i	2.989i	25.864i
	Ω	5.905	9.412	13.594	14.712	15.522	19.267	20.952	21.735
0.20	$\alpha a/\pi$	1	1	2	2	1	2	3	3
	$\beta_1 b/\pi$	1.344	2.217	1.325	2.211	3.128	3.126	1.305	2.205
	$\beta_2 b/\pi$	9.383i	9.313i	16.652i	16.612i	9.169i	16.531i	24.342i	24.314i
	$\beta_3 b/\pi$	1.227i	1.546i	1.276i	1.584i	1.643i	1.678i	1.339i	1.635i
	Ω	4.165	6.411	7.828	9.238	9.476	11.583	11.664	12.665

TABLE 7.9
Frequency Parameters $\Omega = (\omega b^2/\pi^2)\sqrt{\rho h/D_0}$ for Three-Ply Laminated (0°, 90°, 0°) Square SFSF Plates

		Mode Sequence Number							
h/b		1	2	3	4	5	6	7	8
0.05	$\alpha a/\pi$	1	1	1	1	2	2	1	2
	$\beta_1 b/\pi$	0.115i	0.732	1.591	2.543	0.168i	0.807	3.531	1.669
	$\beta_2 b/\pi$	21.541i	21.541i	21.539i	21.532i	25.594i	25.594i	21.513i	25.592i
	$\beta_3 b/\pi$	0.992i	1.234i	1.845i	2.614i	1.642i	1.828i	3.381i	2.302i
	Ω	5.756	5.929	7.357	11.864	19.339	19.479	19.519	20.244
0.10	$\alpha a/\pi$	1	1	1	1	2	2	2	2
	$\beta_1 b/\pi$	0.105i	0.689	1.574	2.547	0.132i	0.751	1.615	2.570
	$\beta_2 b/\pi$	12.797i	12.796i	12.788i	12.752i	18.804i	18.803i	18.797i	18.771i
	$\beta_3 b/\pi$	0.819i	1.061i	1.664i	2.286i	1.116i	1.330i	1.837i	2.393i
	Ω	4.834	4.968	6.324	10.311	13.118	13.210	13.898	16.233
0.20	$\alpha a/\pi$	1	1	1	2	2	1	2	2
	$\beta_1 b/\pi$	0.077i	0.609	1.574	0.075i	0.686	2.570	1.602	2.587
	$\beta_2 b/\pi$	9.402i	9.400i	9.370i	16.663i	16.661i	9.266i	16.643i	16.584i
	$\beta_3 b/\pi$	0.556i	0.796i	1.335i	0.706i	0.940i	1.609i	1.397i	1.648i
	Ω	3.279	3.365	4.645	7.385	7.451	7.552	8.154	10.117

TABLE 7.10
Frequency Parameters $\Omega = (\omega b^2/\pi^2)\sqrt{\rho h/D_0}$ for Three-Ply Laminated (0°, 90°, 0°) Square SSSF Plates

		Mode Sequence Number							
h/b		1	2	3	4	5	6	7	8
0.05	$\alpha a/\pi$	1	1	1	1	2	2	2	2
	$\beta_1 b/\pi$	0.418	1.310	2.275	3.266	0.427	1.356	2.305	3.284
	$\beta_2 b/\pi$	21.541i	21.540i	21.535i	21.520i	25.594i	25.593i	25.588i	25.574i
	$\beta_3 b/\pi$	1.081i	1.628i	2.398i	3.182i	1.702i	2.109i	2.734i	3.423i
	Ω	5.801	6.650	10.278	17.223	19.377	19.848	21.681	26.076
0.10	$\alpha a/\pi$	1	1	1	2	2	1	2	2
	$\beta_1 b/\pi$	0.403	1.296	2.275	0.419	1.322	3.274	2.288	3.282
	$\beta_2 b/\pi$	12.797i	12.792i	12.766i	18.804i	18.800i	12.696i	18.781i	18.732i
	$\beta_3 b/\pi$	0.914i	1.468i	2.126i	1.193i	1.656i	2.649i	2.240i	2.725i
	Ω	4.870	5.669	8.965	13.143	13.549	14.516	15.341	19.208
0.20	$\alpha a/\pi$	1	1	1	2	2	2	1	3
	$\beta_1 b/\pi$	0.375	1.285	2.285	0.410	1.304	2.294	3.291	0.439
	$\beta_2 b/\pi$	9.402i	9.385i	9.305i	16.662i	1.266i	16.606i	9.135i	24.349i
	$\beta_3 b/\pi$	0.665i	1.196i	1.561i	0.805i	16.652i	1.601i	1.639i	0.944i
	Ω	3.303	4.058	6.624	7.403	7.808	9.419	10.056	11.381

TABLE 7.11
Frequency Parameters $\Omega = (\omega b^2/\pi^2)\sqrt{\rho h/D_0}$ for Three-Ply Laminated (0°, 90°, 0°) Square SSSC Plates

		Mode Sequence Number							
h/b		1	2	3	4	5	6	7	8
0.05	$\alpha a/\pi$	1	1	1	2	2	2	1	2
	$\beta_1 b/\pi$	1.203	2.221	3.215	1.158	2.198	3.201	4.201	4.192
	$\beta_2 b/\pi$	21.540i	21.536i	21.521i	25.593i	25.589i	25.575i	21.489i	25.547i
	$\beta_3 b/\pi$	1.548i	2.354i	3.143i	1.997i	2.659i	3.365i	3.858i	4.030i
	Ω	6.450	9.983	16.796	19.675	21.369	25.594	26.012	32.676
0.10	$\alpha a/\pi$	1	1	2	1	2	2	1	3
	$\beta_1 b/\pi$	1.200	2.190	1.177	3.158	2.179	3.152	4.124	1.165
	$\beta_2 b/\pi$	12.793i	2.074i	18.801i	12.707i	2.178i	18.741i	12.590i	1.670i
	$\beta_3 b/\pi$	1.400i	12.770i	1.567i	2.598i	18.784i	2.671i	2.956i	25.865i
	Ω	5.496	8.577	13.426	13.796	15.041	18.593	20.145	21.632
0.20	$\alpha a/\pi$	1	1	2	2	1	2	3	3
	$\beta_1 b/\pi$	1.175	2.116	1.164	2.113	3.066	3.065	1.153	2.109
	$\beta_2 b/\pi$	9.390i	9.324i	16.655i	16.618i	9.181i	16.538i	24.344i	24.318i
	$\beta_3 b/\pi$	1.137i	1.521i	1.197i	1.561i	1.643i	1.678i	1.270i	1.613i
	Ω	3.880	6.103	7.687	9.031	9.256	11.408	11.574	12.518

TABLE 7.12
Frequency Parameters $\Omega = (\omega b^2/\pi^2)\sqrt{\rho h/D_0}$ for Three-Ply Laminated (0°, 90°, 0°) Square SCSF Plates

		Mode Sequence Number							
h/b		1	2	3	4	5	6	7	8
0.05	$\alpha a/\pi$	1	1	1	1	2	2	2	2
	$\beta_1 b/\pi$	0.552	1.509	2.492	3.477	0.515	1.522	2.499	3.480
	$\beta_2 b/\pi$	21.541i	21.539i	21.533i	21.515i	25.594i	25.592i	25.586i	25.569i
	$\beta_3 b/\pi$	1.139i	1.781i	2.574i	3.341i	1.725i	2.209i	2.871i	3.559i
	Ω	5.842	7.124	11.548	19.041	19.393	20.037	22.329	27.309
0.10	$\alpha a/\pi$	1	1	1	2	2	1	2	2
	$\beta_1 b/\pi$	0.554	1.486	2.456	0.535	1.494	3.423	2.459	3.425
	$\beta_2 b/\pi$	0.985i	12.790i	12.757i	18.804i	18.798i	12.681i	18.775i	18.722i
	$\beta_3 b/\pi$	12.797i	1.602i	2.234i	1.234i	1.762i	2.712i	2.334i	2.782i
	Ω	4.910	6.091	9.847	13.160	13.736	15.458	15.862	19.912
0.20	$\alpha a/\pi$	1	1	1	2	2	2	1	3
	$\beta_1 b/\pi$	0.563	1.433	2.389	0.563	1.443	2.394	3.344	0.567
	$\beta_2 b/\pi$	9.401i	9.379i	9.292i	0.876i	1.330i	16.599i	9.123i	0.995i
	$\beta_3 b/\pi$	0.7686i	1.271i	1.581i	16.662i	16.648i	1.620i	1.636i	24.349i
	Ω	3.348	4.339	6.956	7.424	7.954	9.649	10.246	11.395

7.2.6 Circular Polar Orthotropic Plates

For polar orthotropic circular plates, exact fundamental frequency is possible only for a special case of circumferential rigidity to radial rigidity ratio, and the plate edge must be simply supported.

The governing differential equation for the vibration of a polar orthotropic circular plate is given by (Elishakoff and Pentaras 2006)

$$rD_r\frac{d^4w}{dr^4}+\left(2D_r+2r\frac{dD_r}{dr}+\nu_\theta D_r-\nu_r D_\theta\right)\frac{d^3w}{dr^3}$$
$$-\left[\frac{d}{dr}\left(-D_r-r\frac{dD_r}{dr}-\nu_\theta D_r+\nu_r D_\theta\right)-\nu_\theta\frac{dD_r}{dr}+\frac{1}{r}D_\theta\right]\frac{d^2w}{dr^2} \quad (7.80)$$
$$-\left[\frac{d}{dr}\left(-\nu_\theta\frac{dD_r}{dr}+\frac{1}{r}D_\theta\right)\right]\frac{dw}{dr}-r\rho h\omega^2 w=0$$

where w is the deflection, D_r and D_θ are, respectively, the radial and circumferential rigidities, which vary with respect to radial coordinate r; ν_r, and ν_θ are the Poisson ratios in the radial and circumferential direction, respectively; ρ is the material density; h is the thickness; and ω is the angular frequency.

By using a semi-inverse method, Elishakoff and Pentaras (2006) showed that an exact fundamental frequency solution is possible for a simply supported, circular, polar orthotropic plate by assuming that

$$w=\frac{1}{8}\left[\frac{(5+\nu_\theta)}{(1+2\nu_\theta)}-\frac{2(3+\nu_\theta)}{(1+\nu_\theta)}\left(\frac{r}{R}\right)^3+\left(\frac{r}{R}\right)^4\right] \quad (7.81)$$

which satisfies Equation (7.80). The fundamental frequency of vibration is given by

$$\omega^2=\frac{24\alpha}{\rho h}\left(21+7\nu_\theta-3\nu_r k^2-21k^2\right) \quad (7.82)$$

where α is coefficient, $k=\sqrt{D_\theta/D_r}=\sqrt{\nu_\theta/\nu_r}$, and the radial flexural rigidity takes on the following expression:

$$D_r=\frac{\alpha}{\left(1+k^2\nu_r\right)^2\left(2k^2\nu_r-k^2+15\right)\left(k^2-9\right)}\left[\left(-135+50k^4\nu_r+28k^6\nu_r^2-k^4+24k^2\right.\right.$$
$$-171k^4\nu_r^2-288k^2\nu_r-18k^6\nu_r^3+2k^8\nu_r^3-k^8\nu_r^2-2k^6\nu_r\Big)+\Big(6k^4-108k^6\nu_r^2$$
$$+2304k^8\nu_r^2+72k^6\nu_r^3-8k^8\nu_r^3+1566+8k^6\nu_r+810k^4\nu_r^2-328k^4\nu_r-288k^2\Big)R^2r^2$$
$$+\Big(-k^8\nu_r^2-4896k^2\nu_r-1734k^4\nu_r^2-3159+246k^4\nu_r-8k^8\nu_r^4-214k^6\nu_r^3+80k^6\nu_r^2$$
$$\left.+2k^4-36k^2+6k^8\nu_r^3-6k^6\nu_r\Big)R^4r^4\right] \quad (7.83)$$

A few other variations of circumferential rigidity to radial rigidity ratio also give exact fundamental frequencies, and the associated complicated variations of radial rigidity are presented in the paper by Elishakoff and Pentaras (2006).

7.3 SANDWICH PLATES

Exact vibration solutions are available for a simply supported polygonal plate of a sandwich construction with core thickness h_c, core modulus of elasticity E_c, core shear modulus G_c, core Poisson ratio ν_c, core mass density ρ_c, facing thickness h_f, facing modulus of elasticity E_f, facing shear modulus G_f, facing Poisson ratio ν_f, and facing mass density ρ_f, as shown in Figure 7.9.

Wang (1996) showed that an exact relationship exists between the frequencies ω_s of such a sandwich plate based on the Mindlin plate theory and the corresponding frequencies ω of the isotropic plate based on the Kirchhoff (classical thin) plate theory. This relationship is given by

$$\omega_s^2 = \chi_1 \left\{ \left[1+(1+\chi_2)\chi_3\omega\sqrt{\frac{\rho h}{D}}\right] - \sqrt{\left[1+(1+\chi_2)\chi_3\omega\sqrt{\frac{\rho h}{D}}\right]^2 - 4\chi_2\chi_3^2\omega^2\frac{\rho h}{D}} \right\} \tag{7.84}$$

where h is the plate thickness of the isotropic thin plate, D is the flexural rigidity of the isotropic thin plate, ρ is the mass density of the isotropic thin plate, and

$$\chi_1 = \frac{1}{2}\left[\frac{\kappa^2(G_c h_c + 2G_f h_f)}{\rho_c I_c + \rho_f I_f}\right] \tag{7.85a}$$

$$\chi_2 = \left[\frac{\rho_c h_c + 2\rho_f h_f}{\kappa^2(G_c h_c + 2G_f h_f)}\right]\left[\frac{D_c + D_f}{\rho_c I_c + \rho_f I_f}\right] \tag{7.85b}$$

$$\chi_3 = \left[\frac{\rho_c I_c + \rho_f I_f}{\rho_c h_c + 2\rho_f h_f}\right] \tag{7.85c}$$

$$D_c = \frac{E_c I_c}{1-\nu_c^2},\ D_f = \frac{E_f I_f}{1-\nu_f^2},\ I_c = \frac{h_c^3}{12},\ I_f = \frac{2}{3}h_f\left(h_f^2 + \frac{3}{4}h_c^2 + \frac{3}{2}h_c h_f\right) \tag{7.86d}$$

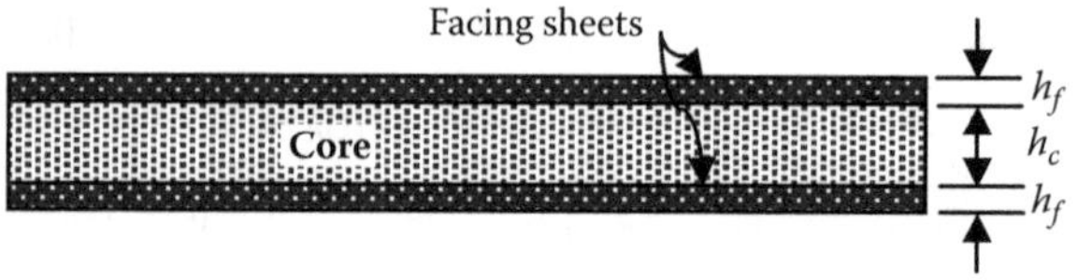

FIGURE 7.9 Sandwich plate.

By providing exact frequencies of the simply supported, polygonal, isotropic thin plate, such as those given in Tables 5.35 to 5.38, Equation (7.84) furnishes the corresponding exact frequencies of the sandwich plate.

7.4 LAMINATED PLATES

Consider a laminate of total thickness h comprising N orthotropic layers, as shown in Figure 7.10. A typical lamina, say the kth layer ($k = 1, 2, \ldots, N$), has a uniform thickness h_k; material properties E_1^k, E_2^k; and principal material coordinates oriented at an angle $\theta_1 = k$ with respect to the laminate (global) coordinate x.

In the development of the classical thin-plate theory for laminated plates, the following assumptions are made:

- Layers are perfectly bonded together.
- Material of each layer is linearly elastic and orthotropic.
- Each layer has a uniform thickness.
- Strains are small.
- The Kirchhoff hypothesis is adopted.

In view of the foregoing assumptions, the governing equations of motion for a laminated plate are given by (Reddy and Miravete 1995)

$$\frac{\partial}{\partial x}\left[A_{11}\frac{\partial u}{\partial x}+A_{12}\frac{\partial v}{\partial y}+A_{16}\left(\frac{\partial u}{\partial y}+\frac{\partial v}{\partial x}\right)-B_{11}\frac{\partial^2 w}{\partial x^2}-B_{12}\frac{\partial^2 w}{\partial y^2}-2B_{16}\frac{\partial^2 w}{\partial x\partial y}\right]$$
$$+\frac{\partial}{\partial y}\left[A_{16}\frac{\partial v}{\partial x}+A_{26}\frac{\partial u}{\partial y}+A_{66}\left(\frac{\partial u}{\partial y}+\frac{\partial v}{\partial x}\right)-B_{16}\frac{\partial^2 w}{\partial x^2}-B_{26}\frac{\partial^2 w}{\partial y^2}-2B_{66}\frac{\partial^2 w}{\partial x\partial y}\right]$$
$$=I_0\frac{\partial^2 u}{\partial t^2}-I_1\frac{\partial^3 w}{\partial x\partial t^2} \qquad (7.87)$$

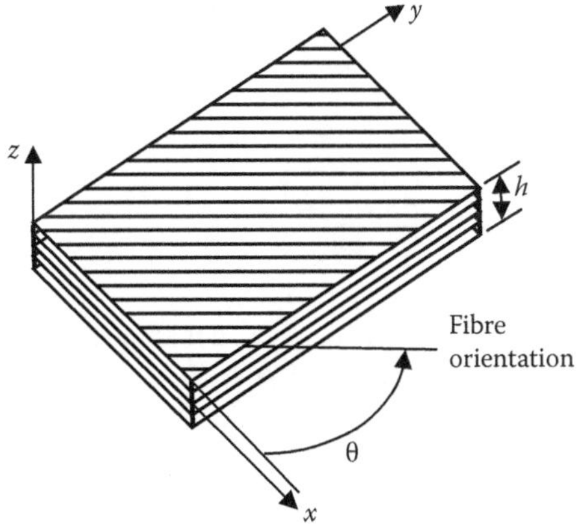

FIGURE 7.10 Laminate comprising N orthotropic layers.

$$\frac{\partial}{\partial x}\left[A_{16}\frac{\partial u}{\partial x}+A_{26}\frac{\partial v}{\partial y}+A_{66}\left(\frac{\partial u}{\partial y}+\frac{\partial v}{\partial x}\right)-B_{16}\frac{\partial^2 w}{\partial x^2}-B_{26}\frac{\partial^2 w}{\partial y^2}-2B_{66}\frac{\partial^2 w}{\partial x\partial y}\right]$$

$$+\frac{\partial}{\partial y}\left[A_{12}\frac{\partial u}{\partial x}+A_{22}\frac{\partial v}{\partial y}+A_{26}\left(\frac{\partial u}{\partial y}+\frac{\partial v}{\partial x}\right)-B_{12}\frac{\partial^2 w}{\partial x^2}-B_{22}\frac{\partial^2 w}{\partial y^2}-2B_{26}\frac{\partial^2 w}{\partial x\partial y}\right]$$

$$=I_0\frac{\partial^2 v}{\partial t^2}-I_1\frac{\partial^2 w}{\partial y\partial t^2} \tag{7.88}$$

$$\frac{\partial^2}{\partial x^2}\left[B_{11}\frac{\partial u}{\partial x}+B_{12}\frac{\partial v}{\partial y}+B_{16}\left(\frac{\partial u}{\partial y}+\frac{\partial v}{\partial x}\right)-D_{11}\frac{\partial^2 w}{\partial x^2}-D_{12}\frac{\partial^2 w}{\partial y^2}-2D_{16}\frac{\partial^2 w}{\partial x\partial y}\right]$$

$$+2\frac{\partial^2}{\partial x\partial y}\left[B_{16}\frac{\partial u}{\partial x}+B_{26}\frac{\partial v}{\partial y}+B_{66}\left(\frac{\partial u}{\partial y}+\frac{\partial v}{\partial x}\right)-D_{16}\frac{\partial^2 w}{\partial x^2}-D_{26}\frac{\partial^2 w}{\partial y^2}-2D_{66}\frac{\partial^2 w}{\partial x\partial y}\right]$$

$$+\frac{\partial^2}{\partial y^2}\left[B_{12}\frac{\partial u}{\partial x}+B_{22}\frac{\partial v}{\partial y}+B_{26}\left(\frac{\partial u}{\partial y}+\frac{\partial v}{\partial x}\right)-D_{12}\frac{\partial^2 w}{\partial x^2}-D_{22}\frac{\partial^2 w}{\partial y^2}-2D_{26}\frac{\partial^2 w}{\partial x\partial y}\right]$$

$$=I_0\frac{\partial^2 w}{\partial t^2}+I_1\frac{\partial^2}{\partial t^2}\left(\frac{\partial u}{\partial x}+\frac{\partial v}{\partial y}\right)-I_2\frac{\partial^2}{\partial t^2}\left(\frac{\partial^2 w}{\partial x^2}+\frac{\partial^2 w}{\partial y^2}\right) \tag{7.89}$$

where t is the time; u, v, and w are the displacements in the x-, y-, and z-directions, respectively; and I_i are the mass inertias given by

$$I_0=\sum_{k=1}^{N}\rho^{(k)}(z_{k+1}-z_k) \tag{7.90a}$$

$$I_1=\frac{1}{2}\sum_{k=1}^{N}\rho^{(k)}\left(z_{k+1}^2-z_k^2\right) \tag{7.90b}$$

$$I_2=\frac{1}{3}\sum_{k=1}^{N}\rho^{(k)}\left(z_{k+1}^3-z_k^3\right) \tag{7.90c}$$

A_{ij} denotes the extensional stiffnesses, D_{ij} denotes the bending stiffnesses, and B_{ij} denotes the bending-extensional coupling stiffnesses of a laminate. Note that for symmetric laminates, the coupling stiffnesses vanish. These stiffnesses are given by

$$A_{ij}=\sum_{k=1}^{N}\breve{Q}_{ij}^{(k)}(z_{k+1}-z_k) \tag{7.91a}$$

$$B_{ij} = \frac{1}{2}\sum_{k=1}^{N} \breve{Q}_{ij}^{(k)}\left(z_{k+1}^2 - z_k^2\right) \tag{7.91b}$$

$$D_{ij} = \frac{1}{3}\sum_{k=1}^{N} \breve{Q}_{ij}^{(k)}\left(z_{k+1}^3 - z_k^3\right) \tag{7.91c}$$

where $\breve{Q}_{ij}^{(k)}$ denotes the material stiffnesses of the kth layer given by

$$\breve{Q}_{11} = Q_{11}\cos^4\theta + 2(Q_{12} + 2Q_{66})\sin^2\theta\cos^2\theta + Q_{22}\sin^4\theta \tag{7.92a}$$

$$\breve{Q}_{12} = (Q_{11} + Q_{22} - 4Q_{66})\sin^2\theta\cos^2\theta + Q_{12}\left(\sin^4\theta + \cos^4\theta\right) \tag{7.92b}$$

$$\breve{Q}_{22} = Q_{11}\sin^4\theta + 2(Q_{12} + 2Q_{66})\sin^2\theta\cos^2\theta + Q_{22}\cos^4\theta \tag{7.92c}$$

$$\breve{Q}_{16} = (Q_{11} - Q_{12} - 2Q_{66})\sin\theta\cos^3\theta + (Q_{12} - Q_{22} + 2Q_{66})\sin^3\theta\cos\theta \tag{7.92d}$$

$$\breve{Q}_{26} = (Q_{11} - Q_{12} - 2Q_{66})\sin^3\theta\cos\theta + (Q_{12} - Q_{22} + 2Q_{66})\sin\theta\cos^3\theta \tag{7.92e}$$

$$\breve{Q}_{66} = (Q_{11} + Q_{22} - 2Q_{12} - 2Q_{66})\sin^2\theta\cos^2\theta + Q_{66}(\sin^4\theta + \cos^4\theta) \tag{7.92f}$$

and the plane stress-reduced stiffnesses Q_{ij} involve four independent material constants—E_1, E_2, ν_{12}, and G_{12}—shown as follows:

$$Q_{11} = \frac{E_1}{1 - \nu_{12}\nu_{21}} \tag{7.93a}$$

$$Q_{12} = \frac{\nu_{12}E_2}{1 - \nu_{12}\nu_{21}} = \frac{\nu_{21}E_1}{1 - \nu_{12}\nu_{21}} \tag{7.93b}$$

$$Q_{22} = \frac{E_2}{1 - \nu_{12}\nu_{21}} \tag{7.93c}$$

$$Q_{66} = G_{12} \tag{7.93d}$$

Exact vibration solutions are possible for rectangular plates with simply supported edges of the type that satisfy the following boundary conditions:

$$\begin{aligned}
&u(x,0,t) = 0, \quad \frac{\partial w}{\partial x}(x,0,t) = 0, \quad u(x,b,t) = 0, \quad \frac{\partial w}{\partial x}(x,b,t) = 0 \\
&v(0,y,t) = 0, \quad \frac{\partial w}{\partial y}(0,y,t) = 0, \quad v(a,y,t) = 0, \quad \frac{\partial w}{\partial y}(a,y,t) = 0 \\
&w(x,0,t) = 0, \quad w(x,b,t) = 0, \quad w(0,y,t) = 0, \quad w(a,y,t) = 0
\end{aligned} \tag{7.94}$$

The Navier solutions for rectangular laminated plates with these simply supported edges exist only when the laminate stacking sequences are such that

$$A_{16}=0,\ A_{26}=0,\ B_{16}=0,\ B_{26}=0,\ D_{16}=0,\ D_{26}=0 \tag{7.95}$$

This means that plates with a single, generally orthotropic layer; symmetrically laminated plates with multiple specially orthotropic layers; and antisymmetric cross-ply laminated plates admit the Navier solution for the simply supported boundary conditions given by Equation (7.94).

By assuming a harmonic solution of the form

$$u(x,y,t)=\left(\sum_{n=1}^{\infty}\sum_{m=1}^{\infty} u_{mn}\cos\frac{m\pi x}{a}\sin\frac{n\pi y}{b}\right)e^{i\omega t}$$

$$v(x,y,t)=\left(\sum_{n=1}^{\infty}\sum_{m=1}^{\infty} v_{mn}\sin\frac{m\pi x}{a}\cos\frac{n\pi y}{b}\right)e^{i\omega t} \tag{7.96}$$

$$w_{mn}(x,y,t)=\left(\sum_{n=1}^{\infty}\sum_{m=1}^{\infty} w_{mn}\sin\frac{m\pi x}{a}\cos\frac{n\pi y}{b}\right)e^{i\omega t}$$

where $i=\sqrt{-1}$ and ω is the frequency of natural vibration, Equations (7.87) to (7.89) reduce to

$$\left(\begin{bmatrix} C_{11} & C_{12} & C_{13} \\ C_{12} & C_{22} & C_{23} \\ C_{13} & C_{23} & C_{33} \end{bmatrix}-\omega^2\begin{bmatrix} I_0 & 0 & 0 \\ 0 & I_0 & 0 \\ 0 & 0 & I_0+I_2\left[\left(\frac{m\pi}{a}\right)^2+\left(\frac{n\pi}{b}\right)^2\right] \end{bmatrix}\right)\begin{Bmatrix} u_{mn} \\ v_{mn} \\ w_{mn} \end{Bmatrix}=\begin{Bmatrix} 0 \\ 0 \\ 0 \end{Bmatrix} \tag{7.97}$$

where

$$C_{11}=A_{11}\left(\frac{m\pi}{a}\right)^2+A_{66}\left(\frac{n\pi}{b}\right)^2 \tag{7.98a}$$

$$C_{12}=(A_{12}+A_{66})\left(\frac{m\pi}{a}\right)\left(\frac{n\pi}{b}\right) \tag{7.98b}$$

$$C_{13}=-A_{11}\left(\frac{m\pi}{a}\right)^3-(B_{12}+2B_{66})\left(\frac{m\pi}{a}\right)\left(\frac{n\pi}{b}\right)^2 \tag{7.98c}$$

$$C_{22} = A_{66}\left(\frac{m\pi}{a}\right)^2 + A_{22}\left(\frac{n\pi}{b}\right)^2 \tag{7.98d}$$

$$C_{23} = -(B_{11} + 2B_{66})\left(\frac{m\pi}{a}\right)^2\left(\frac{n\pi}{b}\right) - B_{22}\left(\frac{n\pi}{b}\right)^3 \tag{7.98e}$$

$$C_{33} = D_{11}\left(\frac{m\pi}{a}\right)^4 + 2(D_{12} + 2D_{66})\left(\frac{m\pi}{a}\right)^2\left(\frac{n\pi}{b}\right)^2 + D_{22}\left(\frac{n\pi}{b}\right)^4 \tag{7.98f}$$

For nontrivial solution, the determinant of the coefficient matrix in Equation (7.97) should vanish, thereby furnishing the characteristic equation. The frequencies are obtained by solving the characteristic equation.

In the case of specially orthotropic laminated plates, the only nonzero stiffnesses are A_{11}, A_{12}, A_{22}, A_{66}, D_{11}, D_{12}, D_{22}, and D_{66}. Therefore, $C_{13} = 0$ and $C_{23} = 0$, and so Equation (7.97) yields the following characteristic equation

$$\left[C_{33} - \omega^2\left(I_0 + I_2\left\{\left(\frac{m\pi}{a}\right)^2 + \left(\frac{n\pi}{b}\right)^2\right\}\right)\right]\left[(C_{11} - \omega^2 I_0)(C_{22} - \omega^2 I_0) - C_{12}C_{12}\right] = 0 \tag{7.99}$$

The three roots of the characteristic equation are

$$\left(\omega_{mn}^2\right)_1 = \frac{C_{33}}{I_0 + I_2\left\{\left(\frac{m\pi}{a}\right)^2 + \left(\frac{n\pi}{b}\right)^2\right\}} \tag{7.100}$$

$$\left(\omega_{mn}^2\right)_2 = \frac{1}{2}\left[C_{11} + C_{22} - \sqrt{(C_{11} - C_{22})^2 + 4C_{12}C_{12}}\right] \tag{7.101}$$

$$\left(\omega_{mn}^2\right)_3 = \frac{1}{2}\left[C_{11} + C_{22} + \sqrt{(C_{11} - C_{22})^2 + 4C_{12}C_{12}}\right] \tag{7.102}$$

The frequencies given by Equation (7.100) correspond to flexural vibration, whereas the frequencies furnished by Equations (7.101) and (7.102) correspond to in-plane vibration, which has higher values than those of the flexural vibration. Equation (7.100) gives the frequencies of flexural vibration for specially orthotropic laminated, simply supported rectangular plates

$$\omega_{mn}^2 = \frac{\pi^4}{\left(I_0 + I_2\left[\left(\frac{m\pi}{a}\right)^2 + \left(\frac{n\pi}{b}\right)^2\right]\right)b^4}\left[D_{11}m^4\left(\frac{b}{a}\right)^4 + 2(D_{12} + 2D_{66})m^2n^2\left(\frac{b}{a}\right)^2 + D_{22}n^4\right] \tag{7.103}$$

7.5 FUNCTIONALLY GRADED PLATES

Functionally graded materials (FGMs) are new composite materials where the material property changes smoothly from one surface to the other as a result of gradually changing the volume fraction of the constituent materials, usually in the thickness direction. An example of FGMs is a mixture of ceramics and metal, as shown in Figure 7.11.

Consider the Young's modulus E and density ρ per unit volume varying continuously through the plate thickness according to the power-law distribution given by

$$E(z) = (E_1 - E_2)V_f(z) + E_2 \tag{7.104}$$

$$\rho(z) = (\rho_1 - \rho_2)V_f(z) + \rho_2 \tag{7.105}$$

in which the subscripts 1 and 2 refer to constituent 1 and constituent 2, respectively, and the volume fraction $V_f(z)$ may be given by

$$V_f(z) = \left(\frac{z}{h} + \frac{1}{2}\right)^{\alpha} \tag{7.106}$$

where α is the gradient index and takes only positive values. Poisson ratio ν may be assumed to be the same for the two constituents in order to simplify the formulation.

The governing equations of motion of functionally graded rectangular plates, according to the first-order shear deformable plate theory, are given by (Hosseini-Hashem et al. 2010)

$$\begin{gathered}\frac{\partial^2 \varphi_x}{\partial x^2} + \left(\frac{a}{b}\right)^2 \frac{\partial^2 \varphi_x}{\partial y^2} + \frac{1+\nu}{1-\nu}\left(\frac{\partial^2 \varphi_x}{\partial x^2} + \left(\frac{a}{b}\right)\frac{\partial^2 \varphi_y}{\partial x \partial y}\right) \\ -\frac{2\kappa^2 Bha^2}{A(1-\nu)}\left(\varphi_x - \frac{\partial w}{\partial x}\right) + \omega^2 \frac{Ch^3 a^2}{6A(1-\nu)}\varphi_x = 0\end{gathered} \tag{7.107}$$

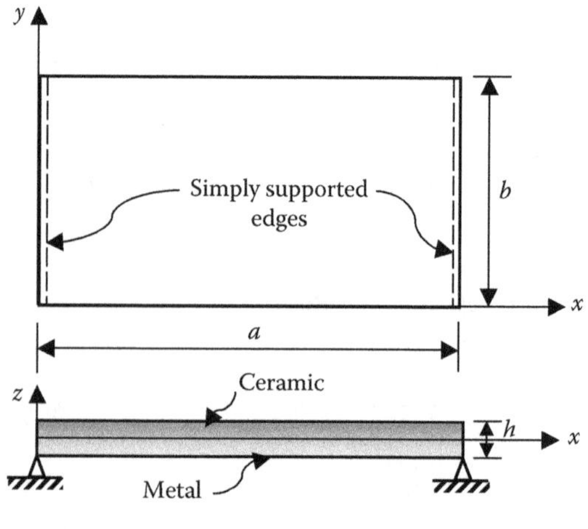

FIGURE 7.11 Rectangular FG plate with coordinate axes.

$$\frac{\partial^2 \varphi_y}{\partial x^2}+\left(\frac{a}{b}\right)^2\frac{\partial^2 \varphi_y}{\partial y^2}+\frac{1+\nu}{1-\nu}\left(\frac{a}{b}\right)\left(\frac{\partial^2 \varphi_y}{\partial x\,\partial y}+\left(\frac{a}{b}\right)\frac{\partial^2 \varphi_y}{\partial y^2}\right)$$
$$-\frac{2\kappa^2 Bha^2}{A(1-\nu)}\left(\varphi_y-\left(\frac{a}{b}\right)\frac{\partial w}{\partial y}\right)+\omega^2\frac{Ch^3a^2}{6A(1-\nu)}\varphi_y=0 \tag{7.108}$$

$$\frac{\partial^2 w}{\partial x^2}+\left(\frac{a}{b}\right)^2\frac{\partial^2 w}{\partial y^2}-\left(\frac{\partial \varphi_x}{\partial x}+\left(\frac{a}{b}\right)\frac{\partial \varphi_y}{\partial y}\right)-\omega^2\frac{Da^2}{B\kappa^2}w=0 \tag{7.109}$$

where

$$A=\frac{h^3}{(1-\nu^2)}\left[\frac{3(2+\alpha+\alpha^2)E_1+\alpha(8+3\alpha+\alpha^2)E_2}{12(1+\alpha)(2+\alpha)(3+\alpha)}\right] \tag{7.110a}$$

$$B=\frac{1}{2(1+\nu)}\left[\frac{E_1+\alpha E_2}{1+\alpha}\right] \tag{7.110b}$$

$$C=\frac{3(2+\alpha+\alpha^2)\rho_1+\alpha(8+3\alpha+\alpha^2)\rho_2}{(1+\alpha)(2+\alpha)(3+\alpha)} \tag{7.110c}$$

$$D=\frac{\rho_1+\alpha\rho_2}{1+\alpha} \tag{7.110d}$$

For a rectangular plate where the edges $x = 0$ and $x = a$ are simply supported as shown in Figure 7.11, the boundary conditions at these two edges are

$$M_{xx}=0,\quad \varphi_y=0,\quad w=0 \tag{7.111}$$

The boundary conditions along the edges $y = 0$ and $y = b$ are given by

$$M_{xx}=0,\; M_{xy}=0,\; Q_y=0 \quad \text{for a free edge} \tag{7.112}$$

$$M_{yy}=0,\quad \varphi_x=0,\quad w=0 \quad \text{for a simply supported edge} \tag{7.113}$$

$$\varphi_x=0,\quad \varphi_y=0,\quad w=0 \quad \text{for a clamped edge} \tag{7.114}$$

The general solutions to Equations (7.107) to (7.109) in terms of three nondimensional potentials Θ_1, Θ_2, Θ_3 may be expressed as

$$\varphi_x=C_1\frac{\partial \Theta_1}{\partial x}+C_2\frac{\partial \Theta_2}{\partial x}-\frac{a}{b}\frac{\partial \Theta_3}{\partial y} \tag{7.115}$$

$$\varphi_y = C_1 \frac{a}{b}\frac{\partial \Theta_1}{\partial y} + C_2 \frac{a}{b}\frac{\partial \Theta_2}{\partial y} - \frac{\partial \Theta_3}{\partial x} \tag{7.116}$$

$$w = \Theta_1 + \Theta_2 \tag{7.117}$$

where

$$C_1 = \frac{B_2}{\alpha_1^2 - B_1}, \quad C_2 = \frac{B_3}{\alpha_2^2 - B_1} \tag{7.118}$$

$$B_1 = \frac{2B\kappa^2 ha^2}{A(1-\nu)} - \omega^2 \frac{Ch^3 a^2}{6A(1-\nu)} \tag{7.119}$$

$$B_2 = \frac{(1+\nu)}{(1-\nu)}\left[\alpha_1^2 + \omega^2 \frac{Da^2}{\kappa^2 B}\right] + \frac{2\kappa^2 Bha^2}{A(1-\nu)} \tag{7.120}$$

$$B_3 = \frac{(1+\nu)}{(1-\nu)}\left[\alpha_2^2 + \omega^2 \frac{Da^2}{\kappa^2 B}\right] + \frac{2\kappa^2 Bha^2}{A(1-\nu)} \tag{7.121}$$

$$\alpha_1^2, \alpha_2^2 = \omega^2 \frac{Ca^2}{24B\kappa^2}\left[12\frac{D}{C} + \frac{\kappa^2 Bh^3}{A}\right]$$
$$\pm \sqrt{\left\{\omega^2 \frac{Ca^2}{24B\kappa^2}\left[12\frac{D}{C} + \frac{\kappa^2 Bh^3}{A}\right]\right\}^2 - \omega^2 \frac{\kappa^2 Bh^2 a^6 D}{12A^2}\left(\omega^2 \frac{a^2 Ch^3}{A} - 12\frac{\kappa^2 Bha^2}{A}\right)} \tag{7.122}$$

In view of the three nondimensional potentials, the governing equations (7.107) to (7.109) may be written as

$$\frac{\partial^2 \Theta_1}{\partial x^2} + \left(\frac{a}{b}\right)^2 \frac{\partial^2 \Theta_1}{\partial y^2} = -\alpha_1^2 \Theta_1 \tag{7.123}$$

$$\frac{\partial^2 \Theta_2}{\partial x^2} + \left(\frac{a}{b}\right)^2 \frac{\partial^2 \Theta_2}{\partial y^2} = -\alpha_2^2 \Theta_2 \tag{7.124}$$

$$\frac{\partial^2 w}{\partial x^2} + \left(\frac{a}{b}\right)^2 \frac{\partial^2 w}{\partial y^2} = B_1 \Theta_3 \tag{7.125}$$

One set of solutions to Equations (7.123) to (7.125) is

$$\Theta_1 = \left(A_1 \sin\frac{\lambda_1 y}{a} + A_2 \cos\frac{\lambda_1 y}{a}\right)\sin\frac{\mu_1 x}{a} + \left(A_3 \sin\frac{\lambda_1 y}{a} + A_4 \cos\frac{\lambda_1 y}{a}\right)\cos\frac{\mu_1 x}{a} \tag{7.126}$$

$$\Theta_2 = \left(A_5 \sinh \frac{\lambda_2 y}{a} + A_6 \cosh \frac{\lambda_2 y}{a} \right) \sin \frac{\mu_2 x}{a} + \left(A_7 \sinh \frac{\lambda_2 y}{a} + A_8 \cosh \frac{\lambda_2 y}{a} \right) \cos \frac{\mu_2 x}{a} \tag{7.127}$$

$$\Theta_3 = \left(A_9 \sinh \frac{\lambda_3 y}{a} + A_{10} \cosh \frac{\lambda_3 y}{a} \right) \sin \frac{\mu_3 x}{a} + \left(A_{11} \sinh \frac{\lambda_3 y}{a} + A_{12} \cosh \frac{\lambda_3 y}{a} \right) \cos \frac{\mu_3 x}{a} \tag{7.128}$$

where A_i, $i = 1, 2, \ldots, 12$ are unknown coefficients, λ_i, $i = 1, 2, 3$, and μ_i, $i = 1, 2, 3$ are related by

$$\alpha_1^2 = \mu_1^2 + \left(\frac{a}{b} \right)^2 \lambda_1^2, \quad \alpha_2^2 = \mu_2^2 - \left(\frac{a}{b} \right)^2 \lambda_2^2, \quad -B_1 = \mu_3^2 - \left(\frac{a}{b} \right)^2 \lambda_3^2 \tag{7.129}$$

In view that the edges $x = 0$ and $x = a$ are simply supported, Equations (7.126) to (7.128) reduce to

$$\Theta_1 = \left(A_1 \sin \frac{\lambda_1 y}{a} + A_2 \cos \frac{\lambda_1 y}{a} \right) \sin \frac{m\pi x}{a} \tag{7.130}$$

$$\Theta_2 = \left(A_5 \sinh \frac{\lambda_2 y}{a} + A_6 \cosh \frac{\lambda_2 y}{a} \right) \sin \frac{m\pi x}{a} \tag{7.131}$$

$$\Theta_3 = \left(A_9 \sinh \frac{\lambda_3 y}{a} + A_{10} \cosh \frac{\lambda_3 y}{a} \right) \sin \frac{m\pi x}{a} \tag{7.132}$$

By substituting Equations (7.130) to (7.132) into Equations (7.115) to (7.117), and then into the appropriate boundary conditions for edges $y = 0$ and $y = b$, one obtains six homogeneous equations. To obtain nontrivial solutions, the determinant of the coefficient matrix must vanish, which furnishes the characteristic equations for rectangular FGM plates. Some vibration frequencies are computed for two FGMs (i.e., Al/Al_2O_3 and Al/ZrO_2) by Hosseini-Hashemi et al. (2010). In the aforementioned paper, the effect of a two-parameter elastic foundation is also considered.

7.6 CONCLUDING REMARKS

Approximate and numerical vibration results of nonisotropic plates may be obtained from books by Reddy (2007), Qatu (2004), Reddy and Miravete (1995), Yu (1996), and Leissa (1969) and from papers published in journals such as *Journal of Sound and Vibration*, *Journal of Vibration and Acoustics*, *Composite Structures*, *International Journal of Solids and Structures*, *Journal of Applied Mechanics*, and the *International Journal of Structural Stability and Dynamics*.

REFERENCES

Cope, R. J., and L. A. Clark. 1984. *Concrete slabs: Analysis and design*. New York: Elsevier Science.

Elishakoff, I., and D. Pentaras. 2006. Lekhnitskii's classic formula serving as an exact mode shape of simply supported polar orthotropic inhomogenous circular plates. *Journal of Sound and Vibration* 291 (3–5): 1239–54.

Hosseini-Hashem, Sh., H. Rokni Damavandi Taher, H. Akhavan, and M. Omidi. 2010. Free vibration of functionally graded rectangular plates using first order shear deformation plate theory. *Applied Mathematical Modelling* 34:1276–91.

Leissa, A. W. 1969. *Vibration of plates*. NASA SP-160. Washington, DC: U.S. Government Printing Office. Repr. Sewickley, PA: Acoustical Society of America, 1993.

Levy, M. 1899. Sur l'equlibre elastique d'une plaque rectangulaire. *C.R. Acad. Sci.* 129:535–39.

Liew, K.M. (1996). Solving the vibration of thick symmetric laminates by Reissner/Mindlin plate theory and the p-Ritz method, *Journal of Sound and Vibration*, 198(3), 343–360.

Liu, B., and Y. F. Xing. 2011. Exact solutions for free vibrations of orthotropic rectangular Mindlin plates. *Composite Structures* 93:1664–72.

Mindlin, R. D., and H. Deresiewicz. 1955. Thickness-shear and flexural vibrations of rectangular crystal plates. *Journal of Applied Physics* 26 (12): 1435–42.

Qatu, M. S. 2004. *Vibration of laminated shells and plates*. Waltham, MA: Elsevier, Academic Press.

Reddy, J. N. 2007. *Theory and analysis of elastic plates*. 2nd ed. Boca Raton, FL: Taylor & Francis.

Reddy, J. N., and A. Miravete. 1995. *Practical analysis of composite laminates*. Boca Raton, FL: CRC Press.

Szilard, R. D. 1974. *Theory and analysis of plates: Classical and numerical methods*. Englewood Cliffs, NJ: Prentice-Hall.

Timoshenko, S. P., and S. Woinowsky-Krieger. 1959. *Theory of plates and shells*. New York: McGraw-Hill.

Voigt, W. 1893. Bemerkungen zu dem problem der transversalen schwingungen rechteckiger platten. *Nachr. Ges. Wiss*. (Göttingen) 6:225–30.

Wang, C. M. 1996. Vibration frequencies of simply supported polygonal sandwich plates via Kirchhoff solutions. *Journal of Sound and Vibration* 190 (2): 255–60.

Yu, Y. Y. 1996. *Vibration of elastic plates*. New York: Springer.

Index

L

M

N

O

P

CPSIA information can be obtained
at www.ICGtesting.com
Printed in the USA
BVHW04*0556230918
527994BV00003B/9/P

9 781466 576841